THE BEGINNER'S HANDBOOK OF AMATEUR RADIO

THE BEGINNER'S HANDBOOK OF AMATEUR RADIO

Clay Laster, W5ZPV

FOURTH EDITION

McGraw-Hill

New York   San Francisco   Washington, D.C.   Auckland   Bogotá
Caracas   Lisbon   London   Madrid   Mexico City   Milan
Montreal   New Delhi   San Juan   Singapore
Sydney   Tokyo   Toronto

Cataloging-in-Publication Data is on file with the Library of Congress

McGraw-Hill

A Division of The McGraw·Hill Companies

2 3 4 5 6 7 8 9 0   DOC/DOC   0 9 8 7 6 5 4 3 2 1

ISBN 0-07-136187-1

The sponsoring editor for this book was Scott Grillo, the editing supervisor was Steven Melvin, and the production supervisor was Pamela Pelton. It was set in Melior per the CMS design by Joanne Morbit of McGraw-Hill's Hightstown, N.J., Professional Book Group composition unit.

Printed and bound by R. R. Donnelley & Sons Company.

McGraw-Hill books are available at special quantity discounts to use as premiums and sales promotions, or for use in corporate training programs. For more information, please write to the Director of Special Sales, Professional Publishing, McGraw-Hill, Two Penn Plaza, New York, NY 10121-2298. Or contact your local bookstore.

This book is printed on recycled, acid-free paper containing a minimum of 50% recycled de-inked fiber.

CONTENTS

Welcome to the world of amateur radio! We (the term "we" applies not only to myself, but to all the pioneering hams who helped to make amateur radio available to everyone) prepared this book with two major objectives in mind. First, we want to introduce you to the exciting world of amateur radio, one of the most interesting and challenging hobbies available today. Once you're hooked, our second objective is help prepare you for the FCC Technician Class license examinations, including both theory and Morse code. There is also another important objective: enticing the many dedicated electronics and computer experimenters, technicians, engineers, and serious CB enthusiasts throughout the country to join the ranks of amateur radio operators.

In no other hobby can one find such a diversity of individual and group endeavors involving scientific study and experimentation. The fields of amateur radio range from basic electronics, voice and data communications on a local and world-wide basis, and space and satellite communications to public service (especially during emergencies) and close camaraderie or friendships that last a lifetime. Truly, amateur radio is one of the largest fraternities in the world, open to persons of all ages and nationalities. The No-Code Technician Class license—yes, at last an amateur radio license with no Morse code examination—makes it easy for you to become a ham operator in a minimum of time. Then to earn hf Morse code and single-sideband voice privileges in selected hf bands, you need only to pass a 5 words-per-minute Morse code test. We'll show you how to study for and pass both the Technician written examination and the Morse code test.

This book is designed for either self-study or group use in Technician Class code and theory classes. The only additional material needed is a source of Morse code practice signals (cassettes or CDs) and a telegraph key/code oscillator for generating Morse code signals. Also, excellent PC and other home computer educational programs for both Morse code and theory instruction are available.

Chapters 1 and 2 provide an introduction to amateur radio, the FCC Part 97 Rules and Regulations for amateur radio, and how to prepare for and take the Technician Class license code and theory examinations. Chapters 3 through 10 form the basis for a "mini-course" in electronics technology, with emphasis on radio communications. Chapter 11 provides specific and technical information on amateur radio operating procedures, rf radiation safety requirements, and advanced amateur radio communications systems involving single sideband, frequency modulation, repeaters, digital and packet communications, and space and satellite communications. The appendices contain the FCC Technician question pool, as well as answers for each question. Sounds easy, doesn't it? All it takes on your part is a concentrated study effort to prepare for and pass the FCC license examinations.

We suggest that you review this book, chapter by chapter, before starting a detailed study of the course material. This will give you an overall perspective on amateur radio and help you develop a study schedule. Finally, review the question pool in Appendix A. One proven study technique is to write each FCC question on a 3-inch reference card, and your answer on the back. Check your answer against the correct answer given in the appendix. The writing of the questions and digging out the correct answers is part of the learning process. As you write each question on the card, you also help to write the question in your memory. To complete this process, review the set of cards until you learn the correct answers to all questions for each required license class examination. Remember, passing the Morse code examination requires a code speed proficiency of at least five words per minute. Build up this code speed before the end of your studies. With these factors in mind, you can develop a realistic schedule for preparing for the FCC Technician code and theory examinations.

One final recommendation: contact the local amateur radio club or a neighborhood ham and make arrangements for a volunteer examiner (VE) to administer the FCC Technician Class license examinations. Finding a ham in the neighborhood is easy—just check with the local radio club or the volunteer examiner before taking the examinations. Good luck!

Clay Laster

INTRODUCTION

Do you want to learn all about amateur radio? How to prepare for the FCC license examinations, all about radio communications and electronics principles, how personal computers can be used to enhance the learning, as well as the operation, of amateur radio comunications, how to assemble and operate an amateur radio "ham" station? This is one of the most comprehensive books for the beginner in amateur radio. It combines theory and practice in a clear and easy-to-understand manner. The many illustrations and photographs make learning all about amateur radio easy and exciting. The book contains complete information on how to study for the Technician Class license examinations and includes *all* of the FCC test questions and answers.

A variety of construction projects, transmitters, receivers, and other electronic equipment are examined to give the beginner practical experience in designing and building amateur radio projects. The book also provides parts and source information for radio components, equipment, and antennas, all typical of current-day technology. This book is unique in that technical information concerning vacuum tubes and associated circuits are covered. Why are these obsolete devices covered in a modern book? Many excellent vacuum-tube amateur receivers, transmitters, and transceivers are to be found at ham swapfests and conventions. Also, many kilowatt final power amplifiers for amateur hf service are marketed by amateur suppliers today.

Anyone interested in amateur radio—student, electronics or CB enthusiast, personal computer buff, doctor, lawyer, or engineer—will want to explore this worldwide hobby. This book offers you the opportunity to get in on the ground floor of amateur radio.

ACKNOWLEDGMENTS

The author is grateful for the cooperation and assistance extended by individuals, companies, and amateur organizations during the preparation of this book. Space does not permit listing these entities here. However, an attempt is made to acknowledge these many contributions as each particular item—circuit, device, or equipment—is covered. This approach also provides the beginner in amateur radio with specific information and source(s) concerning amateur parts and equipment.

Many thanks are due the author's wife, Irma, who provided invaluable assistance and encouragement during the preparation of the manuscript. Finally, the publisher's cooperation and patience for extending the deadline, as well as superb technical expertise in publishing this book, is greatly appreciated.

Introduction to Amateur Radio

Have you ever listened to an AM or FM radio receiver and wondered how radio waves travel through space? Have you gazed at a television set and asked how pictures generated on the other side of the earth can be transmitted to your TV receiver? Have you operated a citizens band radio and wanted to extend your range of communications worldwide? If you answered yes to any of these questions, you may be on the road to amateur radio, an exciting and rewarding hobby extending into almost every phase of electronics. Just pass the required Federal Communications Commission (FCC) exams given by authorized amateur radio operators in your neighborhood and you're ready to become an amateur operator (or "ham") yourself.

Who Can Become a Ham Radio Operator?

Anyone, regardless of age, is eligible to become a ham radio operator in the United States. Persons from less than 10 years to over 80 years of age have become hams.

Many disabled individuals have entered the ranks of amateur radio. Operators can apply for special testing provisions if they have physical disabilities. Learning materials are available on video or audio tape or in books to assist in the learning processes. After becoming an amateur radio operator, the ham has many radio equipment options. Many modern amateur radios (receivers, transmitters, etc.) provide voice outputs and can be controlled by computers using voice commands. Organizations such as the Courage HANDIHAM System (see Chap. 2) provide a wide variety of services to aid those with disabilities.

Aliens, except representatives of foreign governments, are encouraged to apply for a U.S. ham license if they plan to be in the country for an extended period of time. We'll tell you how to prepare and apply for your ham license in Chap. 2.

New Avenues into Amateur Radio

The FCC No-Code Technician License will allow you to enter the ranks of amateur radio and get on the air with a minimum of effort. Yes, that's right—you can skip the Morse code tests and pass a written test for the new No-Code Tech class license. This license allows you to operate on all amateur bands above 50 megahertz (MHz), including the popular 6-meter and 2-meter bands with exciting local repeater and packet radio operation, 440 MHz for amateur television, satellite and long-range "radio link" operation, and advanced microwave systems.

In case the megahertz and meter terms confuse you, just remember that they refer to the bands of frequency for amateur operation. You probably already know that the AM broadcast band covers 540 to 1650 kilohertz (kHz) and the FM broadcast band covers 88 to 108 MHz. However, we'll explain the meaning of these and other terms later. The important thing is that you can talk to other hams, using single sideband (SSB), frequency modulation (FM), or packet radio modes of operation. Then later, if you want to get on some of the amateur radio high-frequency bands (i.e., below 50 MHz) for direct long-distance voice or Morse code contacts, you can pass an easy five words per minute (WPM) Morse code test to qualify for the Technician Class license. Finally, you can earn even more amateur operating privileges by passing General and Extra Class license tests. Remember, you can take any of these tests from qualified hams in your area at a time convenient to you and the ham volunteer examiner.

About Amateur Radio

We live in a world of fantastic technological developments that affect almost every phase of our daily lives. Space exploration, high-speed jet aircraft, and advanced developments in medicine, digital computers, and worldwide electronic communications are but a few of these achievements. Amateur radio offers a challenging entry into exploring much of the technology involved in many of these fields. Amateur radio operators have made many significant contributions to radio communications and electronics technology. In fact, many scientists, engineers, and even astronauts pursue amateur radio as a rewarding hobby. Many of NASA's space shuttle craft flights include Shuttle Amateur Radio Experiment (SAREX) contacts with amateur radio operators.

There are many fascinating aspects to amateur radio—talking to fellow hams, participating in emergency communications during disasters, studying electronics technology, or building ham gear and antennas using state-of-the-art electronic components. Many hams use amateur radio as a steppingstone to a rewarding career in electronics.

But wait—this marvelous technology is only about 100 years old. Before we get into the details of amateur radio and how to pass the FCC exam for an amateur radio license, let's look into the fascinating history of amateur radio.

A History of Amateur Radio

A history of amateur radio would not be complete without acknowledging some of the early discoveries in the field of electricity and magnetism. From the dawn of history, man has been fascinated by the effects of electricity and magnetism produced by nature such as lightning, static electricity, and magnets.

Early pioneers

One of the first major scientific breakthroughs was made by Michael Faraday, an English physicist and the son of a blacksmith, in the early nineteenth century. Although he made many discoveries in several scientific fields, Faraday's major achievement was the discovery of electromagnetic induction and the formulation of the laws of induction. Today we know this as the process where electrons move in a conductor when the conductor is moved through a magnetic field. This discovery led to the development of the electric generator and motor.

In 1873 James Clerk Maxwell, a Scottish physicist-astronomer, mathematically proved the existence of electromagnetic (or radio) waves traveling at the speed of light. Now the stage was set to prove the existence of radio waves in the laboratory.

About 15 years after Maxwell's investigation (and 5 years after his death), Heinrich Hertz, a German physicist, demonstrated that radio waves could be generated and transmitted over short distances of up to about 60 feet (or about 20 meters). With his crude laboratory apparatus, Hertz was able to measure the wavelength of the waves he generated and show that these waves could be reflected, refracted, and polarized just as light waves are. Working in the 150-MHz-and-above radio spectrum, he designed and built spark-gap transmitters, resonator circuits for receiving radio waves, and directional antennas. As you will see later, 1 MHz is the expression of radio frequency for 1 million Hertz, or 1 million cycles per second. This frequency of 150 MHz is just above the amateur 2-meter band at 144–148 MHz.

In the early 1890s Guglielmo Marconi, an Italian inventor, began to experiment with radio waves using equipment similar to that developed by Hertz and other scientists of the era (Fig. 1.1). Marconi made many improvements and inventions that resulted in extending the range of radio transmissions. He conceived the concept of a vertical radiating antenna, and the Marconi (or vertical) antenna is one of his major accomplishments.

Marconi's first crude equipment was capable of a range of about one-half mile (about 800 meters). In 1896 he moved from Bologna, Italy, to England, where he made substantial improvements that increased the operating range of his equipment to about 4 miles (about 6.5 kilometers). By 1898, he succeeded in transmitting "wireless" signals across the English Channel. A major milestone was reached in 1901 when Marconi and his English associates transmitted radio waves across the Atlantic Ocean from Poldu, England, to Halifax, Newfoundland. Thus born, long-range radio communications would have an impact upon the lives of all people regardless of nationality or position in life.

The first major use of the new wireless telegraph sets was to provide for maritime communications. By 1905, spark-gap transmitters and coherer detector receivers were installed on many of the merchant ships and naval vessels on the high seas. For the first time, instant

FIGURE 1.1

Guglielmo Marconi with one of his early wireless sets. This photograph, taken in 1902, illustrates the induction spark coil transmitter (*right*), the "grasshopper" handkey (*center*), and the receiving apparatus (*left*).

communications between remote ships and land-based communications centers were feasible. All of this was accomplished without the use of vacuum tubes or transistors—they would be invented later to provide for amplification of the weak radio signals intercepted by the antennas.

Beginning of amateur radio

The introduction of commercial wireless telegraph equipment after the turn of the century aroused the imagination and interest of people around the world. Some experimenters were content to build simple crystal detector receivers and monitor the raspy code signals transmitted from fixed or rotary spark transmitters at marine or government communications stations. Other experimenters, particularly the restless and aggressive youngsters, assembled spark-gap transmitters, as well as crystal detector receivers, and began sending "dots and dashes" between home separated by a few miles (Fig. 1.2). The range of these early "amateur" stations was increased by continual improvements in equipment and higher power output. Amateur wireless organizations

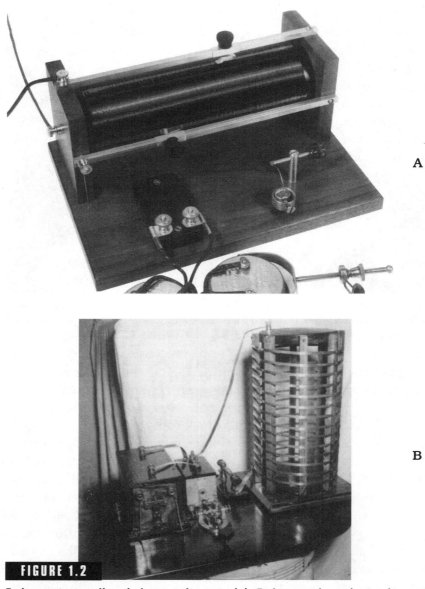

A

B

FIGURE 1.2

Early amateur radio wireless equipment. (*a*) Early crystal receiver using a 2-side antenna tuning inductor, "Cat Whisper" detector, phone blocking capacitor, and head-phones. (*b*) 1914 spark-gap transmitter, employing a vibrator-type spark coil, Dubilier fixed capacitor, Bunnel telegraph key (*in front*), Murdock straight spark gap, and home-made helix coil. These are representative of amateur radios used around 1914. Under favorable conditions, the transmitter could be heard for distances up to about 30 miles.

were formed across the country, beginning with the Junior Wireless Club of New York in January 1909. Amateur radio had come of age!

During radio's infancy, no government rules or regulations were in effect to govern the use of wireless operations. Wavelengths, or operating frequencies, in the vicinity of 300 to 1000 meters were selected by the users on the basis of available equipment. The inevitable conflict between commercial and amateur users abruptly surfaced when interference from amateur transmissions threatened the reliability of commercial radio communications. The U.S. Navy, which was quickly developing radio communications facilities, addressed some of the administrative problems and began issuing "certificates of skill in radio communications" in lieu of licenses. By late 1910, the Navy had issued some 500 of these certificates, many to amateur operators.

By this time, the number of individuals interested in or participating in amateur radio had grown to an estimated 10,000 or more. Amateur transmitters with power outputs of several kilowatts could be heard up to 400 miles (about 650 kilometers) away. However, most amateurs could not afford such luxury and had to be content with ranges of about 5 miles (8 kilometers), with occasional contacts up to 100 miles (160 kilometers). Many wireless equipment stores had appeared by this time, selling crystal detectors, spark-gaps, induction coils, and tuners—the basic ingredients of wireless stations.

First licenses for amateur radio operators

Beginning in 1902, the U.S. Congress recognized the need for regulating the use of wireless operation. However, it was not until 1910 that the first bill (the Act of June 24, 1910) was passed. This bill required the mandatory use of wireless equipment on certain ocean steamers and did not apply to amateur radio. Some bills were introduced in the Congress that would have given all authority for radio communications to the government. If passed, any of these bills would have abolished amateur radio. Fortunately, these bills were defeated and amateur experimentation increased at a rapid rate.

Finally in 1912, Congress passed an all-encompassing bill covering all phases of radio communications in the United States. Amateur operation was restricted to wavelengths below 200 meters and maximum power levels of 1 kilowatt. The lawmakers believed that wavelengths below 200 meters were useless and that this restriction would eventually eliminate the troublesome amateur radio society.

This action by the government proved to be a gold mine for the amateur operators. Refinements in electronics technology and the introduction of the new deForest vacuum tube allowed the amateurs to build short-wave equipment capable of spanning distances of hundreds of miles with low power. Radio clubs were established in all parts of the country. Amateurs began to send personal messages for other individuals to distant cities and remote locations. Emergency communications during periods of disasters were now emerging as a major contribution by the amateurs.

By 1914, the American Radio Relay League (ARRL) was formed to promote the concept of national relaying of amateur traffic across the country. Through the dynamic leadership of Hiram Percy Maxim, the ARRL grew to become the largest amateur radio organization in the United States. By late 1914, efficient relay networks were organized over most of the eastern United States and many stations were dedicated to handling traffic as a public service. By 1915, the ARRL introduced *QST*, a radio amateur journal devoted solely to the pursuits of amateur radio. Today, the ARRL membership includes over 120,000 in North America and some 12,000 foreign and unlicensed associate members. You'll want to consider joining the ARRL to support amateur radio and to receive the monthly *QST*. This magazine features construction articles, technical information, news concerning amateurs and amateur meetings, etc. For more information, you can contact the American Radio Relay League at ARRL, 225 Main Street, Newington, CT 06111-1494; telephone: (860) 594-0200; or on the World Wide Web at: *http://www.arrl.org/*.

The death knell for amateur radio

All amateur radio operations were suspended when the United States entered World War I in 1917. The government order suspending all forms of amateur radio required that the aerial wires be lowered to the ground and all radio apparatus for transmitting and receiving be disconnected from the antennas and rendered inoperative.

The U.S. military forces were faced, at the beginning of the war, with virtually no radio operators, instructors, or technical specialists. The ARRL assisted the government in locating volunteers among the amateur ranks to meet immediate needs in the military. In addition, some of the equipment from the more elaborate amateur stations was released to the government for use by the military. Altogether, an estimated 3500 to 4000 amateur radio operators served in the U.S. Armed Forces, contributing their knowledge and expertise in radio communications to the war effort. When the armistice ending World War I was signed in November 1918, the amateurs were impatient to resume operation. However, the U.S. Congress was considering legislation that would eliminate all forms of amateur radio activity. This legislation would have granted control of all forms of radio communications to the Navy Department. Fortunately, amateur operators, led by ARRL President Maxim, descended upon Congress and mounted sufficient opposition to defeat the proposed bill. Control and regulation of radio communications was left to the Commerce Department. The wartime ban on amateur radio activity was finally lifted on November 1, 1919.

After 2½ years of silence, the amateurs were finally permitted to resume operation. There was a rush to obtain licenses and get back on the air. Some of the amateurs dusted off their old spark-gap transmitters and crystal-detector receivers for immediate operation. Others began to adapt to the new vacuum tubes for both transmitter and receiver use. Radio parts houses sprang up in all major urban areas and mail-order firms began to advertise regularly in the few available radio publications, such as *QST*.

The golden age of amateur radio

Beginning in the late 1920s, amateur radio experienced phenomenal growth in the numbers of operators and in radio communications technology. This growth continued until the United States entered World War II in 1941 and the amateurs were forced to cease operation for the duration of the war. During this period, the number of amateurs had increased from almost 17,000 in 1929 to over 54,000 in 1941.

The typical ham of the 1930s was a young high-school or trade-school graduate with little or no extra money with which to buy radio equipment. Fortunately, the coming of age of the commercial radio broadcasting field brought a bonanza of radio parts that could be adapted to amateur radio uses. Used AM broadcast receiving tubes were readily available at little cost for the amateur-built receivers and transmitters. The typical amateur transmitter consisted of a single-tube oscillator circuit. The power output of this simple rig varied from about 1 to 2 watts, depending on the power supply and the type of tube. A two-tube receiver with a regenerative detector and audio amplifier was used by most amateurs in the early 1930s. Dry-cell batteries or a junk-box "battery-eliminator" power supply normally served to power the transmitter and receiver. Today, you can build and use simple short-wave receivers and QRP (low-power) transmitters for contacts with other hams over distances of thousands of miles (Fig. 1.3). These easy-to-build radios are described in Chaps. 8 and 9.

This period of amateur radio saw many important technological advances. In fact, most of the techniques and tools used by the modern amateur were developed and tested during this time period. These developments included the superheterodyne receiver, single-sideband

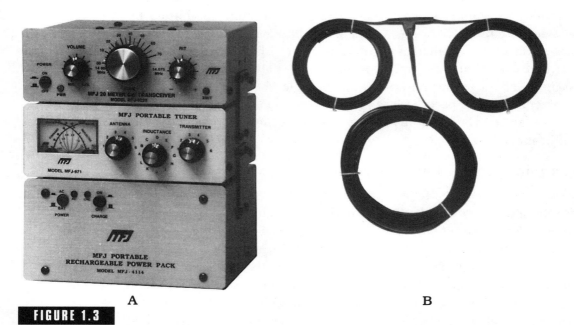

A   B

FIGURE 1.3

The MFJ enterprised low-power CW "QRP" 20-meter amateur radio station. (*a*) The MFJ-0120 20-meter transceiver, MFJ-971 antenna tuner and MFJ-4114 power pack. (*b*) The MFJ-1772-20 meter folded dipole antenna for fixed or portable operation. This compact ham station is capable of 4 watts output for virtually worldwide contacts. Other similar MFJ transceivers and antennas are available for 40-, 30-, and 17-, and 15-meter operations. [*MFJ Enterprises, Inc., Box 494, Mississippi State, MS 39762, Telephone (601) 323-5869*] modulation, frequency modulation, high-gain beam antennas, and vhf communications techniques. Amateur experimentation ranged from improving simple transmitters and receivers to pioneering developments in radio astronomy.

The old raspy sounds of the spark-gap transmitters had been replaced by the clear, crisp signals of continuous-wave (cw) transmitters and voice-modulated rigs. Amateur radio had become a permanent part of the American way of life.

Amateur Radio—A Scientific Hobby

Amateur radio today is a highly specialized hobby involving almost all phases of electronics. There are about 500,000 ham radio operators in the United States. Worldwide, there are some 700,000 amateurs, and almost all countries authorize some form of amateur radio. Hams come from all walks of life: for example, students, salesmen, plumbers, doctors, lawyers, engineers, seamen, and the handicapped.

The FCC, which administers and regulates all radio activity in the United States, defines the Amateur Radio Service as one of self-training, intercommunication, and technical investigation carried on by amateur radio operators.

The great appeal of amateur radio is centered on the incredible ability to communicate person-to-person worldwide and a fascination with the marvels of electronics. Hams constantly keep abreast of technical advances in electronics, and in some cases it is the ham who pioneers the development of some phase of radio or electronics.

Amateur radio activity today ranges from high-frequency (80 to 100 meters) communications [involving cw, single-sideband voice, data (computer-to-computer and teletypewriter), and slow-scan television] to amateur communications satellites operating on frequencies from 28 MHz to 435 MHz and microwave frequencies (Figs. 1.4 and 1.5). Another space-age development is that of moon-bounce communications using the moon as a passive reflector for communications signals in the 144-MHz to 2450-MHz frequency bands. A recent development in the electronics field—digital microelectronics—is being adapted to amateur radio with increasing applications. The integrated-circuit (IC) chips containing logic gates, flip-flops, and even a complete "computer-on-a-chip" or digital processor, are

FIGURE 1.4

The OSCAR 10 radio communications satellite. One of a series of amateur satellites, OSCAR 10 was launched in June 1983. Amateurs on a worldwide basis are able to extend their communications using these satellites. Additional information can be obtained from AMSAT-NA, 850 Sligo Avenue, Silver Spring, MD 20910-4703; telephone: (301) 589-6062.

available to hams at bargain-basement prices. A typical IC chip—containing up to a hundred or more transistors, diodes, and resistors and forming many digital circuits—will cost the ham a mere $0.50 to several dollars each. Hams have adapted these chips to digital circuits ranging from electronic keyers to Morse code translators. Many ham "shacks" include a personal computer (PC) for maintaining station log records, propagation analysis, packet radio, locations of amateur satellites, and even automatic tracking of the station antennas for satellite communications. By the way, don't be alarmed by the subjects mentioned above—we'll cover all of them in this book.

Another exciting aspect of microelectronic circuits is the operational amplifier (op-amp) and related types of "analog" IC chips. The availability of the op-amp has allowed the amateur to design and build active filters for cw (Morse code) applications with bandwidths of less than 100 Hz. Other linear-circuit ICs containing many transistors are used as basic building blocks for receivers, modulators, and low-power transmitters. We will examine some of these devices in subsequent chapters and show you how you can build a complete ham station with a minimum of cost and effort.

Amateur Radio Public Service

One of the most important aspects of amateur radio is that of public service. The FCC states that one justification for the Amateur Radio Service is the recognition and enhancement of the value of the amateur service to the public as a voluntary noncommercial communi-

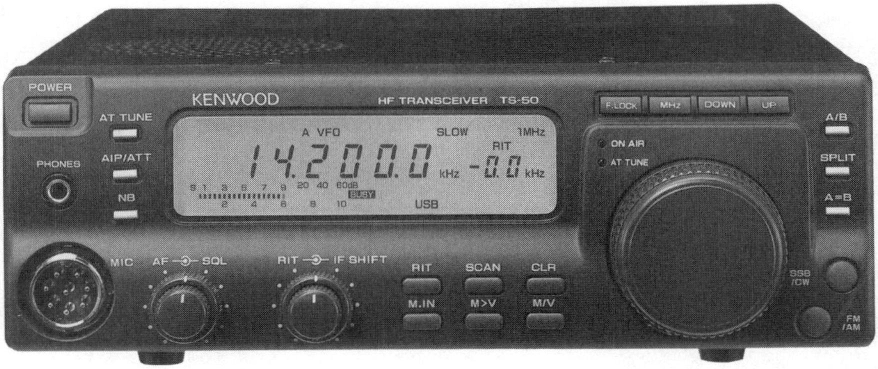

FIGURE 1.5

The Kenwood TS-50S/60S hf transceiver with 6-meter capability. The TS-50S/60 is one of the smallest and hottest hf mobile radios available today. Featuring most of the capabilities of the "big brother" hf radios, this compact mobile rig provides 100 watts power output, 100 memory channels, dual VFOs, split operation and a power menu system. [*Kenwood Communications Corporation, Amateur Radio Products, P.O. Box 22745, 2201 E. Dominguez St., Long Beach, CA 90801-5745; telephone: (310) 639-5300; Internet address: http://www.kenwood.net/.*]

cations service, particularly with respect to providing emergency communications.

One of the first disasters for which amateurs provided emergency communications was the heavy sleet storm that hit western and northern New York State in December 1929. The storm tore down telephone and power lines and isolated many cities in the area. Hams assisted telephone and power companies and the railroads in establishing emergency communications.

Since that time, hams have been instrumental in establishing emergency communications during natural disasters such as floods, earthquakes, fires, storms, explosions, train wrecks, and plane crashes. Although space does not permit a complete summary of these activities, some of the major disasters during which hams provided emergency communications were Hurricane Carla in Texas (September 1961), Hurricane Hugo in the Caribbean and Puerto Rico (September 1989), and the Loma Prieta earthquake (October 1989).

In addition to major disasters, hams have continually been in the forefront in providing communications in the event of personal emergencies such as a lost child, an automobile accident, or participation in an eye transplant bank.

Hams are organized in almost all communities to provide communications service to the public. Each year, hams participate in Field Day exercises during which they set up simulated emergency field communications centers powered by portable gasoline or diesel generators (Fig. 1.6). In addition to Field Day, hams across the country practice sending emergency traffic on Simulated Emergency Traffic (SET) Day.

The radio amateur's contribution to the country during war has been outstanding. In World War I, some 4000 hams served in the Armed Forces. Again in World War II, more than 24,000 hams served in the Army, Air Force, Navy, Coast Guard, and Marine Corps, contributing their skills and talent to the war effort. Countless numbers of other hams were engaged in electronics research and manufacturing activities supporting the war effort. During the Korean and Vietnam

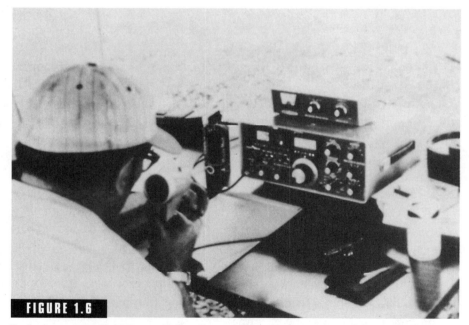

FIGURE 1.6

Typical amateur Field Day activity. Hams across the country participate in Field Day, held each year during the summer. The purpose is to develop and test emergency communications techniques for disasters such as floods, storms, or power blackouts.

conflicts, amateurs relayed many messages from service personnel overseas to their families and friends in the States. These traffic nets handled personal messages via relay stations located mostly on the West Coast. In many cases, the service personnel were able to talk directly to their loved ones in the United States using telephone patches established by the amateurs.

Another important area of public service by the amateur is in providing communications for expeditions to remote areas such as the Arctic and Antarctic. In 1923, Con Mix, 1TS (early call signs did not have a W, K, A, or N prefix) accompanied MacMillan to the Arctic aboard the schooner *Bowdoin*. Hams in Canada and the United States provided home contacts for the explorers. Since that time, amateurs have assisted some 200 expeditions. At present, amateurs provide much of the personal communications for the personnel at the Antarctic research stations.

The Path into Amateur Radio—The Technician Class License

Prior to April 15, 2000, the beginner in amateur radio could select the Novice Class, Technician (Tech-Plus) Class, or the No-Code Technician Class License. Today, the entry-level license is the Technician Class, which requires only a 35-question written examination. The operating privileges for the Technician cover all amateur bands above 50 MHz. This includes the 6-meter band (50–54 MHz), the 2-meter band (144–148 MHz), and the 70-cm band (420–450 MHz). Additional hf operating frequencies are available when the Technician Class amateur operator passes the 5-WPM Morse code test. Tables 1.1 and 1.2 present the key information on these Technician operating bands.

With this preliminary background, let's get started on the basics of amateur radio. We want you to learn the theory of ham radio as well as a good proficiency in Morse code as soon as possible. In this way, you can schedule and pass the Technician examination in a minimum of time. Good luck and good contacts!

TABLE 1.1 Technician Operating Bands, Privileges, and Normal Communications Capability

Operating Band	Frequencies and Operating Privileges	Communications Capability
6 meters	50–54 MHz: Morse code (cw), single sideband voice (SSB), frequency modulation voice (SSB), radio printer (RTTY), FM repeaters, radio control (RC), experimental modes and signal beacons	This band features almost all modes of propagation as it is situated between the hf and vhf frequency bands. During times of high sunspot activity, daytime contacts of up to 3200 km (2000 miles) are possible via skywave propagation. Routine contacts of up to 500 km (300 miles) are possible with high-power transmitters and high-gain antennas. Sporadic E-layer propagation is a favorite of 6-meter hams, providing short band openings at distances of up to about 3000 km (1200 miles).
2 meters	144–148 MHz: Most of the above privileges plus: amateur satellite operation, earth-moon-earth (EME or "moon bounce"), packet (digital) radio	Most hams use the 2-meter band for local repeater operation. Most metropolitan areas will have literally dozens of 2-meter repeaters. The higher frequencies permit smaller antennas and inexpensive repeater equipment. The 2-meter mobile for automobiles (and even airplanes and boats) and hand-held (HT, or walkie-talkies) permit new hams to get on the air with a minimum of equipment and small, inexpensive antennas. Some hams do build elaborate 2-meter stations that can provide long-range contacts of up to about 2000 km (1200 miles) via sporadic E-layer propagation and tropospheric ducting. Finally, the "line of sight" propagation mode of 2-meter frequency signals is used extensively in amateur ratio satellite and earth-moon-earth (EME) operations.

How to Prepare for the FCC Technician Class Examination

The Entrance to the World of Amateur Radio Is Easy

The FCC has established the Technician license for the beginner in amateur radio. Furthermore, the FCC authorizes local volunteer examiners to administer the examinations for the Technician Class license and higher class licenses such as the General Class and Extra Class licenses. As stated in Chap. 1, the Technician Class license provides for all amateur radio operating privileges above 50 MHz. To qualify for selected operating privileges in the amateur HF bands, you must pass a 5 words per minute (WPM) Morse code test.

The Technician Class license examination has 35 questions, and passing requires a minimum of 75 percent (or a total of 27) correct answers. This written examination covers radio theory, FCC rules and regulations, and operating procedures—all the knowledge you will need to set up and operate your ham station. As a general rule, you can master the requirements for the Technician Class license with two or three months of dedicated study. Considerable less time (six to eight weeks) will be sufficient if you are fortunate to enroll in local "Code and Theory" classes handled by amateur radio clubs or civic organizations.

To prepare for the Technician examination, the beginner in amateur radio can obtain any necessary help from a variety of sources. One of the more successful approaches is to enroll in a Technician class sponsored by a local ham club or other organizations, such as the YMCA or a church group. These courses give you the advantage of studying with your peers in an educational environment. Most of these courses will run for about 7 to 10 weeks with one or two class periods per week. Each period may last for about 2 to 4 hours. The final class period is usually a test session whereby the volunteer examiner (VE) team is scheduled to hold an open test session. Classes sponsored by the local ham club are usually very inexpensive, sometimes only the cost of the course text and a year's club membership dues. If no local ham course is available, the neighborhood ham (or "Elmer" as they are affectionately called) will be pleased to help you prepare for your Technician exam.

Self-study is another approach for learning the material required for the Technician examination. This book covers all of the material contained in the Technician examination—namely, FCC rules and regulations, electronics theory, ham receivers and transmitters, antennas and propagation, and radio communications practices and operating procedures. Furthermore, this book contains the actual FCC question pool from which the 35 Technician examination questions are taken. Yes, you actually have access to the Technician question pool, and answers to each question are available at the end of this book (see Appendix A).

Additional study guides for amateur radio examinations are available from various sources. For example, the amateur radio organizations listed below offer study texts, videotapes, and computer software packages designed to make learning the required material more efficient (Fig. 2.1).

QST. A monthly magazine published as the official journal of the American Radio Relay League, 225 Main Street, Newington, CT 06111-1494. For more information call (800) 326-3942, or e-mail: newham@arrl.org. Web page: *http://www.arrl.org*

73 Amateur Radio Today. A monthly magazine published by 73 Magazine, 70 N202, Peterborough, NH 03458-1107. For more information call (603) 924-0058. Web page: *www.waynegreen.com*

The W5YI Group, Inc. Offers ham test prep tapes, books, software, and videos. Contact the W5YI Group at P.O. Box 565101, Dallas, TX 75356, or call toll-free 1-800-669-9594. Web page: *www.w5yl.org*

The HANDIHAM World. Published three times annually by the Courage HANDIHAM System, an organization devoted since 1967 to amateur radio for persons with physical disabilities and sensory impairments. 3915 Golden Valley Road, Golden Valley, MN 55422. Telephone:

(763) 520-0511 (voice), (763) 520-0577 (fax), or e-mail: handiham@mtn.org. Additional information may be obtained from the HANDIHAM Web site: *http://www.handiham.org*

When you are ready to take the Technician (or higher) Class license examination, a local VE team consisting primarily of EXTRA Class hams, will be available to schedule the examination. Most VE teams regularly schedule VE examinations on a monthly or semimonthly basis. For further information on the VE test teams, contact the local amateur radio club or ARRL at the above address. Practically everyone in the United States, regardless of age, is eligible to take the FCC amateur radio examinations. This includes aliens who have a valid U.S. address and are not representatives of a foreign government.

The Federal Communications Commission

The FCC, established by the Communications Act of 1934, regulates and licenses all

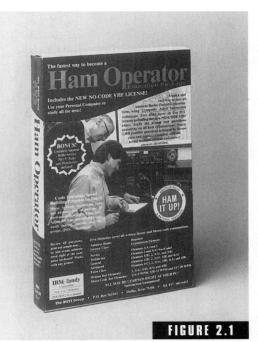

FIGURE 2.1

The W5YI Computer Aided Instruction Software Package for all Amateur Radio Morse code and written elements, including Technician and up to Extra Class. Probably the fastest way to learn Morse code, and study all of the questions that appear on the various ham radio written tests, is with a personal computer. This Ham Operator software package contains material that covers everything, including teaching code and all questions in every question pool.

radio transmitters in the United States and aboard all U.S.-registered ships and aircraft. Unlike many other countries, the U.S. does not license the possession of radio receivers. The "bible" of radio laws that governs all FCC activity contains some 40 parts. The Amateur Radio Service is covered in Part 97, Rules and Regulations. A copy of Part 97 can be obtained from the ARRL, local ham radio and book stores, or the FCC. Also, Part 97 can be downloaded from the Government Printing Office at www.gpo.gov.

The structure of the FCC includes a headquarters located in Washington, DC, a license and call-sign processing facility in Gettysburg, Pennsylvania, some 16 radio districts or field offices, and some 13 FCC monitoring stations. For more information concerning any FCC radio district or monitoring station, contact the Public Service Division, Federal Communications Commission, 445 12th Street SW, Washington, DC 20554 [telephone: (202) 418-0200, fax: (202) 418-2555].

The FCC Volunteer Examiner Program

The FCC has "deregulated" the amateur radio service in terms of amateur radio license examinations. In effect, the responsibility for administering all examinations for amateur radio licenses has been transferred to the amateur radio community. The chances are good that any local ham club will be affiliated with the ARRL and will be authorized by the ARRL to administer all amateur radio license examinations, including the Technician Class exam.

Examinations for the Technician and higher-class (General and Extra) amateur radio licenses are administered by amateur radio VE teams. Each VE team consist of a minimum of three qualified volunteer examiners, certified by a volunteer examiner coordinator (VEC) authorized by the FCC. The ARRL and the W5YI Group are representative of the VECs. Each VEC must maintain a question pool (available to the public), prepare and administer license examinations, make public announcements for examination schedules, and qualify VE personnel.

The VEC entity must be organized, at least partially, for the purpose of furthering amateur radio, and agree not to accept any compensation from any source for its services. However, in order to defray the expense of preparing and administering tests, the VEC is authorized to charge a maximum fee (on the order of $6) for each applicant.

FCC Amateur Radio Operating Classes

There are currently five classes of amateur radio licenses, which provide increasing privileges at each step forward. However, the recent FCC changes to the amateur license structure call for the eventual elimination of the Novice and Advanced Class Licenses. Current licensed Novice and Advanced amateur operators may continue to operate on the authorized amateur frequency bands or elect to upgrade to higher-class licenses. No new FCC examinations for the Novice and Advanced class licenses will be permitted. The Technician Class license is now the entry level into amateur radio. As a Technician

Class amateur operator, you will enjoy all amateur privileges above 50 MHz. No Morse code test is required for the Technician Class license. However, you must pass a 35-question written examination (Element 2) to receive this license. Technician Class privileges can be expanded to selected portions of the HF bands by passing the Element 1, a 5-WPM Morse code test.

The second step in amateur radio is the General Class license. This license requires that you pass a second 35-question written exam (Element 3) and have credit for the 5-WPM Morse code test (Element 1). In return, you will receive operating privileges for all or portions of all amateur bands. This includes the exciting hf bands between 3 and 30 MHz.

The top amateur license, the Amateur Extra Class, requires passing a 50-question written exam (Element 4) on advanced topics. This "top prize" in amateur radio gives you all the operating privileges in all amateur bands. Keep this goal in mind as you progress through the ranks of amateur radio.

A summary of the amateur radio license requirements and privileges is given in Table 2.1. Note that the Technician Class license permits the beginner to obtain valuable experience in virtually all types of amateur radio communications.

The Amateur Radio Frequency Spectrum

The amateur radio frequency begins just above the AM broadcast band (540–1650 kHz) at 1800 kHz and contains small "chunks" of bandwidth or bands of frequencies up to 300 GHz and beyond. At this time, the amateur portion of the radio frequency spectrum is probably the largest of any single user organization. However, as electronic techniques and development of communications equipment for the higher frequency bands advance, commercial and government organizations will find ways to use these bands.

One of the strengths of amateur radio that attracts so many people is the abundance and diverse nature of the frequency spectrum allocated to amateur radio operators. Amateurs can build and/or experiment with a wide variety of antennas, radio receivers, radio transmitters, computers, and other communications systems that operate from the 1800–2000 kHz band to the microwave frequency bands (tens of thousands of megahertz near the optical frequencies of light).

TABLE 2.1 FCC Amateur Radio Licenses, Examination Requirements, and Operating Privileges

Operator license class	Morse code examination	Written examination	Operating privileges
Technician (entry level)	None	Element 2 (35 questions)	All amateur privileges above 50 MHz. This includes specific frequency bands up to 300 GHz and beyond
Technician with Morse code	Element 1 5-WPM test	Element 2 (35 questions)	All privileges above 50 MHz and privileges on portions of selected amateur HF bands below 30 MHz.
General	Element 1 5-WPM test	Element 3 (35 questions)	All privileges above 50 MHz and privileges on portions of all amateur HF bands below 30 MHz.
Amateur Extra	Note 1	Element 4 (50 questions)	All privileges on all amateur bands.

Each amateur radio band exhibits a unique set of characteristics, requiring special techniques for designing and building antennas, transmission lines, transmitters, and receivers. Some amateur radio bands permit reliable communications on a worldwide basis, while others are limited primarily to "line-of-sight" propagation. Some amateurs specialize in the HF bands (3–30 MHz) for long-range DX contacts involving single sideband (SSB) voice, Morse code (CW) communications, or amateur slow-scan television. Some of these amateurs use low-power "QRP rigs" of only a few watts to work with other amateurs over long distances. Many of these QRP rigs are easily built and tested. Other amateurs experiment with digital data transmissions involving computer-to-computer operation and slow-scan television systems in the VHF and UHF frequency bands. You will probably want to investigate many of these exciting aspects of ham radio.

Figure 2.2 shows the total amateur radio frequency spectrum, including the Technician Class license operating privileges. Many of the amateur bands are assigned on approximate octave intervals throughout the spectrum. For example, the 40-meter band is approximately twice the frequency of the 80-meter band. In many instances, this octave relationship aids in the design and construction of multi-band radio transmitters, receivers, and even antennas. Additional information concerning the methods of modulation and propagation characteristics is discussed in Chap. 3. For a more detailed and up-to-date listing of amateur frequencies and operating privileges, refer to the current edition of the *ARRL Handbook for the Radio Amateur.*

FCC Rules and Regulations

Part 97: Amateur Radio Service

Basic law comprising rules and regulations essential to the operation of amateur radio stations is contained in the Rules and Regulations of the FCC, Part 97: Amateur Radio Service. As stated earlier, a copy of Part 97 may be obtained from the U.S. Printing Office or directly from the FCC. You are encouraged to obtain a copy of this document.

We will review Part 97 in terms of the Technician Class license examination and subsequent Technician station operation, and provide selected portions of Part 97. You should become completely familiar with these rules and regulations prior to taking the written examination. Virtually all the FCC rules presented here relate to questions that will appear on the Element 2 exam.

BASIS AND PURPOSE.

This section expresses the justification for the existence of the Amateur Radio Service. You may have wondered why the Amateur Radio Service controls so much of the valuable radio frequency spectrum. This justification is based on the following five functions of amateur radio.

97.1 Basis and purpose. The rules and regulations in this part are designed to provide an amateur radio service having a fundamental purpose as expressed in the following principles:

(a) Recognition and enhancement of the value of the amateur radio service to the public as a voluntary noncommercial service particularly with respect to providing emergency communications.

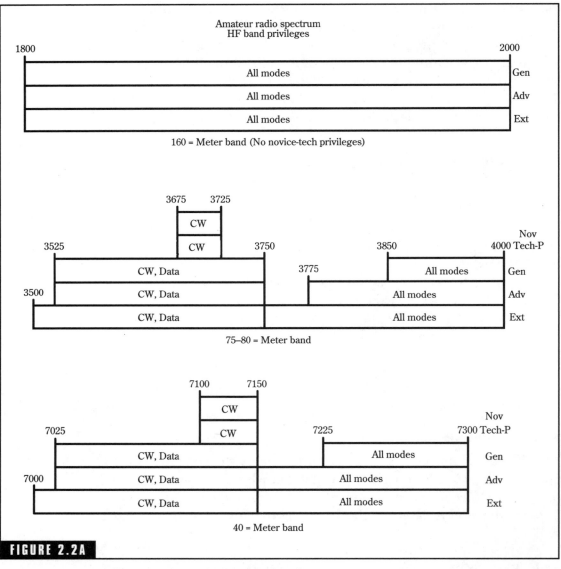

FIGURE 2.2A

Amateur radio frequency spectrum and operating privileges.

(b) Continuation and extension of the amateur's proven ability to contribute to the advancement of the radio art.

(c) Encouragement and improvement of the amateur radio service through rules which provide for advancing skills in both the communication and technical phases of the art.

(d) Expansion of the existing reservoir within the amateur radio service of trained operators, technicians, and electronics experts.

(e) Continuation and extension of the amateur's unique ability to enhance international good will.

DEFINITIONS

The definitions of terms essential to the operation of amateur radio stations are contained in Part 97.3. Almost all Technician examinations will include one or more of these definitions.

97.3 Definitions.

(a) The definitions of terms used in Part 97 are:
   (1) Amateur operator. A person holding a written authorization to be the control operator of an amateur station.
   (2) Amateur radio service. The amateur service, the amateur-satellite service and the radio amateur civil emergency service (RACES).
   (3) Amateur-satellite service. A radiocommunications service using stations on Earth satellites for the same purpose as those of the amateur service.
   (4) Amateur service. A radiocommunications service for the purpose of self training, intercommunication and technical investigations carried out by amateurs, that is, duly authorized persons interested in radio technique solely with a personal aim and without pecuniary interest.
   (5) Amateur station. A station in an amateur radio service consisting of the apparatus necessary for carrying on radio communications.
   (6) Automatic control. The use of devices and procedures for control of a station when it is transmitting so that compliance with the FCC Rules is achieved without the control operator being present at a control point.

(7) Auxiliary station. An amateur station, other than in a message forwarding system, that is transmitting communications point-to-point within a system of cooperating amateur stations.

(8) Bandwidth. The width of a frequency band outside of which the mean power of the transmitted signal is attenuated at least 26 dB below the mean power of the transmitted signal within the band.

(9) Beacon. An amateur station transmitting communications for the observation of propagation and reception or other related experimental activities.

(10) Broadcasting. Transmissions intended for reception by the general public, either direct or relayed.

(12) Control Operator. An amateur operator designated by the licensee of a station to be responsible for the transmissions from that station to assure compliance with the FCC Rules.

(13) Control Point. The location at which the control operator function is performed.

(14) CSCE. Certificate of successful completion of an examination.

(16) Earth Station. An amateur station located on or within 50 km of the Earth's surface intended for communications with space stations or with other Earth stations by means of one or more objects in space.

(17) EIC. Engineer in Charge of an FCC Field Facility.

(20) FAA. Federal Aviation Administration

(22) Frequency coordinator. An entity, recognized in a local or regional area by amateur operators whose stations are eligible to be auxiliary or repeater stations, that recommends transmitter/receive channels and associated operating and technical parameters for such stations in order to avoid or minimize potential interference.

(23) Harmful interference. Interference which endangers the functioning of a radionavigation service or other safety services or seriously degrades, obstructs, or repeatedly interrupts a radiocommunication service operating in accordance with the Radio Regulations.

(24) Indicator. Words, letters or numerals appended to and separated from the call sign during the station identification.

(25) Information bulletin. A message directed only to amateur operators consisting solely of subject matter of direct interest to the amateur service.

(26) International Morse Code. A dot-dash code as defined in International Telegraph and Telephone Consultative Committee (CCITT) Recommendation F.1 (1984), Division B.I. Morse Code.

(27) IARP. International Amateur Radio Permit
_____

(30) Local control. The use of a control operator who directly manipulates the operating adjustments in the station to achieve compliance with the FCC Rules.
_____

(37) RACES (radio amateur civil emergency service). A radio service using amateur stations for civil defense communications during periods of local, regional, or national civil emergencies.

(38) Remote control. The use of a control operator who indirectly manipulates the operating adjustments in the station through a control link to achieve compliance with the FCC Rules.

(39) Repeater. An amateur station that simultaneously retransmits the transmission of another amateur station on a different channel or channels.

(40) Space station. An amateur station located more than 50 km above the Earth's surface.
_____

(42) Spurious emission. An emission, on frequencies outside the necessary bandwidth of a transmission, the level of which may be reduced without affecting the information being transmitted.
_____

(46) Third-party communications. A message from the control operator (first party) of an amateur station to another amateur station control operator (second party) on behalf of another person (third party).

(47) ULS (Universal Licensing System). The (FCC) consolidated database, application filing system, and processing system for all Wireless Telecommunications Services.

(48) VE. Volunteer examiner.

(49) VEC. Volunteer-examiner coordinator.

(b) The definition of technical symbols used in this Part are:

(1) EHF (extremely high frequency). The frequency range of 30–300 GHz.[1]

(2) HF (high frequency). The frequency range 3–30 MHz

(3) Hz. Hertz

(4) m. Meters

(5) MF. (medium frequency). The frequency range 300–3000 kHz.

(6) PEP (peak envelope power). The average power supplied to the antenna transmission line by a transmitter during one RF cycle at the crest of the modulation envelope taken under normal operating conditions.

(7) RF. Radio frequency

(8) SHF (super-high frequency). The frequency range is 3–30 GHz.

(9) UHF. (ultra-high frequency). The frequency range is 300–3000 MHz.

(10) VHF (very high frequency). The frequency range is 30–300 MHz.

(11) W. (watt)[2]

---

[1]The frequency of 1 Hertz (Hz) is defined as one cycle per second. Voice, music, radio transmission, AC power lines, and other signals are usually given in terms of operating frequencies. For example, voice frequencies normally range from about 300 Hz to 3000 Hz. Power-line frequency in the United States is 60 Hz while those in countries such as England and France are 50 Hz. You need to learn the frequency range definitions to help you pass the FCC written examinations as well as their use during amateur operations. You will find that the frequencies of amateur bands can be expressed in kilohertz (kHz), megahertz (MHz), or gigahertz (GHz), depending on the particular amateur band.

1 kHz = 1000 Hz.

1 MHz = 1000 KHz = 1,000,000 Hz.

1 GHz = 1000 MHz = 1,000,000,000 Hz.

[2]The watt is a unit of electrical power.

Part 97.3(c) defines the standard terms used to describe signal emission types from amateur radio transmitters. The signal types authorized for Technicians are given as follows:

1. **CW.** Sometimes referred to as "continuous wave," CW is defined as the International Morse Code telegraphy, usually transmitted from a radio transmitter via an operator using a hand key or automatic keying device, sometimes generated by a computer.

2. **RTTY.** RTTY or radioteletypewriter direct-printing telegraph emissions provide for transmission of mostly textual information. Older mechanical teletypewriter equipment has a typewriter-like keyboard to generate messages composed primarily of alpha and numeric characters along with generation of punctuation symbols and end-of-line and line-feed control characters. RTTY electrical signals, generated by the keyboards and connected to a printer to print messages, consist of digital codes representing characters such as letters or numbers. Only FCC specified codes (such as the old 5-unit Baudot or the 7-unit ASCII codes) can be used for transmitting RTTY signals. The newer personal computers can be used to generate and receive RTTY signals. We'll cover more on this subject in Chap. 11.

3. **Data.** Data represents a broad class of telemetry, telecommand, or computer communications emissions. Here's where the Technician operators can use their PCs for such exciting transmission modes as Packet and AMTOR. In general, the amateur operator is restricted to the use of FCC-approved digital codes.

4. **Phone.** Phone, or voice communications, allows the Technician operator to talk to other hams via local or long-distance contacts. The Technician with 5 WPM certification, for example, can use single-sideband voice on specified frequencies in the 10-meter band and all voice modes on all amateur bands above 50 MHz.

5. **Image.** Image emission types include facsimile and television signals. Technician operators can experiment with these modes of emission on all amateur bands above 50 MHz.

6. **Spread spectrum.** Spread-spectrum emissions involving "frequency hopping" over a wide band of frequencies are authorized on

some of the higher UHF bands. This is a relatively new technology for amateur operation.

AMATEUR STATION/OPERATOR LICENSE REGULATIONS

Part 97 contains the following FCC regulations concerning amateur operator and station license requirements. You will find that the FCC Amateur Radio License actually covers the operator privileges and the station authorization—two separate licenses on one slip of paper! The operator portion authorizes you to be the "control operator" of your amateur station or another amateur's station provided the station owner gives you permission.

The combination FCC amateur license also authorizes you to have your own amateur radio station.

Part 97.5 Station license required.

(a) The station apparatus must be under the physical control of a person named in an amateur station license grant on the ULS consolidated license database or a person authorized for alien reciprocal operation by Part 97.107 of this part, before the station may transmit on any amateur service frequency from any place that is:
  (1) Within 50 km of the Earth's surface and at a place where the amateur service is regulated by the FCC;
  (2) Within 50 km of the Earth's surface and aboard any vessel or craft that is documented or registered in the United States; or
  (3) More than 50 km above the Earth's surface aboard any craft that is documented or registered in the United States.

(b) The types of station license grants are:
  (1) An operator/primary station license grant. One, but only one, operator/primary station license grant may be held by any one person. The primary station license is granted together with the amateur operator license. Except for a representative of a foreign government, any person who qualifies by examination is eligible to apply for an operator/primary station license grant.
  (2) A club station license grant. A club station license grant may be held only by the person who is the license trustee desig-

nated by an officer of the club. The trustee must be a person who holds an Amateur Extra, Advanced, General, Technician Plus, or Technician operator license grant. The club must be composed of at least four persons and must have a name, a document of organization, management, and a primary purpose devoted to amateur service activities consistent with this part.

(3)   A military recreation station license grant. A military recreation station grant may be held only by the person who is the license custodian designated by the official in charge of the United States military recreation premises where the station is situated. The person must not be a representative of a foreign government. The person need not hold an amateur license grant.

(4)   A RACES station grant license. A RACES station license grant may be held only by the person who is the license custodian designated by the official responsible for the government agency served by that civil defense organization. The custodian must be the civil defense official responsible for coordination of all civil defense activities in the area concerned. The custodian must not be a representative of a foreign government. The custodian need not hold an amateur operator license grant.

Part 97.7   Control operation required.   When transmitting, each amateur station must have a control operator. The control operator must be a person:

(a)   For whom the amateur operator/primary station license grant appears on the ULS consolidated license database, or

(b)   Who is authorized for alien reciprocal operation by Part 97.107 of this part.

Part 97.9   Operator license.

(a)   The classes of amateur operation license grants are: Novice, Technician, Technician Plus (until such licenses expire, a Technician Class license before February 14, 1991, is considered a Technician Plus Class license), General, Advanced, and Amateur Extra. The person named in the operator license grant is authorized to be the control operator of an amateur station with the privileges authorized to the operator class specified on the license grant.

(b) The person named in an operator license grant of Novice, Technician, Technician Plus, General, or Advanced Class who has properly submitted to the administering VEs an FCC Form 605 document requesting examination for an operator license grant to a higher class, and who holds a CSCE indicating that the person has completed the necessary examinations within the previous 365 days, is authorized to exercise the rights and privileges of the higher operator class until final disposition of the application or until 365 days following the passing of the examination, whichever comes first.

Part 97.13   Restrictions on station location.

(a) Before placing an amateur station on land of environmental importance or that is significant in American history, architecture, or culture, the licensee may be requested to take certain actions prescribed by Para 1.1305-1.1319 of the Communications Act of 1934, as amended.

(b) A station within 1600 meters (1 mile) of an FCC monitoring facility must protect that facility from harmful interference.

(c) Before causing or allowing an amateur station to transmit from any place where the operation of the station could cause human exposure to RF electromagnetic field levels in excess of those allowed under Part 1.1310 of this chapter, the licensee is required to take certain actions:

   (1) The licensee must perform the routine RF environmental evaluation prescribed by Para 1.1307(b) of this chapter, if the power of the licensee's station exceeds the limits given in the following table:

Part 97.21 Application for a modified or renewed license.

(a) A person holding a valid amateur station license:

(1) Must apply to the FCC for a modification of the license grant as necessary to show the correct mailing address, licensee name, club name, license trustee name or license custodian name in accordance with Part 1.913 of this chapter. For a club, military recreation, or RACES station license grant, it must be presented in document form to a Club Station Call Sign Administrator who must submit the information thereon to the FCC in an electronic batch file. The Club Station Call Sign Administrator must retain the collected information for at least 15 months and make it available to the FCC upon request.

(2) May apply to the FCC for a modification of the operator/ primary station license grant to show a higher operator class. Applicants must present the administering VEs with all information required by the rules prior to the examination. The VEs may collect all necessary information in any manner of their choosing, including creating their own forms.

(3) May apply to the FCC for renewal of the license grant for another term in accordance with Part 1.913 of this chapter. Application for renewal of a Technical Plus operator/primary station license will be processed as an application for renewal of a Technician Class operator/primary station license.

(b) A person whose amateur station license grant has expired may apply to the FCC for renewal of the license grant for another term during a two-year filing grace period. The application must be received at the address specified above prior to the end of the grace period. Unless and until the license grant is renewed, no privileges in this Part are conferred.

Part 97.23 Mailing address. Each license grant must show the grantee's correct name and mailing address. The mailing address must be in an area where the amateur service is regulated by the FCC and where the grantee can receive mail delivery by the United States Postal Service. Revocation of the station license or suspension of the operator license may result when correspondence from the FCC is returned as undeliverable because the grantee failed to provide the correct address.

Part 97.25 License term. An amateur service license is normally granted for a 10-year term.

Part 97.29 Replacement license grant document. Each grantee whose amateur station license grant document is lost, mutilated, or destroyed may apply to the FCC for a replacement in accordance with Para 1.913 of this chapter.

STATION OPERATING STANDARDS.

This section of Part 97 covers amateur radio station operating standards. You will need to master these standards to help you pass the FCC examinations and for effective operation of your amateur station.

Part 97.101 General standards.

(a) In all respects not specifically covered by FCC Rules each amateur station must be operated in accordance with good engineering and good amateur practice.

(b) Each station licensee and each control operator must cooperate in selecting transmitting channels and in making the most effective use of the amateur service frequencies. No frequency will be assigned for the exclusive use of any station.

(c) At all times and on all frequencies, each control operator must give priority to stations providing emergency communications, except to stations transmitting communications for training drills and tests in RACES.

(d) No amateur operator shall willfully or maliciously interfere with or cause interference to any radio communication or signal.

Part 97.103 Station licensee responsibilities.

(a) The station licensee is responsible for the proper operation of the station in accordance with FCC Rules. When the control operator is a different amateur than the station licensee, both persons are equally responsible for proper operation of the station.

(b)   The station licensee must designate the station control operator. The FCC will presume that the station licensee is also the control operator, unless documentation to the contrary is in the station records.

(c)   The station licensee must make the station and the station records available for inspection upon request by an FCC representative. When deemed necessary by an EIC to assure compliance with the FCC Rules, the station licensee must maintain a record of station operation containing such items of information as the EIC may require with Part 0.314(x) of the FCC Rules.

Part 97.105   Control operator duties.

(a)   The control operator must ensure the immediate proper operation of the station, regardless of the type of control.

(b)   A station may only be operated in the manner and to the extent permitted by the privileges authorized for the class of operator license held by the control operator.

Part 97.109   Station control.

(a)   Each amateur station must have at least one control point.

(b)   When a station is being locally controlled, the control operator must be at the control point. Any station may be locally controlled.

(c)   When a station is being remotely controlled, the control operator must be at the control point. Any station may be remotely controlled.

(d)   When a station is being automatically controlled, the control operator need not be at the control point. Only stations specifically designated elsewhere in this Part may be automatically controlled. Automatic control must cease upon notification by an EIC that the station is transmitting improperly or causing harmful interference to other stations. Automatic control must not be resumed without prior approval of the FCC.

(e) No station may be automatically controlled while transmitting third party communications, except a station transmitting a RTTY or data emission. All messages that are retransmitted must originate at a station that is being locally or remotely controlled.

Part 97.111   Authorized transmissions.

(a) An amateur station may transmit the following types of two-way communications:

  (1) Transmissions necessary to exchange messages with other stations in the amateur service, except those in any country whose administration has given public notice to such communications. The FCC will issue public notices of current arrangements for international communications.

  (2) Transmissions necessary to exchange messages with a station in another FCC-regulated service, while providing emergency communications.

  (3) Transmissions necessary to exchange messages with a United States Government station, necessary to providing communications with RACES.

  (4) Transmissions necessary to exchange messages with a station not regulated by the FCC, but authorized by the FCC to communicate with amateur stations. An amateur station may exchange messages with a participating United States military station during Armed Forces Day Communications Test.

(b) In addition to one-way transmissions specifically authorized elsewhere in this Part, an amateur station may transmit the following types of one-way communications:

  (1) Brief transmissions necessary to make adjustments to the station.

  (2) Brief transmissions necessary to establishing two-way communications with other stations.

  (3) Telecommand.

  (4) Transmissions necessary to providing emergency communications.

  (5) Transmissions necessary to assisting persons learning, or improving proficiency in, the international Morse code.

(6) Transmissions necessary to disseminate information bulletins.

(7) Transmission of telemetry.

Part 97.113 Prohibited transmissions.

(a) No amateur station shall transmit:

(1) Communications specifically prohibited elsewhere in this Part.

(2) Communications for hire or for material compensation, direct or indirect, paid or promised, except as otherwise provided in these rules.

(3) Communications in which the station licensee or control operator has a pecuniary interest, including communications on behalf of an employer. Amateur operators may, however, notify other operators of the availability for sale or trade of apparatus normally used in an amateur station, provided that such activity is not conducted on a regular basis.

(4) Music using a phone emission except as specifically provided elsewhere in this Section; communications intended to facilitate a criminal act; messages in codes or ciphers intended to obscure the meaning thereof, except as otherwise provided herein; obscene or indecent words or language; of false or deceptive messages, signals, or identification.

(5) Communications on a regular basis which could reasonably be furnished alternately through other radio services.

(b) An amateur station shall not engage in any form of broadcasting, nor may an amateur station transmit one-way communications except as specifically provided in these rules; nor shall an amateur station engage in any activity related to program production or news gathering for broadcasting purposes, except that communications directly related to the immediate safety of human life or the protection of property may be provided by amateur stations to broadcasters, for dissemination to the public where no other means of communication is reasonably available before or at the time of the event.

(c) A control operator of a club station may accept compensation as an incident of a teaching position during periods of time when an amateur station is used by that teacher as a part of classroom instruction at an educational institution.

(d) The control operator of a club station may accept compensation for the periods of time when the station is transmitting telegraphy practice or information bulletins, provided that the station transmits such telegraphy practice and bulletins for at least 40 hours per week and schedules operations on at least six amateur service MF and HF bands using reasonable measures to maximize coverage, where the schedule of normal operating times and frequencies is published at least 30 days in advance of the actual transmissions and where the control operator does not accept any direct or indirect compensation for any other service as a control operator.

(e) No station shall transmit programs or signals emanating from any type of radio station other than an amateur station, except propagation and weather forecast information intended for use by the general public and originating from United States Government stations and communications, including incidental music, originating on United States Government frequencies between a space shuttle and its associated Earth stations. Prior approval for shuttle retransmissions must be obtained from the National Aeronautics and Space Administration. Such retransmissions must be for the exclusive use of amateur operators. Propagation, weather forecasts, and shuttle retransmissions may not be conducted on a regular basis, but only occasionally, as an incident of normal amateur radio communications.

(f) No amateur station, except an auxiliary, repeater, or space station, may automatically retransmit the radio signals of other amateur stations.

Part 97.115 Third party communications.

(a) An amateur station may transmit messages for a third party to:
    (1) Any station within the jurisdiction of the United States.
    (2) Any station within the jurisdiction of a foreign government whose administration has made arrangements with the

United States to allow amateur stations to be used for transmitting international communications on behalf of third parties. No station shall transmit messages for a third party to any station within the jurisdiction of any foreign government whose administration has not made such an arrangement. This prohibition does not apply to a message for any third party who is eligible to be a control operator of the station.

(b) The third party may participate in stating the message where:
    (1)   The control party is present at the control point and is continuously monitoring and supervising the third party's participation.
    (2)   The third party is not a prior amateur service licensee whose license was revoked; suspended for less than the balance of the license term and the suspension is still in effect; suspended for the balance of the license term and relicensing has not taken place; or surrendered for cancellation following notice of revocation, suspension, or monetary forfeiture proceedings. The third party may not be the subject of a cease and desist order which relates to amateur service operation and which is still in effect.

(c) At the end of an exchange of international third party communications, the station must also transmit in the station identification procedure the call sign of the station with which a third party message was exchanged.

Part 97.119   Station identification.

(a) Each amateur station, except a space station or telecommand station, must transmit its assigned call station on its transmitting channel at the end of each communication, and at least every 10 minutes during a transmission, for the purpose of clearly making the source of the transmission from the station known to those receiving the transmissions. No station may transmit unidentified communications or signals, or transmit as the station call sign, any call sign not authorized to the station.

(b) The call sign must be transmitted with an emission authorized for the transmitting channel in one of the following ways:

(1) By a CW emission. When keyed by an automatic device used only for identification, the speed must not exceed 20 words per minute;

(2) By a phone emission in the English language. Use of a standard phonetic alphabet as an aid for correct station identification is encouraged;

(3) By a RTTY emission using a specified digital code when all or part of the communications are transmitted by a RTTY or data emission;

(4) By an image emission conforming to the applicable transmission standards, either color or monochrome, of Part 73.682(a) of the FCC Rules, when all or part of the communications are transmitted in the same image emission.

(c) One or more indicators may be included with the call sign. Each indicator must be separated from the call sign by the slant mark (/) or by any suitable word that denotes the slant mark. If an indicator is self-assigned, it must be included before, after, or both before and after the call sign. No self-assigned indicator may conflict with any other indicator specified by the FCC Rules or with any prefix assigned to another country.

(d) When transmitting in conjunction with an event of special significance, a station may substitute for its assigned call sign a special event call sign as shown for that station for that period of time on the common database, maintained and disseminated by the special event call sign database coordinators. Additionally, the station must transmit its assigned call sign at least once per hour during such transmissions.

(e) When the operator license class held by the control operator exceeds that of the station licensee, an indicator consisting of the call sign assigned to the control operator's station must be included after the call sign.

(f) When the control operator is exercising the rights and privileges authorized by Part 97.9(b), an indicator must be included after the call sign as follows:

(1) For a control operator who has requested a license modification from Novice to Technician Class: KT

(2) For a control operator who has requested a license modification from Novice or Technician Class to General Class: AG

(3) For a control operator who has requested a license modification from Novice, Technician, or General Class operator to Advanced Class: AA; or

(4) For a control operator who has requested a license modification from Novice, Technician, General, or Advanced Class operator to Amateur Extra Class: AE

Part 97.121   Restricted operation.

(a) If the operation of an amateur station causes general interference to the reception of transmissions from stations operating in the domestic broadcast service where receivers of good engineering design, including adequate selectivity characteristics, are used to receive such transmissions, and this fact is made known to the amateur station licensee, the amateur station shall not be operated during the hours from 8 p.m. to 10:30 p.m., local time, and on Sunday for the additional period from 10:30 a.m. until 1 p.m., local time, upon the frequency or frequencies used when the interference is created.

(b) In general, such steps as may be necessary to minimize interference to stations operating in other services may be required after investigation by the FCC.

Part 97.203   Beacon station.

(a) Any amateur station licensed to a holder of a Technician, Technician Plus, General, Advanced, or Amateur Extra operator license may be a beacon. A holder of a Technician, Technician Plus, General, Advanced, or Amateur Extra Class operator license may be the control operator of a beacon, subject to the privileges of the class of operator license held.

(b) A beacon must not currently transmit on more than one channel in the same amateur service frequency band, from the same station location.

(c) The transmitting power of a beacon must not exceed 100 W.

(d) A beacon may be automatically controlled while it is transmitting on the 28.20–28.30 MHz, 50.06–50.08 MHz, 144.275–144.300 MHz, 222.05–222.06 MHz, and 432.300—432.400 MHz segments, or on the 33 cm and shorter wavelength bands.

(g) A beacon may transmit one-way communications.

Part 97.205   Repeater stations.

(a) Any amateur station licensed to a holder of a Technician, General, Advanced, or Amateur Extra Class operator license may be a repeater. A holder of a Technician, General, Advanced, or Amateur Extra Class operator license may be the control operator of a repeater, subject to the privileges of the class of operator license held.

(b) A repeater may receive and transmit only on the 10 m and shorter wavelength frequency bands except the 28.0–29.5 MHz, 50.0–51.0 MHz, 144.0–144.5 MHz, 145.5–146.0 MHz, 222.00–222.15 MHz, 431.0–433.0 MHz, and 435.0–438.0 MHz segments.

(c) When the transmissions of a repeater cause harmful interference to another repeater, the two station licensees are equally and fully responsible for resolving the interference unless the operation of one station is recommended by a frequency coordinator and the operation of the other is not. In that case, the licensee of the noncoordinated repeater has primary responsibility to resolve the interference.

(d) A repeater may be automatically controlled.

(e) Ancillary functions of a repeater that are available to users on the input channel are not considered remotely controlled functions of that station. Limiting the use of a repeater to only certain user stations is permissible.

(g) The control operator of a repeater that retransmits inadvertently communications that violate the rules in this Part is not accountable for the violative communications.

Part 97.207   Space station.

(a) Any amateur station may be a space station. A holder of any class operator license may be the control operator of a space station, subject to the privileges of the class of operator license held by the control operator.

(b) A space station must be capable of effecting a cessation of transmissions by telecommand whenever such cessation is ordered by the FCC.

(c) The following frequency bands and segments are authorized to space stations:
(1) The 17 m, 15 m, 12 m, and 10 m bands, 6 mm, 2 mm, and 1 mm bands; and
(2) The 7.0–7.1 MHz, 14.00–14.25 MHz, 144–146 MHz, 435–438 MHz, 1260–1270 MHz, and 2400–2450 MHz; and 3.40–3.41 GHz, 5.83–5.85 GHz, 10.45–10.50 GHz, and 2400–2405 GHz segments.

(d) A space station may automatically retransmit the radio signals of Earth stations and other space stations.

(e) A space station may transmit one-way communications.

Part 97.215 Telecommand of model craft. An amateur station transmitting signals to control a model craft may be operated as follows:

(a) The station identification procedure is not required for transmissions directed only to the model craft, provided that a label indicating the station call sign and the station licensee's name and address is affixed to the station transmitter.

(b) The control signals are not considered codes or ciphers intended to obscure the meaning of the communication.

(c) The transmitter power must not exceed 1 W.

Part 97.307 Emission standards.

(a) No amateur station transmission shall occupy more bandwidth than necessary for the information rate and emission type being transmitted, in accordance with good amateur practices.

(b) Emissions resulting from modulation must be confined to the band or segment available to the control operator. Emissions outside the necessary bandwidth must not cause splatter or keyclick interference to operations on adjacent frequencies.

(c) All spurious emissions from a station transmitter must be reduced to the greatest extent practicable. If any spurious emission, including chassis or power line radiation, causes harmful interference to the reception of another radio station, the licensee of the interfering amateur station is required to take steps to eliminate the interference, in accordance with good engineering practice.

(d) The mean power of any spurious emission from a station transmitter or external RF power amplifier transmitting on a frequency below 30 MHz must not exceed 50 MW and must be at least 40 dB below the mean power of the fundamental emission. For a transmitter of mean power less than 5 W, the attenuation must be at least 30 dB. A transmitter built before April 15, 1977, or first marketed before January 1, 1978, is exempt from this requirement.

(e) The mean power of any spurious emission from a station transmitter or external RF power amplifier transmitting on a frequency between 30–225 MHz must be at least 60 dB below the mean power of the fundamental. For a transmitter having a mean power of 25 W or less, the mean power of any spurious emission supplied to the antenna transmission line must not exceed 25 μW (25 microwatts) and must be at least 40 dB below the mean power of the fundamental emission, but need not be reduced below the power of 10 μW. A transmitter built before April 15, 1977, or first marketed before January 1, 1978, is exempt from this requirement.

(f) The following standards and limitations apply to transmissions on the frequencies specified in Part 97.305 of this Part:
  (1) No angle-modulation (equivalent to frequency modulation) emission may have a modulation index greater that 1 at the highest modulating frequency.
  (2) No non-phone emission shall exceed the bandwidth of a communications quality phone emission of the same modulation type. The total bandwidth of an independent sideband emission (having B as the first symbol), or a multiplexed image and phone emission, shall not exceed that of a communications quality A3E emission.

(3)   Only a RTTY or data emission having a specified digital code listed in Part 97.309(a) of this part may be transmitted. The symbol rate must not exceed 300 bauds, or for frequency-shift keying, the frequency shift between mark and space must not exceed 1 kHz.

Notes:

1. This section includes the FCC's technical specifications for modulation characteristics involving voice (phone), RTTY (radio teletypewriter) and data, and image emissions. The defining lines between RTTY and data forms of communications radio have become somewhat blurred since RTTY, data, and some imaging (TV) signals represent digital signals. For example, one manufacturer markets a multimode data controller that can be connected between your ham radio and personal computer. This controller processes the following signals: Packet, PACTOR, AMTOR, RTTY, Color SSTV TV, FAX, Weather, ASCII, Navtex, CW, and Memory Keyer. Don't worry about all these complex abbreviations—you'll learn all of them as you progress in amateur radio. We'll cover the ones you need to know for the Technician Examination.

2. RTTY and digital code transmission rates are measured in either bauds or bits per second. A baud may be thought of as a signal transition rate, e.g., 300 baud represents 300 signal transitions per second. One or more digital bits (usually designated as "ones" or "zeros," or 1s and 0s) may be encoded in each signal transition. Thus 1200 baud may represent 1200 bits per second for one communications system or 4800 bits per second in another communications system. The second system provides for 4 bits being transmitted during each baud transition.

3. The FCC Rules restrict the baud rates in the lower amateur bands (mostly HF bands) while permitting higher baud rates in the higher bands (VHF and UHF). However, Technician Class Amateur operators are permitted RTTY and data privileges *only* above 50 MHz. The maximum baud (or symbol) rate for the 2-meter and 6-meter bands is 19.6 kilobauds.

Part 97.313   Transmitter power standards.

(a)   An amateur station must use the minimum transmitter power necessary to carry out the desired communications.

(b) No station may transmit with a power exceeding 1.5 kW PEP.

(c) No station may transmit with a power exceeding 200 W PEP on:

   (1) The 3.675–3.725 MHz, 7.10–7.15 MHz, 10.10–10.15 MHz, and 21.1–21.2 MHz segments;

   (2) The 28.1–28.5 MHz segment when the control operator is a Novice Class operator or a Technician Class operator who has received credit for proficiency in telegraphy in accordance with the international requirements; or

   (3) The 7.050–7.075 MHz segment when the station is within ITU Regions 1 and 3.

EMERGENCY COMMUNICATIONS.

The FCC is very specific when it comes to emergency communications. As you may remember, Part 97.1(1) clearly states that one major justification for the existence of the Amateur Service is its ability to provide communications during emergency conditions and public service events.

Part 97.401   Operation during a disaster.

(a) When normal communications are overloaded, damaged, or disrupted because a disaster has occurred, or is likely to occur, in an area where the amateur service is regulated by the FCC, an amateur station may make transmissions necessary to meet essential communications needs and facilitate relief actions.

(b) When normal communications systems are overloaded, damaged, or disrupted because a natural disaster has occurred, or is likely to occur, in an area where the amateur service is not regulated by the FCC, a station assisting in meeting essential communications needs and facilitating relief actions may do so only in accordance with ITU[3] Resolution No. 640 (Geneva, 1979). The 80 m, 75 m, 40

---

[3]The International Telecommunications Union (ITU), an agency of the United Nations, coordinates and establishes radio frequency allocations on a worldwide basis. Region 1 covers Europe, the African Continent, and all of Russia; Region 2 covers the Western Hemisphere, including North and South America, Greenland, and Hawaii; and Region 3 covers the Far East, including China, Japan, Indonesia, Australia, India, and Pakistan.

m, 30 m, 20 m, 17 m, 15 m, 12 m, and 2 m bands may be used for these purposes.

(c) When a disaster disrupts normal communications systems in a particular area, the FCC may declare a temporary state of communications emergency. The declaration will set forth any special conditions and special rules to be observed by stations during the communications emergency. A request for a declaration of a temporary state of emergency should be directed to the RIC in the area concerned.

(d) A station in or within 92.5 km of Alaska may transmit emissions J3E and R3E on the channel of 5.1675 MHz for emergency communications. The channel must be shared with stations licensed in the Alaska private fixed service. The transmitter power must not exceed 150 watts.

Part 97.405   Station in distress.

(a) No provisions of these rules prevent the use by an amateur station in distress of any means at its disposal to attract attention, make known its condition and location, and obtain assistance.

(b) No provisions of these rules prevent the use of a station, in the exceptional circumstances described in paragraph (a), of any means of radiocommunications at its disposal to assist a station in distress.

HOW TO QUALIFY FOR AN FCC AMATEUR RADIO OPERATOR LICENSE

Part2 97.501 through 97.527 describes the processes involved in applying for the entry- and advanced-level amateur licenses. The specific telegraphy (or Morse code) tests and written examinations, known as "elements," were summarized in Table 2.1. If you are planning to become a ham, you'll want to review this table carefully to become familiar with the entry-level class license—i.e., the Technician Class—requirements. You may elect to start with the No-Code Technician Class license and enjoy some activity with 2-meter and 440-centimeter repeaters, as well as attending local ham club meetings.

However, if you have an opportunity to also prepare for the Element 1 Morse code test and the Element 2 written examination for the Technician Class license, this is the best of both worlds. For example, code and theory ham classes offered by the local ham club present an ideal approach to preparing for Elements 1 and 2 examinations.

Prior to 1984, the FCC administered amateur radio exam at selected FCC facilities throughout the country. Since that time, the FCC has delegated the administration of all amateur radio examinations to qualifying (or accredited) amateurs and amateur organizations such as the ARRL and the W5YI Group. The amateurs who perform this testing function are known as volunteer examiners (VEs). The following sections of Part 97 govern the amateur testing sessions for all Technician (and higher) Class license examinations.

Part 97.501 Qualifying for an amateur operator license. Each applicant must pass an examination for a new amateur operator license grant and for each change in operator class. Each applicant for the class of operator license grant specified below must pass, or otherwise receive examination credit for, the following examination elements:

(a)   Amateur Extra Class operator: Elements 1, 2, 3, and 4;

(b)   General Class operator: Elements 1, 2, and 3;

(c)   Technician Class operator: Element 2.

Part 97.503   Element standards.

(a)   A telegraphy examination must be sufficient to prove that the examinee has the ability to send correctly by hand and to receive correctly by ear texts in the international Morse code at not less than the prescribed speed, using the letters of the alphabet, numerals 0–9, period comma, question mark, slant mark, and prosigns AR, BT, and SK:
(1)   Element 1: 5 words per minute.

(b)   A written examination must be sufficient to prove that the examinee possesses the operational and technical qualifications required to perform properly the duties of an amateur service licensee. Each written examination must be comprised of a question set as follows:

(1) Element 2: 35 questions concerning the privileges of a Technician Class operator license. The minimum passing score is 26 questions answered correctly.

(2) Element 3: 35 questions concerning the privileges of a General Class operator license. The minimum passing score is 26 questions answered correctly.

(3) Element 4: 50 questions concerning the privileges of a Amateur Extra Class operator license. The minimum passing score is 37 questions answered correctly.

Amateur Call-Sign Allocations

The FCC assigns a unique amateur radio call sign to each licensed station. Once you receive your call sign, you can retain it for life if you want to. However, call signs are issued based on the class of license and the geographical location. When upgrading to a higher license class such as the Amateur Extra, you may want to request the prestigious call sign assigned to that class of licenses!

The amateur call sign consists of three separate sections: a prefix of one or two letters (the first one or two letters), a number ranging from 0 to 9 in the center of the call sign, and a suffix of one to three letters at the end of the call sign.

The prefix and the suffix provide an indication of the license class—Amateur Extra, Advanced, General, Technician, or Novice. The numbers and specific letters (L, P, H) serving as the second letter in the prefix identify the geographical location of the amateur station. Letters other than L, P, and H, used in the second location of the prefix, permit the numbers 1 through 10 to identify regions within the continental United States. Table 2.2 lists specific numeral designators for amateur call signs. To distinguish between numbers assigned to continental U.S. regions (i.e., within the 48 states), Alaska and overseas U.S. locations, the second letter in the prefix is used to provide an indicator. Here is the key to this designator:

TABLE 2.2 FCC Call-Sign Regions and Designators

Region	Numeral	Geographical area
1	1	Connecticut, Maine, Massachusetts, New Hampshire, Rhode Island, and Vermont
2	2	New Jersey and New York
3	3	Delaware, Maryland, Pennsylvania, and the District of Columbia
4	4	Alabama, Florida, Georgia, Kentucky, North Carolina, South Carolina, Tennessee, and Virginia
5	5	Arkansas, Louisiana, Mississippi, New Mexico, Oklahoma, and Texas
6	6	California
7	7	Arizona, Idaho, Montana, Nevada, Oregon, Utah, Washington, and Wyoming
8	8	Michigan, Ohio, and West Virginia
9	9	Illinois, Indiana, and Wisconsin
10	0	Colorado, Iowa, Kansas, Minnesota, Missouri, Nebraska, North Dakota, and South Dakota
11	7–9	Alaska
12		Caribbean area:
	1	Navassa Island
	2	Virgin Islands
	3–4	Puerto Rico (except Desecheo Island)
	5	Desecheo Island
13		Hawaii and Pacific islands:
	0	Northern Mariana islands
	1	Baker or Howland Island
	2	Guam
	3	Johnston Island
	4	Midway Island
	5	Palmyra or Jarvis Island; followed by suffix letter K: Kingman Reef
	6–7	Hawaii; 7 followed by suffix K: Kure Island
	8	American Samoa
	9	Wake, Pearl, or Wilkes islands

Designator	Region	Geographical location
L	11	Alaska
P	12	Caribbean islands
H	13	Hawaii and Pacific islands

AMATEUR EXTRA CLASS CALL SIGNS

Call signs authorized for Amateur Extra Class stations in the continental United States (Regions 1–10) use a 1 × 2 format using K, N, or W as the prefix and a two letter suffix; a 2 × 1 format using A, N, K, or W as the first letter in the prefix; and a 2 × 2 format with an A as the first letter in the prefix. Examples of call signs for the Amateur Extra Class licensees from Texas are N5RA and AA5KB.

Call signs authorized for Amateur Extra Class stations in areas outside the continental United States use unique letter identifiers in the second position of a two-letter prefix. In Regions 11, 12, and 13, a special identifier is inserted after the first letter of the prefix. Two-letter suffixes are currently being used. Here are examples of Amateur Extra call signs for stations located outside of the continental United States:

Region 11: Prefixes of AL, Kl, NL, or WL
Example: KL7H (Alaska)

Region 12: Prefixes of KP, NP, or WP
Example: WP3X (Puerto Rico)

Region 13: Prefixes of AH, KH, NH, or WH
Example: AH4L (Midway Island)

ADVANCED CLASS CALL SIGNS

Although these call signs, a 2 × 2 format, are no longer issued by the FCC as of May 15, 2000, holders of current Advanced Class licenses may retain their licenses indefinitely. The first letter of the prefix may use K, N, or W. Advanced Class licensees who live outside of the continental United States will use the prefixes AL, KP, or AH to identify locations in Alaska, islands in the Caribbean, and islands in the Pacific, respectively.

GENERAL, TECHNICIAN, AND TECHNICIAN WITH HF PRIVILEGES CALL SIGNS

Call signs for these licensees use a 1×3 format with the first letter of the prefix being a K, N, or W letter. The General or Technician Class licensees located outside of the continental United States will be issued call signs with a 2×3 format with prefixes KL, NL, or WL (Region 11); NP or WP (Region 12); and KH, NH, or WH (Region 13). This latter group applies to locations in Alaska, islands in the Caribbean, and islands in the Pacific, respectively.

NOVICE, CLUB, AND MILITARY RECREATION STATION CALL SIGNS

Although these call signs, a 2×3 format, are no longer issued by the FCC as of May 15, 2000, holders of current Novice Class licenses may retain their licenses indefinitely. Novice Class licensees located outside of the continental United States were issued call signs with a 2×3 format with prefixes KL, NL, or WL (Region 11); NP or WP (Region 12); and KH, NH, or WH (Region 13). This latter group applies to locations in Alaska, islands in the Caribbean, and islands in the Pacific, respectively.

How to Learn International Morse Code

If you plan to enter the exciting realm of amateur radio via the Technician Class with HF privileges, the FCC requires that you develop a 5 word per minute (5 WPM) proficiency in Morse code. This first step can be easy and fun—or it can be difficult and a burden, depending upon your attitude and study habits. You can get off to a good start in amateur radio by developing a sound plan of attack for learning the Morse code. And remember, code (or CW) transmissions will be your means for contacting fellow hams all over the United States and in foreign countries.

One final comment: 5 words per minute is a slow code speed. By the time you have learned the Morse code, you normally have developed a 5 WPM proficiency. To begin learning this new language, let's get started with a complete review of Morse code.

What is Morse Code?

International Morse code, consisting of a series of dits (dots) and dahs, (dashes), is a universal language complete in itself. It affords a means

the rate of about 13 WPM. The advantage to this approach is that the beginner learns the sounds of the 13 WPM characters early in his or her amateur career. For many Novice operators, this is an effective way to progress to the 13 WPM code requirement for the General license.

Most International Morse code signals are transmitted as CW or continuous-wave transmissions. Bursts of CW radiofrequency energy are produced by alternately keying on and off the transmitter carrier frequency according to the coded character combinations of dits and dahs.

Who can learn Morse code?

Although some individuals adapt to the code faster than others, almost everyone can master this fascinating language. The age and educational level of the individual are not significant factors. In fact, some psychological studies indicate that children under the age of 10 will adapt to code faster than adolescents or adults. The ranks of amateur

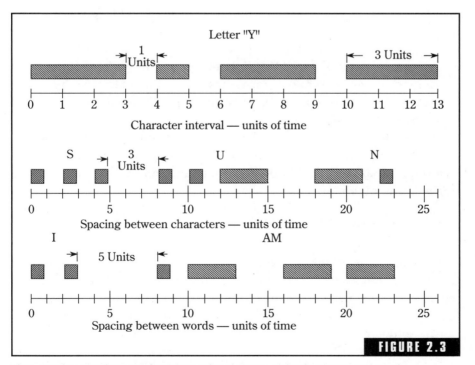

FIGURE 2.3

Time spacing requirement for International Morse code characters and words.

radio range from those less than 10 years of age to those in their 80s and 90s.

Most people can learn the code in about 10 hours. With another 10 to 20 hours of practice, the average individual can achieve a code proficiency adequate to pass the Novice Class 5 WPM code test. Experience has shown that a code proficiency of about 6 to 7 WPM is necessary to ensure that the applicant will pass the Novice code test. The emotional stress encountered in taking a code test tends to reduce the individual's code proficiency by some 2 to 3 WPM, so prepare by practicing at a speed higher than 5 WPM.

LEARNING PLATEAUS

Almost all individuals will experience learning plateaus or periods of time when they cannot seem to progress to higher code speeds. The human mind appears to form mental blocks at code speeds which vary from individual to individual. These plateaus may appear at representative speeds of 6–7, 8–10, 14–16, or 18–20 WPM, or at higher speeds. You may encounter these learning plateaus at several phases in your quest to increase your code speed. However, the average Technician applicant does not normally encounter any plateaus in his/her progression to a speed of 6 to 7 WPM. Later on if you opt for a faster code proficiency, you may experience these learning plateaus. By being aware of this potential problem—common to most amateurs—you will be prepared to cope with it.

Fortunately, there is a solution to the learning-plateau problem—practice and more practice. After several days of effort, you will find that the problem will suddenly disappear and you will be able to copy traffic at the higher speeds.

SCHEDULE PRACTICE SESSIONS

A mandatory requirement for successful code proficiency is a periodic routine practice schedule. You must establish an adequate schedule for code practice sessions and rigidly adhere to this schedule. Experience in teaching numerous code and theory classes has shown that the failure to practice code at routine intervals is the primary factor in failing to achieve the Technician Class with HF privileges license. The secret to learning the code is practice, practice, and more practice.

As a minimum, you should be prepared to devote about one hour per day, seven days per week, to code practice. It is important that this be a continuous schedule. Missing one or two days will probably result in a loss of code proficiency. It is also equally important that you avoid operating fatigue by taking breaks of about 5 to 10 minutes, spaced evenly throughout each session. Research in educational techniques indicates that the average individual's attention span in such activity is about 15 to 20 minutes.

Many excellent code training aids are available from a variety of sources. A code-practice oscillator, a hand key, and a source of code-practice signals are considered essential tools for learning to send and receive International Morse code. Any amateur radio journal will offer sources for all of these training aids. They are also available from most local radio parts stores.

A simple code oscillator consisting of a high-frequency buzzer or an electronic audio oscillator with a loudspeaker are available as kits or completed items. Building and assembling an audio oscillator is an excellent first project for your Technician career. Some amateur equipment suppliers offer combination audio oscillator–hand key units. Figure 2.4 shows the MFJ Enterprises, Inc., Model MFJ-557 Code Practice Oscillator, which can be used for individual as well as group code practice sessions. It is battery powered and has a small internal speaker. An alternate approach is to build a "home-brew" code-practice oscillator using the popular Signetics NE555 integrated-circuit (IC) timing chip. (Other integrated-circuit companies may offer a direct substitute for the Signetics NE555.) Figure 2.5 provides a schematic diagram for this code oscillator. Parts for this device, including a small cabinet or chassis, are available from local radio parts stores. A single 9-volt transistor battery will power this code oscillator for many hours of operation.

A standard radio telegraphy hand key of heavy construction is recommended for code training and later for Technician

FIGURE 2.4

The MFJ-557 Code Practice Oscillator. This versatile device has a Morse key and audio oscillator unit mounted on a heavy steel base. The oscillator section features battery operation, internal speaker with earphone jack for private listening, and volume and tone controls. The key has adjustable contacts and can be used with your CW transmitter after you receive your ham license. *(Courtesy of MFJ Enterprises, Inc., Box 494, Mississippi State, MS 39762; Telephone: (601) 323-5869; Web site: http//www.mfjenterprises.com.)*

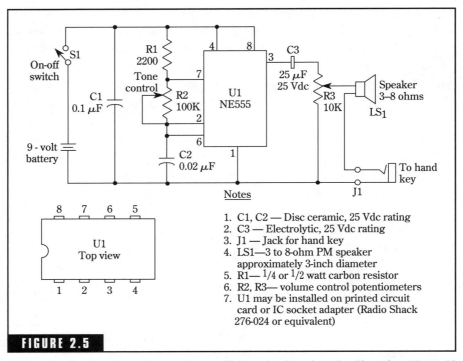

FIGURE 2.5

Schematic diagram for code practice oscillator circuit using the Signetics NE555 IC Timer.

operation. The hand key should be mounted on a wood or plastic base of approximately 4 × 6 in (10 × 15 cm) to provide solid, stable operation.

The source of code practice signals may be cassette tapes, phonograph records, a computer program for generating practice code on a PC or other home computer, or an HF-band amateur radio receiver for on-the-air reception of CW signals. One recommended approach for individual practice sessions is to obtain a small, inexpensive cassette-tape record/playback unit and one or more Technician code training tapes. The cassette tape approach offers the advantages of a stable, uniform source of code-practice signals at specified transmission rates; a capability for recording and playback of the beginner's hand-sent code signals; and if an HF-band receiver is available, an ability to record on-the-air code signals for later playback. Also, the cassette recorder may be taken to code and theory classes to record code-practice sessions and discussions of radio theory and FCC rules and regulations. Some Morse code training tape cassettes are shown in Fig. 2.6.

W1AW, the ARRL amateur station in Newington, Connecticut, transmits daily code-practice sessions on a scheduled basis. Approximate frequencies of transmission are 1.818, 3.5815, 7.0475, 14.0475, 18.0975, 21.0675, 28.0675, 50.0675, and 147.555 MHz. Slow code practice sessions of about 5–15 WPM are scheduled at specific times during the day to late evening hours, seven days a week. On alternating hours, fast code sessions of about 10–35 WPM are transmitted, also on a seven-day basis. However, if variations to this schedule are observed, you can contact the ARRL for the exact schedule. Monthly schedules for W1AW code transmissions are given in *QST*, the ARRL magazine devoted entirely to amateur radio.

BITE-SIZE CODE LESSONS

Most producers of code training tapes or records recognize that it is almost impossible for the beginner to memorize the International Morse code all at once. They divide the code set into "bite-size" lessons for about four or five characters each. Normally, the groups of characters are arranged to contain similar characters, such as those with all dits or dahs, and then progress into the groups of characters containing combinations of dits and dahs. A typical series of bite-size code lessons is given in Table 2.4. You may elect to use this set of lessons if your code training tapes (or records) do not contain such an approach.

FIGURE 2.6

Morse code training tape cassettes for learning the Morse code.

HAND COPY OF CODE SIGNALS

The best approach for fast, efficient copy of code is by longhand writing or using a typewriter. Although hand printing of code messages looks more attractive and

TABLE 2.4 Recommended bite-size lessons for learning the International Morse code.

Lesson number	Character set
1	E, I, S, H
2	T, M, O, R
3	A, W, J, K
4	U, V, F, L
5	N, D, B, G
6	C, Y, Q, X
7	P, Z
8	Numbers 1, 2, 3, 4, 5
9	Numbers 6, 7, 8, 9, 0
10	Punctuation Symbols:
	Period (.), Comma (,)
	Question Mark (?)
	Slant Bar (/), Error
11	Special Symbols:
	Wait ($\overline{\text{AS}}$), Pause ($\overline{\text{T}}$),
	End of Message ($\overline{\text{AR}}$),
	End of Work ($\overline{\text{VA}}$),
	Invitation To Transmit (K)

is used by many CW operators, it is too slow for most individuals at the higher speeds of about 20 WPM and above. Thus, developing a good habit of copying code signals as a Technician will help you when you progress to the higher-class licenses.

MISSED CHARACTERS

One of the most annoying bad habits developed by the beginner in copying code signals is that of being stuck on missed characters. Sometimes the brain will simply refuse to recognize a particular character. Other factors, such as an unexpected noise burst or fatigue from practicing too long, will cause a person to lose the power of concentration and therefore miss characters. It is a natural tendency to stop and try to remember what that particular character was. However, stopping for one character means missing the next two, three, or four.

The best approach is to simply ignore the missed character and concentrate on the following characters. This will require an extra effort on your part. One way to develop this habit is to write in a dash for the missed character. This helps you to proceed to the next character and "turn loose" of the one you missed.

ALWAYS INCREASE CODE SPEED DURING PRACTICE

You should never remain fixed on a code speed that can be copied comfortably and accurately. For example, when you can copy 5 WPM with some 95 to 100 percent accuracy, it is time to move up to a higher code speed of about 7–8 WPM. This forces your mind to adapt and become proficient at the higher speed.

DON'T IGNORE SENDING PRACTICE

Many beginners make the mistake of neglecting sending practice while concentrating on receiving practice. The FCC requires proficiency in both sending and receiving code signals. If available, a cassette tape recorder can be used to record the sending "fist" for subsequent analysis and improvement. Remember also, the first contact on the Novice frequencies may be disappointing if you cannot send an intelligent code message.

Proper sending by a hand key requires a comfortable linkage between the operator and the key. The spring tension and contact spacing should be adjusted to suit the individual's preference. While operating, the arm should be positioned to rest on a flat surface and the key should be loosely grasped by the thumb, forefinger, and index finger. Minimum fatigue and more uniform operating will result.

The FCC Written Examinations

The written examination for the Technician Class license, Element 2, contains 35 questions covering rules and regulations, elementary radio theory, and operating procedures. You must have a passing score of 26 correct answers to achieve the Technician Class license.

The VEC question pool

Appendix A presents the VEC question pool for the Technician Class Element 2 license examinations. The question pool covers 10 specific categories pertinent to the licensing and operation of amateur radio stations. There are some 384 questions in the pool. Each question is a multiple-choice type with one correct and three incorrect answers. Table 2.5 shows the subject areas and number of questions covered by the Technician Class license examinations.

Your Technician exam questions will be taken from this question pool, verbatim. Just think about this for a minute—you have a chance to review and learn every possible question on your Technician Class license written examination.

Some helpful suggestions for successfully passing the Novice or Technician written exam are in order. Most people are apprehensive about taking tests, but knowing some of the pitfalls and reasons for failing will help you to get through with flying colors.

TABLE 2.5   Technician Class Element 2 Written Examination Structure

Topic	Subelement number	Total questions	Number of questions on examination
Commission rules	T1	102	9
Operating procedures	T2	55	5
Radio wave propagation	T3	33	3
Amateur radio practices	T4	44	4
Electrical principles	T5	33	3
Circuit components	T6	22	2
Practical circuits	T7	22	2
Signals and emissions	T8	20	2
Antennas and feed lines	T9	22	2
RF safety fundamentals and definitions	T0	31	3
Totals		384	35

BE PREPARED

Lack of preparation and study is the primary reason for failing amateur license examinations. Since the Technician Class examination is the easiest of all amateur examinations, many beginners do not devote sufficient effort to study and preparation. Being prepared is more than half of the battle—it gives you a high degree of confidence when it is time to take the test.

Develop good study habits and a regular schedule for mastering all of the material anticipated in the exam. A typical schedule may involve one or two hours of study in a quiet room each day for about two or three weeks. The actual time required for each individual will vary, and each person must pace his or her study efforts accordingly. The important thing to remember is that sufficient time must be devoted to studying all of the required areas. For example, it is a certainty that the FCC will stress the rules and regulations on all examinations. As you can see from Table 2.5, the Technician Class examination contains nine questions on FCC Rules and five questions on operating procedures. Again, remember that in Appendix A you have access to the exact questions that will be used.

One suggested study technique used by many beginners in successfully preparing for the Technician examination is to write each question on one side of an index card and the correct answer on the reverse side. The process of writing down the information on the cards results in considerable retention of this material. Then, by constant review, you can begin to separate the cards into two piles—known and unknown. After a while, the stack of cards containing unknown answers will diminish to zero. Oh yes, be sure to go through the entire stack of cards the night before the examination for a final review.

Scheduling the examination

Be sure to check with the local volunteer examiner (VE) for testing requirements and to register for the examination at a time convenient to both parties. You may need to obtain, fill out, and send a form to the VE for preregistration. If you do not know where the nearest volunteer examiner team is in your area, contact the local amateur radio club or the ARRL, Educational Activities Department, 225 Main Street, Newington, CT 06111.

Taking the examination

Plan to arrive early at the examination point. This will ensure no interruption of examinations in progress and will help you locate a comfortable place to sit. At a minimum, you will need to bring the following to the exam:

1. Any current amateur radio license and/or certificate of successful completion of examination.

2. Two documents for identification purposes. This may include a current amateur radio license, a photo ID such as a driver's license, or a birth certificate.

3. Several sharpened No. 2 pencils, an eraser, and two pens (black or blue ink). If a pocket calculator is available, bring it along for possible mathematical calculations. However, if a programmable

calculator is used, be sure it is cleared of all instructions and constants. Ask the VE for blank paper to write out such items as calculations and formulas.

4. Finally, don't forget to bring the testing fee (about $6–7). Check with your VE on this point.

Carefully read the FCC instructions on the examination. Fill out all requested information, including signatures, before proceeding. Check the exam to ensure that all pages are included. If some are missing, report this to the VE immediately. Check the question numbers.

Read each question thoroughly before attempting to answer it. Remember, a multiple-choice exam requires a careful screening of all answers to decide on the most correct one. If the question cannot be answered correctly after the first deliberation, move on to the next one. In this manner, the easy questions can be answered and you can return to the more difficult ones for additional study. In many instances, this delay in tackling the tough questions gives the mind a chance to recall the correct answer. Also, some subsequent questions may relate to the one you are having trouble with. Most volunteer examiners will allow any reasonable length of time for completing the exam.

At the end of the test, carefully check to ensure that all questions have been answered. Go back over each answer to be 100 percent sure that the most correct answer has been given. Some test-takers fail because of carelessness in answering questions, though they may know the required material. When this final check is finished, return the exam to the VE.

If by some chance you fail the examination, chalk this up to experience and try again. The FCC allows you to retake the test at any time convenient to the VE. Good luck!

Radio Communications Theory

The ability to transmit and receive radio waves is a fascinating subject. In this chapter we will take a close look at some of the fundamentals of radio communication. Learning these principles will help you to pass the Technician Class license examination and to understand how amateur radio signals are propagated for long distances.

Definitions

Sky-wave propagation   A type of radio-wave propagation in which radio waves traveling upward are bent or refracted by the ionosphere back to the earth. This is one of the major means of amateur communication in the high-frequency spectrum. Otherwise, these radio waves would be propagated into outer space and lost. Sky-wave propagation provides a capability for long-range or DX communications using one or more "hops" in which the radio waves are reflected back to the earth.

Ground-wave propagation   Another form of propagation in which the ground or surface wave travels along the surface of the ground or water. This mode of propagation is important at the low and medium frequencies. Most commercial AM broadcast stations use ground-wave propagation during daylight hours for local urban and suburban coverage. However, beginning at about 3 MHz, ground-wave propagation at distances greater than 100 miles (or about 160 kilometers) becomes impractical.

Refraction of radio waves Radio waves, like light waves, are refracted or bent when they pass from one medium to another medium with a different density. Since radio waves traveling upward experience less atmospheric density, they may be curved or bent. This is related to sky-wave propagation.

Sunspot cycle The sun exhibits a periodic 11-year cycle of increasing and decreasing sunspots which affect radio communications on earth. Scientists have recorded the number of sunspots appearing on the surface of the sun for the past 300 years and have determined that the number of sunspots reaches a maximum about every 11 years. Also, the number of sunspots will vary during each maximum and minimum cycle. The maximums may range from about 60 to over 200 while minimums may drop to almost zero.

During maximum sunspot activity, excellent amateur communications in the bands up to and including the 10-meter band are possible. However, during minimum sunspot activity, long-range communication is almost nil in the 15- and 10-meter bands. Also, 20-meter operation is restricted primarily to daylight operation.

Skip distance The skip distance is associated with sky-wave propagation for the most part. It is defined as the distance between the transmitter location and the point that the skywave returns to earth after striking the ionosphere. Except for the limited area subject to ground-wave reception, the skip distance or zone does not allow for communications because no radio waves are reflected or bent into this zone.

Wavelength The wavelength of a radio-frequency signal is defined as the length in meters of that signal. More specifically, it is the length of one complete cycle of the radio wave. Thus, a 10-meter signal or wave is approximately 10 meters in length (or about 32.8 feet). For example, a quarter-wave vertical antenna for 10 meters would be approximately 2.5 meters (or about 8.2 feet) of vertical distance.

Frequency Frequency is defined as the number of complete cycles per second that a radio signal exhibits in passing from a value of maximum intensity, through zero intensity, and back to the original value of intensity. Frequency is measured in hertz (or simply Hz), which is directly equivalent to cycles per second; that is, 1 Hz equals 1 cycle per second.

Frequency is related to wavelength and may be used interchangeably to describe the wave motion of a radio wave. Frequency can be converted to wavelength in meters by dividing the frequency in hertz into 300,000,000. For example, 7150 kHz (which is 7,150,000 Hz) corresponds to a wavelength of 41.96 meters.

Ionosphere The ionosphere consists of layers of ionized air at heights above the surface of the earth ranging from about 7 to 250 miles (or about 11 to 402 kilometers). This ionization of the rarefied air particles is caused by the ultraviolet radiation from the sun. The heights of these layers vary from daylight to darkness, depending on the position of the sun with respect to the surface of the earth. A more detailed description of the ionosphere will be given later in this chapter.

The Radio Circuit—Transmitter to Receiver

Radio is defined as the transmission and reception of communications signals through space by means of electromagnetic waves. These communications signals contain intelligence conveyed by such forms as code, voice, television images, teleprinter, and computer-to-computer operation in digital form. Noncommunications electronics systems that use electromagnetic radiation include radar, navigation aids, and identification devices.

Figure 3.1 illustrates a generalized model of a radio communications system. The source of intelligence can be code being generated by a telegraph key or a microphone picking up the sounds of human speech. This intelligence, when applied to the modulator, varies the transmitter's radio-frequency (RF) output energy by the modulation method being used. In the case of code transmissions, the key can be used to switch on and off the carrier frequency of the transmitter, resulting in a series of dits and dahs containing the required information. Voice modulation schemes can employ amplitude or frequency modulation.

The RF power output levels from transmitters can range from a few watts to hundreds or thousands of watts. This RF energy is converted to electromagnetic or radio waves by the transmitting antenna. With proper matching between the transmitter, transmission line, and the antenna, most of the RF output of the transmitter will be used to generate radio waves.

In most communications systems, particularly in the high-frequency (HF) spectrum, the propagation path loss or attenuation is

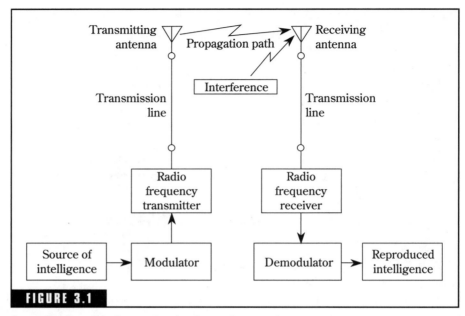

FIGURE 3.1

Generalized model of a communications system.

tremendous. Sky-wave propagation can reduce the transmitted signal power level from many watts to micromicrowatts at the receiving antenna. The radio receiver must have sufficient sensitivity to detect and amplify these weak signals for demodulation and conversion to the required output such as audio speech or code signals.

Radio communications depend on many factors for reliable performance. Some of these factors are the RF power output levels, the antenna directivity and gain characteristics, the radio frequencies being used and the related propagation characteristics at these frequencies, and the receiver performance. An ever-increasing number of transmitters in the already crowded spectrum, as well as interference from nonintentional emitters, also adds to the problems of achieving reliable radio communications. These factors are explored in detail beginning in this chapter.

The Electromagnetic Spectrum

The electromagnetic spectrum, as defined by modern science, ranges from extremely long radio waves to ultrashort cosmic rays. Between

Wavelength is defined as the physical length of one complete cycle of an electromagnetic wave traveling in free space. For example, a 10-meter radio wave in free space exhibits a physical length of 10 meters (or about 32.8 feet) for each complete cycle of the wave.

Because electromagnetic waves travel at the speed of light (300,000,000 meters per second) in free space, the frequency of the wave in complete cycles per second (or Hz) can be determined by dividing 300,000,000 by the wavelength in meters. Also, if the frequency of a wave is known, the wavelength can be determined by dividing 300,000,000 by the frequency in hertz. These relationships or equations can be expressed as follows:

A. To find frequency of a radio wave when wavelength is known:

$$f = \frac{300,000,000}{\lambda \text{ Hz}} \qquad \text{(Eq. 3.1)}$$

where

f is the frequency in Hz,
λ is the wavelength in meters.

B. To find the wavelength when frequency is known:

$$\lambda = \frac{300,000,000}{f \text{ meters}} \qquad \text{(Eq. 3.2)}$$

Two examples are given to show these relationships. Remember, always convert frequency in kilohertz or megahertz to hertz or Equation 3.2 will not give the correct wavelength.

Example 1 Find the wavelength of an electromagnetic wave generated by an amateur transmitter operating at 7.255 MHz.

Solution Step A. 7.255 MHz is converted to 7,255,000 Hz

$$(7.255 \times 1,000,000 \text{ equals } 7,255,000)$$

Step B. By using Equation 3.2,

$$\frac{300,000,000}{7,255,000} = 41.35 \text{ meters (Answer)}$$

Example 2 Find the frequency of a radio signal being received if the wavelength is 20.9 meters.

Solution Step A. Using Equation 3.1

$$f = \frac{300,000,000}{20.9} = 14,354,000 \text{ Hz}$$

Step B. Convert frequency in hertz to megahertz:

$$\frac{14,354,000}{1,000,000} = 14.354000 \text{ MHz}$$

Step C. Round off to the nearest kilohertz: 14.354 MHz (Answer)

This is the first occasion in this book to use a little mathematics, and some comments may be in order to help you become more proficient in the use of "math" in amateur radio computations. The need for mathematical computations in the written examinations for the Technician Class license is extremely limited. However, examinations for the higher class licenses will require more use of mathematical computations. We recommend that you become familiar with the basic equations used in amateur radio. This will help you in constructing antennas for specific operating frequencies, the design of "home-brew" equipment, and other aspects of amateur radio. Another suggestion is to obtain a "scientific" style pocket calculator. While most simple pocket calculators will help with the basic operations of addition, subtraction, multiplication, and division, the more serious amateur will want to make calculations involving squares, square roots, logarithms, and trigonometric functions. Don't let these last math expressions alarm you: much information concerning them can be learned simply by following the instructions furnished with each calculator.

The radio frequency spectrum

Radio frequencies are defined as ranging from about 10,000 Hz (10 kHz) to 300,000,000,000 Hz (300 GHz), with the upper limit approaching infrared and visible-light radiation. The radio spectrum is divided into designated bands, which exhibit individual characteristics such as propagation and methods of generation and detection. Table 3.1 lists these bands and their respective frequency ranges.

No amateur communications are permitted in the VLF band (3 to 30 kHz). Operation in this band is limited primarily to radio navigation aids, time-frequency standard stations, and military, maritime, and aeronautical communications. Limited, mostly experimental, amateur communications is permitted in the 160–190 kHz portion of the LF and ground-wave propagation. Communications on VHF and higher band frequencies use mostly direct- or reflected-wave propagation. Let's examine each of these modes of propagation and see how they apply to amateur radio operation, particularly in the HF bands.

Ground radio waves from the transmitting antenna travel along the surface of the earth. In the HF bands and above, the radio waves are attenuated by the earth's surface and rarely extend beyond 20 to 30 miles (or about 30 to 50 kilometers). Thus amateur radio communications in the HF bands and above cannot rely on ground-wave propagation over long distances.

Virtually all long-range amateur radio communications from 80 to 10 meters are accomplished by sky-wave propagation. The HF radio waves radiated into space from the transmitting antenna, are refracted from the ionosphere back to the earth's surface. Figure 3.6 illustrates ground-wave and sky-wave propagation at the HF frequencies. There is a weak signal area between the point of transmission and the point of return to earth for the refracted waves: this area is referred to as the skip zone or skip distance. Signal reception in the skip zone is virtually impossible,

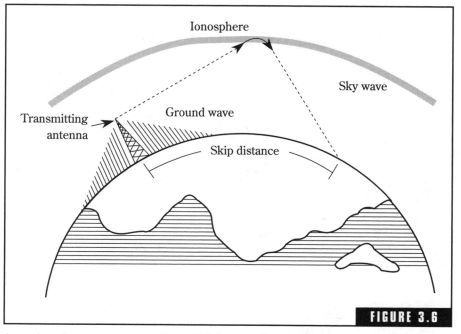

FIGURE 3.6

Ground- and sky-wave propagation of HF frequencies.

particularly in the HF bands above 40-meters. Be sure you are familiar with ground- and sky-wave propagation because the FCC Technician Class examinations will include questions on these subjects.

Beginning with the VHF and UHF frequencies, direct and reflected waves are the major modes of propagation. Figure 3.7 shows an example of a particular transmission with both direct and reflected waves. Most of the reliable communications at these frequencies are often referred to as *line-of-sight* propagation of the direct wave. This simply means that the transmitting antenna must be able to "see" the receiving antenna with no intervening obstructions. Some forms of long-range propagation modes are experienced at these higher frequencies. Occasional skip distances of up to 2500 miles (about 4000 kilometers) or more are encountered in the amateur bands at VHF and UHF frequencies due to abnormal atmospheric conditions.

The Effects of the Ionosphere on Radio Communications

The ionosphere consists of changing layers of ionized air particles in the rarefied regions of the atmosphere beginning about 40 to 50 miles

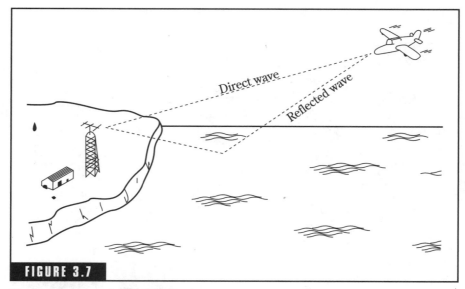

FIGURE 3.7

Direct and reflected wave propagation.

(50 to 80 kilometers) above the earth. The ionosphere differs from other regions of rarefied air in that it contains large numbers of free ions (charged particles) and electrons. The creation of these layers of charged particles and electrons is attributed to the intense radiation of the sun. The layers of the ionosphere extend up to about 250 miles (400 kilometers), depending on the intensity of the radiation of the sun.

The rotation of the earth on its axis, the orbital journey of the earth around the sun, and varying sunspot activity on the surface of the sun will result in a constant changing of the layers of charged particles in the ionosphere. It is the nature of the ionosphere that affects the quality and distance of radio communications.

Ionospheric layers

Up to four distinct layers of ionized or charged particles can be present in the ionosphere at any one time. Each layer consists of varying densities of charged particles ranging from an intense density in the inner region to very little density at the outer regions. Above 250 miles (400 kilometers), the number of air particles in the atmosphere is too small to permit the formation of any sizable concentration of charged particles that would affect radio communications. At altitudes below 40 miles (64 kilometers), the intensity of the radiation of the sun is reduced, or attenuated, by the ionosphere and little potential remains to create more charged particles.

During daylight hours, the radiation of the sun produces four ionospheric layers: the D layer ranging from about 40 to 50 miles (64 to 80 kilometers); the E layer from about 50 to 90 miles (80 to 145 kilometers); the F1 layer at about 140 miles (225 kilometers); and the F2 layer at about 200 miles (320 kilometers). The ionization intensity of these layers rises to a maximum level about midday, local time.

At sunset, a transition of the ionospheric layers begins to take place. The D layer disappears, the F1 and F2 layers combine into a single layer, and the ionization intensity of these two layers is reduced. Figure 3.8 illustrates both the day and night composition of the ionosphere.

Sunspots affect the ionosphere

The presence of sunspots on the surface of the sun has a marked influence on radio communications. Sunspots are dark areas on the surface of the sun, sometimes appearing in groups covering areas of up to

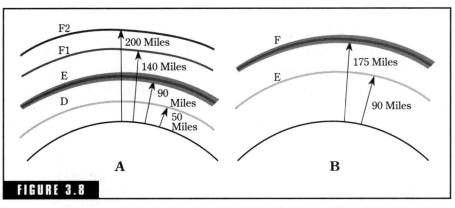

Structure of the ionosphere during day and night. (A) Composition of the ionosphere during daylight hours. (B) Composition of the ionosphere during nighttime hours.

100,000 miles (160,000 kilometers) in diameter. They appear to be solar storms, resembling earth tornadoes. A sunspot can develop in a matter of hours, persist for a few weeks, and then break up into a number of smaller spots. Associated with the sunspots are intense magnetic fields or disturbances.

Sunspots affect the intensity of the radiation of the sun, including ultraviolet waves. During periods of high sunspot activity, the radiation of the sun increases producing more intense ionization of the atmosphere. At approximate 11-year intervals, the number of sunspots builds up to a maximum ranging from about 60 to 200. During minimum sunspot activity, the number of sunspots can decrease to zero for short periods of time.

High sunspot activity

High sunspot activity results in more skywave propagation at the higher HF frequencies. The highest usable frequency is normally referred to as the maximum usable frequency (MUF). During this period, the MUF will extend to the 10-meter band and beyond, providing reliable communications paths. The MUF is reduced to about the 20-meter band during periods of minimum sunspot activity. Only sporadic activity is available at 15 meters, and the 10-meter band is essentially "dead" except for local activity.

Sunspot activity has been observed for some 250 years, with accurate measurements being recorded in the last 150 years. The last two periods of maximum sunspot activity, known as Sunspot Cycles 22 and 23, began in 1986 and 1999, respectively. Current sunspot activity is available from time standard stations WWV at Ft. Collins, Colorado, and WWVH at Kauai, Hawaii—both stations transmit continually on frequencies of 2.5, 5.0, 10.0, 15.0, and 20 MHz. Most amateur receivers will provide reception on 10 MHz to permit monitoring of WWV and WWVH.

Propagation characteristics of the ionosphere

During daylight hours, sky-wave energy of the 160- and 80-meter amateur bands is almost completely absorbed by the D layer. Only high-angle radiation can penetrate the D layer and be reflected back to earth by the E layer. This restricts operation on these bands to about 150 to 250 miles (240 to 400 kilometers). The remaining HF amateur bands experience very little D-layer absorption.

The E layer is probably the best region in the ionosphere for reliable sky-wave propagation. Distances of up to about 1200 miles (2000 kilometers) in one-hop transmissions are common for frequencies up to about 20 MHz. Sporadic E ionization, or occasional erratic regions of E-layer ionization, permits propagation of radio waves with frequencies extending well into the VHF band. Sometimes referred to as "short skip," this phenomenon is most prevalent in the northern hemisphere during summer months and in the southern hemisphere during winter months. Transmission paths range from about 4000 to 1300 miles (650 to 2100 kilometers) for single-hop transmissions and up to 2500 miles (400 kilometers) or more for multihop paths.

Most HF waves penetrating the E layer also pass through the F1 layer and are reflected by the F2 layer, the outermost region of the ionosphere. The F2 layer is the principal reflecting region for long-distance HF communications, providing for international or worldwide transmission paths. Frequencies involved in F2 transmission range from about 20 MHz to as high as 70 MHz during high sunspot activity. At this time, excellent band openings on 10 and 6 meters are available for multihop paths of up to 12,000 miles (20,000 kilometers) or more. However, this capability decreases with diminishing sunspot activity and becomes almost nonexistent during minimum activity.

TABLE 3.2 Propagation Characteristics of Amateur Bands

Amateur bands	Propagation characteristics
160 meters (1.8–2.0 MHz)	■ Ground waves to about 25 miles ■ Day sky waves to about 200 miles ■ Night sky waves to about 2500 miles
80 meters (3.5–4.0 MHz)	■ Ground waves to about 20 miles ■ Day sky waves to about 250 miles ■ Night sky waves to about 2500 miles
40 meters (7.0–7.3 MHz)	■ Ground wave to about 20 miles ■ Day sky waves to about 750 miles ■ Night sky waves to about 10,000 miles
20 meters (14.0–14.35 MHz)	■ Ground waves to about 20 miles ■ Day sky waves with worldwide communications extending from dawn to dusk ■ Excellent night sky waves during high sunspot activity ■ Virtually no sky waves during minimum sunspot activity
15 meters (21.0–21.45 MHz)	■ Ground waves to about 20 miles ■ Worldwide day and night sky waves during high sunspot activity ■ No night sky waves and only occasional day sky waves during minimum sunspot activity
10 meters (28.0–29.7 MHz)	■ Ground waves to about 20 miles ■ Worldwide sky waves in daytime and early evening hours during high sunspot activity ■ Very little day sky waves and virtually no night sky waves during minimum sunspot activity.
VHF frequencies and above (Beginning with 50–54 MHz)	■ No ground wave propagation in sense of MF and HF ground waves. Direct and reflected waves describe line-of-sight radiation for these frequencies. ■ Only 6-meter band exhibits fairly constant propagation at distances of about 75 to 100 miles. Some ionospheric conditions permit propagation paths up to 2500 miles. ■ Frequencies at 144 MHz and above exhibit primarily line-of-sight communications with little ionospheric effects on this propagation. However, unusual propagation modes, such as ducting, will allow limited long-range communications.

FIGURE 3.9

The Yaesu FT-990 All Mode HF Transceiver. This superb amateur HF radio reflects the latest state-of-the-art in electronics technology. The FT-990 features all mode SSB, AM, CW, FSK operation; all amateur bands from 160 meters to 10 meters with 90 memories that store frequency, mode, and bandwidth for dual VFOs; receiver coverage from 100 kHz to 30 MHz, adjustable transmitter power output up to 100 watts (25 watts for AM); built-in high speed antenna tuner with 39 memories; and other advanced capabilities. *(Courtesy of Yaesu USA, 17210 Edwards Road, Cerritos, CA 90701, Telephone: (301) 404-2700)*

Nighttime propagation conditions

The disappearance of the D layer after sunset allows the 160- and 80-meter amateur band waves to pass unobstructed to the E and F layers. Sky-wave propagation at these bands increases to distances of about 2500 miles (4000 kilometers) by F-layer reflections.

The E layer loses most of its ionization properties after sunset and has very little effect on HF propagation. However, occasional sporadic-E regions may be present and contribute to propagation paths well into the VHF frequencies.

Shortly after sunset, the F1 and F2 layers combine into a single layer with slowly decreasing ionization levels. This results in a lowering of the MUF to the 40-meter band for amateur communications.

Table 3.2 lists the primary modes of propagation in the amateur bands. Space does not permit a complete description of all modes of propagation involved in radio communications. Also, for purposes of brevity, the distances are given in miles. To convert to meters, multiply the distance in miles by 1.61.

Principles of Electricity and Magnetism

The study of electricity and magnetism provides a foundation for almost every aspect of radio communications. The basic principles of electrical current flow are used to describe the action of resistors, capacitors, and inductors in ac and dc circuits. The concepts of electromagnetism, for example, can be used to explain the operation of power-line, voice, or radio-frequency transformers.

The material in this chapter is designed to help you learn enough basic theory and practical knowledge for the FCC Technician Class examinations and subsequent operation of your Technician amateur station. You will also find that this theory and practical information will be helpful in understanding the operation of vacuum tubes, transistors, amplifiers, and other radio circuits. Finally, mastering the material in this and subsequent chapters will lay the foundation for your study of more advanced amateur license examinations.

Definitions

Alternating current Alternating current (ac) is a flow of current that alternates or reverses its direction of flow on a periodic basis. The ac power used in most homes in the United States exhibits a complete reversal of current flow 60 times per second. Thus, the frequency of this ac power can be described as 60 cycles per second or 60 hertz (Hz).The unit of current is the ampere.

Capacitance Capacitance is defined as the ability to store electrical energy by means of an electrical field. In a physical sense, a capacitor consists of two conductors (such as two parallel metal plates) separated by an insulator known as a dielectric. The dielectric can be waxed paper, ceramic, air, or other insulating material. The unit of capacitance is the farad.

Direct current Direct current (dc) can be described as a flow of current in one direction only. For example, a battery forces current to flow in an electrical circuit in one direction only.

Electrical power Expressed in units of watts (W), electrical power is the rate at which electricity is performing work. Electrical power can be converted to heat when a current flows in a resistor or to physical motion when current flows in an electric motor.

Electromotive force Electromotive force (emf) can be described as a difference in potential, or voltage, which forces the flow of current in a conductor. It is analogous to pressure in a water pipe that forces water to flow through the pipe. The unit of electromotive force is the volt.

Frequency Frequency is used to describe the number of alternating cycles of an ac current flow in one second, or the number of complete wave cycles per second of an electromagnetic wave. The basic unit of frequency is the cycle per second, usually referred to as a hertz. Related terms are:

1 Kilohertz (kHz) = 1000 Hz.

1 Megahertz (MHz) = 1,000,000 Hz.

INDUCTANCE Inductance implies an ability to store electrical energy by means of a magnetic field. Current passing through an inductor or coil creates magnetic lines of force. The inductance of a coil can be defined in terms of the number of magnetic lines of force generated by a unit of current flowing in the coil. The unit of inductance is the henry.

RESISTANCE Resistance is simply the opposition to the flow of current. Resistance is measured in units of ohms.

Fundamentals of Electricity

The first recorded observation of electricity was made in ancient Greece about 2600 years ago. Thales of Miletus discovered that when amber was rubbed with a wool cloth, it would attract small bits of straw. However, this discovery would go unnoticed until the sixteenth century. William Gilbert, an English physicist, made a detailed study of Thales' observations and found that many substances such as hard rubber and glass would exhibit this force of attraction. Because the Greek word for amber is elektron, Gilbert proposed that this attractive force be called electricity.

By the eighteenth century, scientists had discovered two types of electrical charges, which were designated as positive and negative. This led to the development of a fundamental law of electricity, the *law of charges.* Simply stated, the law of charges governs the behavior of particles of matter possessing electrical charge: *Like Charges Repel—Unlike Charges Attract.*

Figure 4.1 illustrates the basic principles involved in the law of charges. You will find this fundamental law of electricity useful in describing the flow of electrical current in conductors, capacitors, inductors, transistors, and other components.

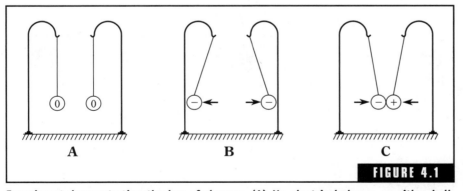

A   B   C

FIGURE 4.1

Experiment demonstrating the law of charges. (A) No electrical charge on either ball. No force of attraction or repulsion is exerted. (B) Both sides possess a negative charge. This causes a force of repulsion which tends to separate the balls. (C) Balls possessing unlike charges are attracted to each other.

Atoms and matter

Near the end of the nineteenth century, most of the basic laws of electricity had been discovered and documented. The scientists involved in this field knew the effects of electricity, but not the cause or origin. The battery, motor, generator, electric lamp, telegraph, telephone, and even the wireless had been invented before science could explain the nature of electricity.

The atomic theory of matter, developed jointly by chemists and physicists, brought about the first plausible explanation of electricity. All matter can be described primarily in terms of elements that are composed of varying numbers of electrons, protons, and neutrons—the basic building blocks of nature. There are 104 different elements known to date, ranging from the simple atoms of hydrogen and helium to the more complex atoms of silicon, copper, and californium. Figure 4.2 shows models of the hydrogen and helium atoms.

Unfortunately, there are still many unanswered questions and incomplete explanations regarding the current theory for the atomic structure of matter. Many other subatomic particles have been discovered or postulated to account for certain inconsistencies observed to date. The dual nature of the electron, acting alternately as a solid particle of matter and as a wave of energy cannot be explained. Fortunately for the electronics field, the present theory of atomic structure provides an adequate and reasonable explanation for the nature of electricity.

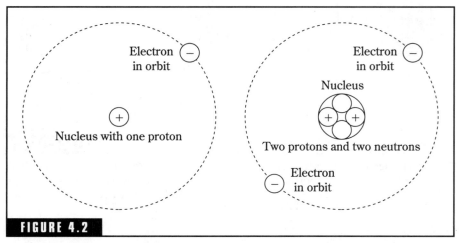

FIGURE 4.2

Models of hydrogen and helium atoms.

Negative and positive charges

Electrons are tiny particles of matter that possess a negative electric charge, sometimes referred to as a "unit of electric charge." Electrons normally travel in specified orbits around the nucleus of an atom. Electrons in the inner orbits are tightly bound to the atoms while electrons in an outer orbit might be loosely held to the atom. Under the influence of external energy such as heat, an outer orbit electron might escape from the parent atom and become a "free electron." This represents a negative charge lost by the atom, making the atom positively charged.

Protons and neutrons are tiny particles contained within the nucleus of the atom. Each proton possesses a positive electric charge, equal and opposite to that of the electron. However, the proton has a mass about 1800 times the mass of the electron. The neutron, which has no electric charge, has a mass approximately equal to the mass of the proton.

Electricity—the flow of electrons

When an atom loses an electron, the free electron becomes mobile and acts as a moving negative charge. In a material such as copper, many free electrons are created and are free to move in random directions throughout the material. These electrons eventually collide with atoms or other electrons, losing or gaining energy in the process. The atom releasing an electron becomes a *positive ion*, which is not free to move about in a solid material.

The movement of the free electrons, each with a negative charge, does not affect the total electric charge on the material. We still have an equal number of positive ions in the material that balances out the number of free electrons. In fact, a free electron colliding with a positive ion can recombine to form a normal or neutral atom.

The type of material and its ambient temperature are the major factors that determine the number of free electrons present at any given time. At normal room temperature, materials such as silver, copper, and gold possess a large number of free electrons. For example, copper has about 1.4×10^{24} free electrons per cubic inch (16.4 cubic centimeters) at normal room temperature. (This extremely large number is equal to 14 followed by 23 zeros!) Figure 4.3A shows the random movement of free electrons in a piece of copper wire.

The availability of large numbers of free electrons allows the flow of negative charge within a material. When we connect the copper wire to a battery as shown in Fig. 4.3B, this creates a potential difference or voltage across the wire. The negatively charged electrons are immediately attracted to the positive pole of the battery while the negative pole repels the electrons. The flow of electrons through the copper wire constitutes a current. In a practical sense, connecting a copper wire across a battery would rapidly drain the battery.

These forces of attraction and repulsion can be thought of as electrical pressure, similar to pressure in a water pipe that forces water through the pipe. Current can be thought of as analogous to the amount of water flowing in the pipe at any one time.

Units of voltage and current

The unit of potential difference or voltage is the volt. Sometimes voltage is referred to as electromotive force or emf. Voltage levels in electronic equipment might vary from microvolts (one microvolt is equal to 0.000001 volt) to the kilovolt level (1000 volts). The instrument used to measure voltage is the voltmeter. Voltmeters are available in many types including the VOM (volt-ohm-milliammeter), VTVM (vacuum-tube voltmeter), and the TVM (transistorized voltmeter).

The flow of electrical charge or current in a conductor is defined in units of amperes. Like voltage levels, current in electronic circuits can

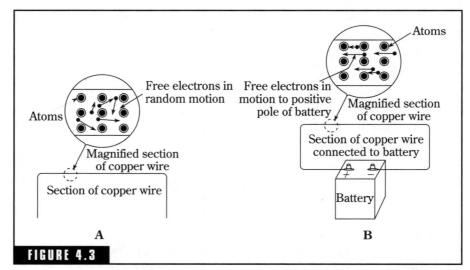

FIGURE 4.3

Flow of free electrons in copper wire. (A) When copper wire is not connected to a circuit, free electrons travel in random directions. (B) When copper wire is connected to a circuit, free electrons are attracted to positive pole of the battery.

vary from microamperes (0.000001 ampere) to many amperes. Current meters, called ammeters, are used to measure current.

Conductors and insulators

In electrical terms, all materials can be divided into two basic categories—*conductors* and *insulators.* Conductors possess large numbers of free electrons and are capable of conducting sizable amounts of electrical current. Silver is the best conductor, followed closely by copper, gold, and aluminum.

Materials that have very few free electrons are known as insulators. Almost all of the electrons in the outer atomic orbits of these materials are tightly bound in their orbits. Glass, wood, plastics, air, and distilled water fall into this category. Electrons can only be dislodged from their orbits when extremely high voltage levels are placed across these materials. For example, air is an efficient insulator until subjected to a voltage level of about 30,000 volts per centimeter (or about 76,000 volts per inch). Mica, one of the best known insulators, will withstand voltage levels of about 2,000,000 volts per centimeter (or about 5,000,000 volts per inch) before it breaks down and acts as a conductor.

An "in-between" group of materials, being neither good conductors nor insulators, includes resistors and semiconductors. Resistors are simply materials with a limited number of free electrons and are used for controlling current in a circuit. The action of semiconductors is covered in Chap. 5.

Resistance and resistors

Resistance can be considered as opposition to the flow of current in a circuit. In the water pipe analogy, the size of the pipe controls the amount of water that can flow for a given pressure. A smaller pipe restricts the flow of water. Similarly, a conductor having fewer free electrons restricts the flow of current for a given voltage.

The flow of free electrons in a material as a result of an applied voltage is opposed by force similar to friction. This opposition is due to the countless number of collisions between the free electrons and atoms in the material. Energy in the form of heat is generated, raising the temperature of the material. If the voltage applied to the material is to a critical level, the amount of heat can raise the temperature past the melting point of the material. At this point, the material will vaporize and the electrical component will be destroyed or damaged. For example, applying too much voltage across a lamp or flashlight bulb will burn out the filament.

The unit of resistance is the *ohm.* Resistors used in electronic circuits can possess values of resistance ranging from less than one ohm to millions of ohms (or megohms). The values of many units in electronics vary over a wide range. A list of standard numerical expressions is given in Table 4.1. These terms apply to voltage, current, resistance, and many other electrical units.

Resistors are available in many types and sizes, and can have either fixed or variable values. Figure 4.4 shows some of the typical resistors used in amateur radio equipment. The resistors shown on the left are fixed power types and are usually wirewound. The fixed carbon composition resistors in the center range from ⅛ watt to 2 watts. Variable resistors and potentiometers are shown on the right. Other fixed resistors include metal film, oxide, and film-chip types with power ratings of 1⁄16 to 1 watt. The size of the resistor usually indicates the power rating, and the color bands or stripes provide a coding for the value of resistance and tolerance factor. Table 4.2 shows the color-code scheme used for these types of resistors. The values of resistances normally vary from 2.7 ohms to 22 megohms. In

Table 4.1. Numerical Expressions Used in Electronics

Expression	Abbreviation	Multiplying factor	Value in powers of ten
giga	G	1,000,000,000	10^9
mega	M	1,000,000	10^6
kilo	k	1000	10^3
milli	m	0.001	10^{-3}
micro	μ	0.000001	10^{-6}
nano	n	0.000000001	10^{-9}
pico	p	0.000000000001	10^{-12}

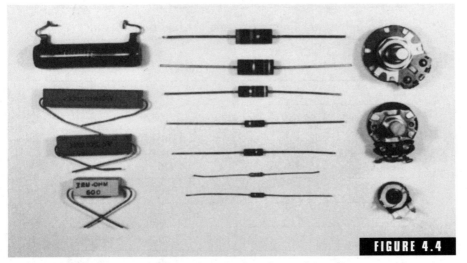

Typical resistors used in amateur radio and other electronic equipment.

some instances, carbon composition resistors with values of one ohm or less are available.

The tolerance factor of a resistor indicates the range that the resistance value can vary due to manufacturing differences. Carbon composition resistors are available with tolerances of 5% (gold band), 10% (silver band), and 20% (no tolerance band). For example, a 1-KΩ (or 1000-ohm) resistor with a tolerance of 10% might vary from 900 ohms to 1100 ohms and still be in tolerance. Resistors with low tolerance ratings cost more and generally are not used unless the circuit requires a more precise value of resistance. Special resistors with tolerance factors of 1% or less are available for special applications.

Variable resistors are available either as *rheostats* (two-terminal devices) or *potentiometers* (three terminals). The rheostat is simply a resistor capable of being adjusted over a specified range of resistance. The potentiometer, as the name implies, is a variable potential device acting as a voltage-divider network. The potentiometer is used to control the level of voltage being applied to a circuit element. For example, a potentiometer might serve as a volume control in a receiver, allowing the level of the audio signals to be varied as desired.

The instrument used to measure the resistance of a resistor or between any two points of an electrical circuit is the ohmmeter.

Table 4-2. Resistor Color Code and Method of Marketing

Color of band	Significant figures	Multiplier	Tolerance, percent (%)
Black	0	1	—
Brown	1	10	—
Red	2	100	—
Orange	3	1000	—
Yellow	4	10,000	—
Green	5	100,000	—
Blue	6	1,000,000	—
Violet	7	10,000,000	—
Gray	8	100,000,000	—
White	9	1,000,000,000	—
Gold	—	0.1	5
Silver	—	0.01	10
No Color	—	—	20

Resistor with color code bands

First significant figure — Tolerance factor

Second significant figure — Multiplier

Example:
Red-red-brown-gold
equals 220 ohms
with a 5% tolerance

Some Basic Electrical Laws

So far we have covered voltage, current, and resistance, the three basic electrical units used in electrical measurements. These three units are interrelated; you can't change one of these units in a particular circuit without changing the others. Figure 4.5 shows a simple circuit consisting of a battery and a resistor. When we apply one volt across a one-ohm resistor, we find that one ampere of current flows in this circuit. George Simon Ohm, a German school teacher, set up a similar circuit in 1828 and discovered this relationship.

Ohm's law

Ohm's law is often referred to as the foundation of electricity. It can be applied to the flow of current in a single resistor or to a complex circuit involving many series- and parallel-connected resistors. The relationship between current, voltage, and resistance can be expressed by the following simple equation:

$$I = \frac{V}{R} \text{ amperes} \qquad \textbf{(Eq. 4.1)}$$

where
    I is the current in amperes
    V is the voltage in volts
    R is the resistance in ohms

This equation is useful for solving for current when the voltage and resistance values are known. For example, if you connect a 10-ohm resistor across a 12-volt battery, the resulting current flow is 12/10, or 1.2 amperes. Ohm's law can be restated in two additional equations when solving for voltage or current:
To find voltage when current and resistance values are given:

$$V = I \times R \text{ volts} \qquad \textbf{(Eq. 4.2)}$$

To find resistance when voltage and current values are given:

$$R = \frac{V}{I} \text{ ohms} \qquad \textbf{(Eq. 4.3)}$$

When you use Ohm's law, you must express voltage, current, and resistance in units of volts, amperes, and ohms, respectively. Current, for example, cannot be expressed in units of milliamperes or microamperes. All such values must be converted to the basic units of measurement. The following example illustrates the use of Ohm's law.

Example 1 If the current through a 2-K resistor is found to be 3 mA, what is the voltage across the resistor?

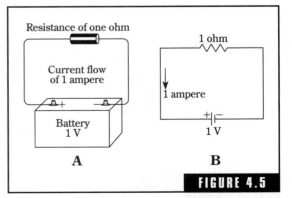

FIGURE 4.5

Electrical circuit illustrating the definition of one ohm of resistance. **(A)** Pictorial representation. **(B)** Schematic symbol representation.

Solution

Step A. Convert the 2-K resistance to ohms:

$$1000 \times 2 = 2000 \text{ ohms}$$

Step B. Convert the 3 mA to amperes:

$$0.001 \times 3 = 0.003 \text{ ampere}$$

Step C. Using Equation 4.2, compute the voltage across the resistor:

$$V = I \times R = 0.003 \times 2000 = 6 \text{ volts (Answer)}$$

Resistors in series

Many electronic circuits have more than one resistor connected in series. Figure 4.6 shows networks with two, three, and four resistors in series. In order to find the total resistance of two or more resistors in series, simply add the values of each individual resistor:

$$R_T = R_1 + R_2 + R_3 + \ldots \text{ ohms} \qquad \textbf{(Eq. 4.4)}$$

where

R_T is the total value of resistance

R_1, R_2, R_3, etc., are the values of each resistor

Resistors in parallel

When two or more resistors are connected in parallel, the resulting resistance of the parallel network is less than the value of the smallest resistor. In the simple case where the value of each resistor is the same, the value

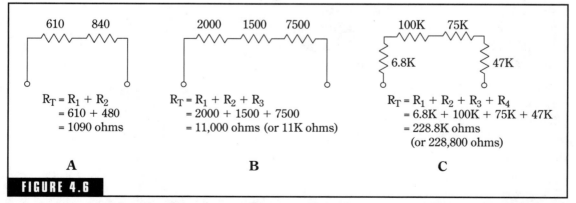

| **A** | **B** | **C** |

FIGURE 4.6

Examples of resistors in series. (A) Two resistors in series. (B) Three resistors in series. (C) Four resistors in series.

of the parallel combination is equal to the value of any one resistor divided by the number of resistors. For example, if four 100-ohm resistors are connected in parallel, the resulting resistance is 100/4, or 25 ohms.

When two resistors of unequal values are connected in parallel, the resulting resistance is found by the following equation:

$$R_T = \frac{R_1 \times R_2}{R_1 + R_2} \text{ ohms} \qquad \textbf{(Eq. 4.5)}$$

where

R_T is the resulting parallel resistance

R1 and R2 are values of each individual resistor

A more complex equation is required to compute the parallel resistance of three or more unequal resistors. A pocket calculator is handy for computing parallel resistances when using this equation or Equation 4.5.

$$R_T = \frac{1}{\dfrac{1}{R_1} + \dfrac{1}{R_2} + \dfrac{1}{R_3} + \ldots} \text{ ohms} \qquad \textbf{(Eq. 4.6)}$$

Figure 4.7 illustrates parallel combinations of resistors and examples for computing these different cases.

Series-parallel combinations

Any complex combination of series and parallel resistors can be reduced to a single equivalent resistance. In general, you determine the equivalent resistance of each parallel combination separately as a first step. Then add up all of these equivalent resistances to find the total circuit resistance. Sometimes it helps to redraw the circuit at each step. Figure 4.8 shows an example of reducing a complex network to a single value of resistance.

Current flow in series dc circuits

The current flowing in a series dc circuit is the same level at any point in the circuit. For example, in Fig. 4.9, the current leaving the positive terminal of the battery is equal to the current flowing through the resistor. In this simple circuit, Ohm's law can be used to compute the circuit current. This type of circuit is referred to as a direct-current or dc circuit because the current flows only in one direction.

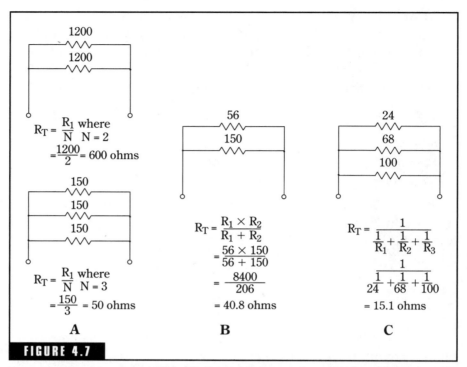

FIGURE 4.7

Examples of resistors in parallel. (A) Parallel resistors of equal value. (B) Two parallel resistors of unequal value. (C) Three parallel resistors of unequal value.

When two or more batteries and resistors are connected in a series circuit, such as Fig. 4.10A, the total circuit voltage and resistance must be determined before the current can be computed. This bring us to a new concept—how battery voltages are added in a series circuit. The following steps are used to analyze this circuit:

- Batteries connected in *series-aiding* have a total voltage equal to the sum of the individual batteries. In this case, batteries A and B are connected in series-aiding and their resulting voltage is 12 + 6, or 18 volts.

- Batteries connected in series-opposing have a total voltage equal to the difference of the individual batteries. The resulting polarity of the combination is equal to the polarity of the higher battery voltage. Thus in Fig. 4.10A, the AB combination of 18 volts is opposing the 6 volts of battery C. The three bat-

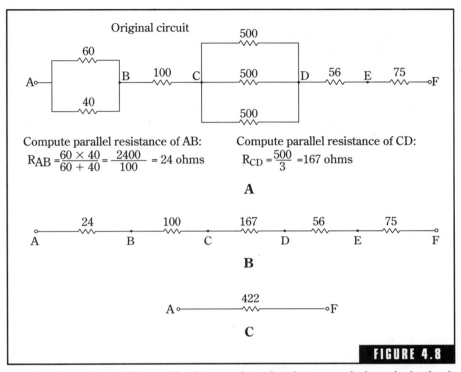

Compute parallel resistance of AB:

$$R_{AB} = \frac{60 \times 40}{60 + 40} = \frac{2400}{100} = 24 \text{ ohms}$$

Compute parallel resistance of CD:

$$R_{CD} = \frac{500}{3} = 167 \text{ ohms}$$

A

B

C

FIGURE 4.8

Resistors in series-parallel combinations can be reduced to an equivalent single circuit resistance. (A) Reduce parallel networks AB and CD to equivalent resistances. (B) Redraw the circuit showing only series resistances. Add values of series resistances to obtain total circuit resistance. (C) Total resistance of series-parallel combination of resistors.

teries connected in series result in a net voltage of 12 + 6 − 6, or 12 volts. In other words, one 6-volt battery connected in one direction cancels out the other 6-volt battery connected in the opposite direction.

■ The total circuit resistance in Fig. 4.10A is equal to the sum of the individual resistors, or 30 + 35 + 5 + 20 + 30, or 120 ohms.

■ Ohm's law can now be used to compute the current flowing in the circuit.

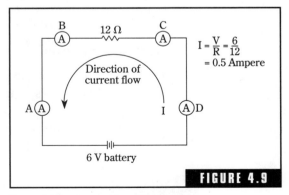

$$I = \frac{V}{R} = \frac{6}{12}$$
$$= 0.5 \text{ Ampere}$$

FIGURE 4.9

Current flow in a series circuit.

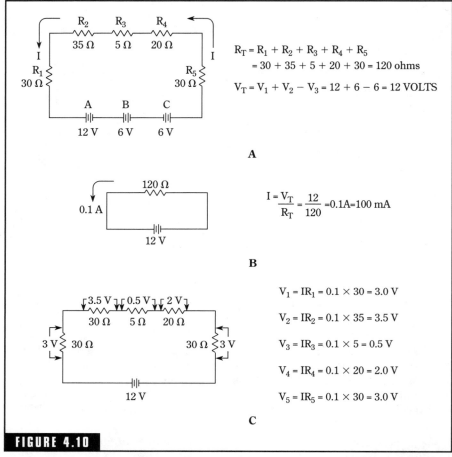

$R_T = R_1 + R_2 + R_3 + R_4 + R_5$
    $= 30 + 35 + 5 + 20 + 30 = 120$ ohms

$V_T = V_1 + V_2 - V_3 = 12 + 6 - 6 = 12$ VOLTS

A

$I = \dfrac{V_T}{R_T} = \dfrac{12}{120} = 0.1A = 100$ mA

B

$V_1 = IR_1 = 0.1 \times 30 = 3.0$ V

$V_2 = IR_2 = 0.1 \times 35 = 3.5$ V

$V_3 = IR_3 = 0.1 \times 5 = 0.5$ V

$V_4 = IR_4 = 0.1 \times 20 = 2.0$ V

$V_5 = IR_5 = 0.1 \times 30 = 3.0$ V

C

FIGURE 4.10

Analysis of a series dc circuit. (A) The original series dc circuit. (B) The circuit reduced to an equivalent circuit. (C) Calculating IR voltage drops.

IR voltage drops

This brings us to another new concept—the voltage across any resistor in a series circuit is equal to the current times the resistance (from Ohm's law). This gives us a powerful tool for computing the voltage at any point in a series circuit. Figure 4.10C shows the voltage across each resistor based on a current of 0.1 ampere.

If you add up the voltage drops in a series circuit, you find that their sum is equal to the source voltage. In Fig. 4.10C, note that the IR drops of 3, 3.5, 0.5, 2, and 3 volts are equal to the effective source voltage of 12 volts.

Current flow in parallel dc circuits

When two resistors are connected in parallel across a battery, the current flowing through each resistor depends on the resistance encountered. Consider the circuit in Fig. 4.11. The battery voltage across resistor R_1 is the same as the voltage across resistor R_2. However, the current through R_1 is 12/24, or 0.5 ampere and the current through R_2 is 12/12 or 1.0 ampere. Therefore, the battery must supply a total of 0.5 and 1.0, or 1.5 amperes to the parallel network.

This simple example demonstrates an important concept in current flow—the current entering a parallel network is equal to the sum of the individual branch currents within the network. At point A in Fig. 4.11, the total current I_T supplied by the battery is equal to the sum of I_1 and I_2, the individual branch currents.

You can use this concept to compute the current flow in any series-parallel network. First, it is necessary to compute the total circuit resistance and the total current flowing in the circuit. The second step is to compute the IR voltage drop across each parallel network. After this is done, the currents in each parallel network can readily be computed. Note that most of these computations use different variations of Ohm's law.

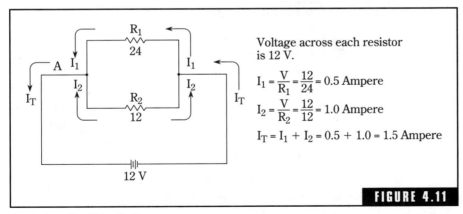

Voltage across each resistor is 12 V.

$$I_1 = \frac{V}{R_1} = \frac{12}{24} = 0.5 \text{ Ampere}$$

$$I_2 = \frac{V}{R_2} = \frac{12}{12} = 1.0 \text{ Ampere}$$

$$I_T = I_1 + I_2 = 0.5 + 1.0 = 1.5 \text{ Ampere}$$

FIGURE 4.11

Parallel dc circuit analysis.

Magnetism

The attraction and repulsion properties of magnets are similar to those described previously for electrical forces. Recorded history of experiments with natural magnets dates back some 2600 years to when the early Greek scientists described the magnetic properties of an ore material. This material, known as *magnetite* or *lodestone,* is the only substance occurring in nature that possesses magnetic qualities. Chinese sailors used crude forms of magnetic compasses as early as the second century for navigation purposes.

It was not until the early 1800s that the relationship between electricity and magnetism was discovered. A Danish physicist, Hans Christian Oersted, found that a compass placed near a conductor carrying an electrical current would swing away from north and point toward the conductor. Later, it was discovered that a current-carrying conductor wound around an iron rod would create a magnetic field and magnetize the rod. You can make a simple magnet by winding an insulated wire around an iron nail and connecting the two ends of the insulated wire to a battery. Figure 4.12 illustrates this simple experiment. The compass is used to detect the presence of a magnetic field. Prior to connecting the battery to the coil, the compass will point toward the magnetic north pole of the earth. When the battery is connected to the circuit, the compass needle will swing from the magnetic north to the direction of the nail magnet.

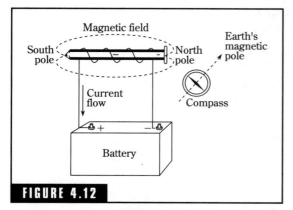

FIGURE 4.12

A simple magnetconsisting of a soft iron nail, a conductor (insulated) wound around the nail, and a battery.

The magnetic field

Magnetic fields exhibit unique forces of attraction and repulsion. For convenience of notation, the two ends of a magnet are designated *north* and *south* poles. All magnets possess a north and a south pole. Magnetic lines of force emerge from the north pole and travel back to the south pole of the magnet. Figure 4.13 illustrates the distribution of magnetic lines of force surrounding a bar magnet. The magnetic field intensity is greatest at each end of the magnet. The total number of magnetic lines of force traveling from the north to the south pole is called the magnetic flux. The flux density of a magnetic field is defined as the number of lines per unit area at a given point. The more lines there are per unit area, the stronger the magnetic field will be.

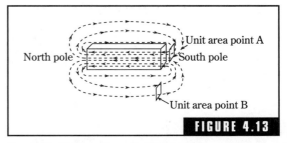

FIGURE 4.13

Magnetic field or flux distribution for a permanent magnet.

If the north pole of a magnet is placed near the south pole of a second magnet, the two magnets will be attracted to each other. However, if two like poles (north-north or south-south) of two magnets are placed close together, there is a repelling force and the two magnets will attempt to move away from each other. This action is known as the law of magnetic poles and can be stated as follows: *Like poles repel; unlike poles attract.*

Although magnetic behavior of attraction and repulsion is similar to that of electrically charged particles, there is one fundamental difference. The electrically charged particle will possess either a negative or positive charge over its entire surface. A magnet, on the other hand, will always possess a north pole at one end and a south pole at the other end. These twin poles can never be separated as individual entities.

Temporary and permanent magnets

Soft iron, such as the nail in Fig. 4.12, will retain magnetism as long as the current flows through the wire. When the current is turned off, the nail will lose most of its magnetism. This type of magnet is called a *temporary* magnet. The remaining magnetism left in the nail is referred to as *residual* magnetism.

When hard iron or steel is placed in a magnetic field, it will retain most of the magnetic property after the magnetic field is removed. When these types of materials are magnetized, they are referred to as *permanent* magnets. Most speakers in amateur receivers contain a powerful permanent magnet placed next to the voice coil. These speakers are designated as permanent-magnet, or pm, speakers.

Electromagnetism

A magnetic field is generated around a conductor when an electrical current flows through the conductor. The intensity of the magnetic

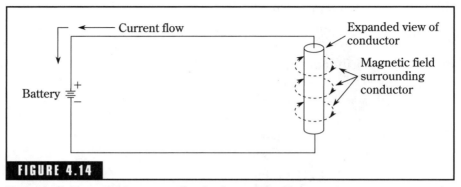

FIGURE 4.14

Magnetic field created by current flowing in a conductor.

field is directly related to the amplitude of the current; any increase in current results in a comparable increase in the magnetic field. Figure 4.14 illustrates this action.

If the conductor is wound in the form of a coil, as shown in Fig. 4.15, the strength or intensity of the magnetic field is increased many times. The magnetic lines of force associated with each turn of the conductor combine in the same direction with the magnetic lines of force from the other turns. Thus, more lines of magnetic force will be concentrated in a small space, all passing through the center of the coil.

A coil designed to create a magnetic field for exerting physical force or performing work in an electrical circuit is called an *electromagnet.* Electric motors, relays, solenoids, and headphones are some of the devices that use electromagnets for performing work functions.

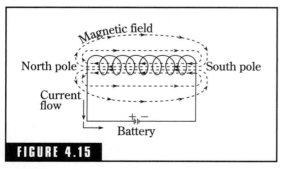

FIGURE 4.15

Magnetic field created by current flowing in a coil of wire.

Magnetomotive force (MMF)

The amount of magnetic force or strength of an electromagnet depends on the amount of current flowing in the coil, the number of turns, and the type of core material used in the coil. The first two factors, current in amperes and number of turns of the conductor, can be expressed in ampere-turns. This represents the magnetomotive force or magnetic potential, similar to the concept of electromotive force

(MMF) or voltage potential difference in electrical circuits.

Electromagnetic induction

When a conductor is moved through a magnetic field, cutting against the lines of magnetic force, a voltage will be induced in the conductor. Figure 4.16 shows the effect of moving a conductor through a magnetic field. This principle of electromagnetic induction is the basis for the operation of electric generators, transformers, and even radio antennas.

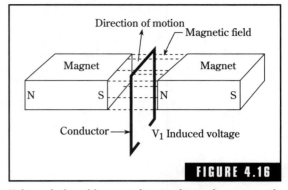

FIGURE 4.16

Voltage induced by a moving conductor in a magnetic field between two permanent magnets.

Alternating current and voltage

At the beginning of this chapter the concepts of direct current (dc) were used to demonstrate the relationships of voltage, current, and resistance. These same concepts can be used to look at the principles of alternating current (ac).

When the direction of current flow is reversed at periodic intervals, the resulting varying current is known as alternating current. The voltage producing an alternating current is, by definition, an alternating voltage that continuously varies in amplitude and periodically reverses its polarity. An ac generator is a source of alternating voltage.

The ac sine wave

Almost all ac waveforms encountered in radio communications, as well as the familiar 120-Vac, 60-Hz power in our homes, are sine waves. Let's take a look at the 60-Hz power and define some basics of ac waveforms. These basics are also applicable at audio and radio frequencies.

Figure 4.17 shows two complete cycles of the standard 120-Vac, 60-Hz voltage waveform. We call this a 60-Hz voltage because the frequency is 60 hertz, or 60 complete cycles per second.

The period of time for each cycle is directly related to the frequency by the following equation:

$$f = \frac{1}{T} \text{ or} \qquad\qquad \text{(Eq. 4.7)}$$

$$T = \frac{1}{f} \qquad\qquad \text{(Eq. 4.8)}$$

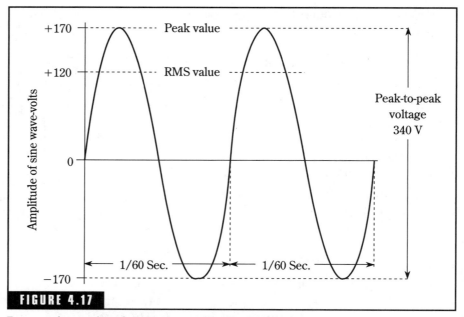

Two complete cycles of a 120-Vac, 60-Hz sine wave.

where

f is the frequency in Hz

T is the period of each cycle in seconds

Thus, the period or the duration of each ac cycle of the 120-Vac, 60-Hz voltage is ¹⁄₆₀, or 0.0167 seconds (16.67 milliseconds).

We are familiar with the expression 120 Vac power. However, Fig. 4.17 shows that the maximum values of voltage range from a positive 170 volts to a negative 170 volts during each ac cycle. This results in a peak-to-peak voltage of 340 Vac.

The ac voltage and current values are often expressed in *rms* terms. The expression rms is an abbreviation for "root mean square," a mathematical expression for describing the effective value of an ac waveform. In practical terms, a 120-Vac, 60-Hz source provides the same amount of power that a 120-Vdc source would provide. This is a convenient way to measure ac power. For example, a 120-volt light bulb will glow with an equal brilliance when connected to either a 120-Vac or 120-Vdc source.

The relationship between peak ac voltage and rms voltage for a sine waveform is given by the following equation. This also applies to peak ac and rms current values.

$$V_p = 1.414 \ V_{rms} \ \text{or} \qquad \textbf{(Eq. 4.9)}$$

$$V_{rms} = 0.707 \ V_P \qquad \textbf{(Eq. 4.10)}$$

where

V_P is the peak value of the ac voltage

V_{rms} is the rms or root-mean-square value of the ac voltage

Each ac sine wave is divided into 360 electrical degrees. This helps us to designate the phase relationships of two or more ac waveforms which are not in phase with each other. Figure 4.18 shows voltage and current waveforms with both in-phase and out-of-phase relationships.

ac circuits

The analyses of ac circuits are more complex when compared to dc circuits. Some of the unusual characteristics of ac circuits that have no equal in dc circuits are:

■ Opposition to ac current flow in an ac circuit is not limited to resistance. The total opposition to ac current flow is known as *impedance,* which includes resistance. Impedance also includes *capacitive reactance* (ac opposition developed by a capacitor) and *inductive reactance* (ac opposition developed by an inductor). Like resistance, impedance, capacitive reactance, and inductive reactance are measured in units of ohms. The symbols used for impedance, capacitive reactance, and inductive reactance are Z, X_C, and X_L, respectively.

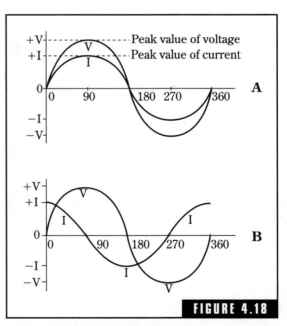

FIGURE 4.18

Phase relationships showing both in-phase and out-of phase voltage and current. (A) Current and voltage in phase. (B) Current leading voltage by 90 degrees.

■ ac current will flow through a capacitor, but a capacitor will block the flow of dc current. As the frequency of the ac current is increased, the capacitive reactance in ohms will decrease. At high frequencies, a capacitor can act as a short circuit for all practical purposes.

■ An inductor or coil possesses both inductance and dc resistance. The inductance produces an inductive reactance that increases as the frequency of operation is increased. In a dc circuit, only the dc resistance of the coil offers opposition to the flow of current. However, in an ac circuit both inductive reactance and dc resistance offer opposition to the flow of current. As the frequency is increased, the ac opposition will increase until finally the inductor acts as an open circuit. For all practical purposes, this reduces the ac current flow to zero.

■ In an ac circuit, energy can be transferred from one circuit to another by the electromagnetic induction action of transformers with no physical connection between the circuits. Also, transformers can be used to step-up or step-down ac voltages or currents. In dc circuits, there are no equivalent transfer characteristics.

■ In an ac circuit, the phase relationship between voltage and current may vary such that the voltage may lead or lag the current by a given phase angle usually expressed in electrical degrees. There is no similar parallel in dc circuits—the voltage is always in phase with the current.

Capacitors and Capacitance

Capacitors, originally called condensers, are used in almost every electronic circuit. A typical amateur transceiver will have almost as many capacitors as resistors, which are probably the most widely used components. Capacitors are available in a wide variety of sizes, capacitances, and working-voltage ratings. They can be divided into two broad categories—fixed and variable. Capacitors perform many functions, which include:

■ Coupling, transferring, and bypassing ac current.

■ Blocking dc voltages and currents.

■ Storing electrical energy and subsequent discharge of the charge stored in the capacitor.

■ Tuning frequency-sensitive circuits.

Figure 4.19 shows a collection of typical capacitors used in amateur radio equipment.

The capacitors shown on the left are variable capacitors, used for tuning radio-frequency circuits. Fixed coupling and bypass capacitors are shown in the center. These include paper, mylar, ceramic, and mica types. On the right are electrolytic capacitors, commonly used as filter capacitors in power supplies and related applications. The bottom-right capacitor is a low-voltage tantalum capacitor available in the range of about 0.1 to 100 microfarads.

We can define capacitance as the ability to store an electrical charge in a dielectric or insulating material. A simple capacitor, shown in Fig. 4.20, consists of two metal plates separated by a dielectric material such as air, glass, mica, or ceramic. The amount of capacitance developed by such a capacitor is related to the surface area of the plates, the

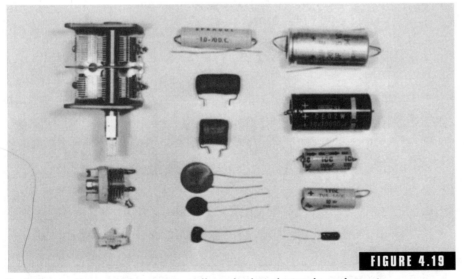

FIGURE 4.19

Typical capacitors used in amateur radio and other electronic equipment.

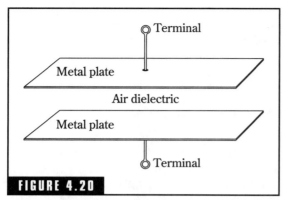

FIGURE 4.20

Simple capacitor consisting of two metal plates placed close to each other.

separation of the plates, and the dielectric constant of the dielectric material.

Capacitance is measured in units of farads (F). Because the farad is usually too large for most electronic applications, you will find the farad microfarads and micromicrofarads (or picofarads).

1 microfarad = 1 μF = 0.000001 F = 1 × 10^{-6} F

1 micromicrofarad = 1 μμF = 0.000000000001 F = 1 × 10^{-12} F

1 picofarad = 1 pF = 1 μμF = 1 micromicrofarad

How capacitors work

In performing its function in a circuit, a capacitor is charged and discharged sometimes as fast as millions to billions of times per second. A simple way to illustrate the action of a capacitor is to connect it across a battery as shown in Fig. 4.21A. The positive pole of the battery attracts the free electrons in the upper plate of the capacitor, leaving a net positive charge on this upper plate. At the same time, the negative pole of the battery repels the free electrons in the lower conductor, forcing them toward the lower plate of the capacitor. This results in a

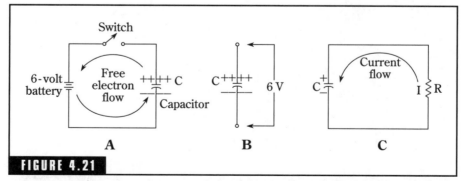

FIGURE 4.21

Charging and discharging action in a capacitor. (A) Charging the capacitor. (B) The capacitor remains charged when removed from the circuit. (C) Discharging the capacitor.

net negative charge on the lower plate. This current flow continues until the voltage across the capacitor plates is equal to the battery voltage, 6 volts in this case. The capacitor is then fully charged.

The positive and negative charges on the capacitor plates create an electric field between the two plates. This field distorts the molecular structure of the dielectric material, producing an electrical stress. In this manner an electric charge is stored in the dielectric.

The charged capacitor can now be disconnected from the battery, retaining an electrical charge that can perform work (Fig. 4.21B). If the capacitor is connected to a load such as a resistor as shown in Fig. 4.21C, the charge within the capacitor causes a current to flow until the electrical energy is dissipated in the resistor in the form of heat. At this point, the voltage across the capacitor drops to zero and the capacitor is discharged.

Capacitor voltage ratings

If the battery voltage in Fig. 4.21 is raised above the specified *working voltage* of the capacitor, the dielectric material will break down and conduct a heavy current. This action will damage or destroy many types of capacitors. The working voltage, usually specified by the manufacturer, is defined as the maximum amount of voltage that can be placed across the capacitor for safe, sustained operation. This voltage rating is usually expressed in terms of dc voltage (Vdc) levels. The working voltage ratings of capacitors can range from as low as 3 Vdc to thousands of volts dc.

If an ac voltage is to be placed across a capacitor, the peak value of the ac voltage must be used in selecting a capacitor with an adequate voltage rating.

Connecting capacitors in parallel

If two or more capacitors are connected in parallel, as shown in Fig. 4.22, the total capacitance is equal to the sum of the individual capacitors. This relationship is expressed as follows:

$$C_T = C_1 + C_2 + C_3 + \ldots \qquad \textbf{(Eq. 4.11)}$$

where

C_T is the total capacitance of the parallel combination in farads
C_1, C_2, C_3, etc., are the values of each individual capacitor in farads

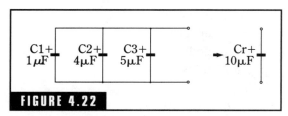

FIGURE 4.22

Parallel connection of capacitors.

Note that some capacitors are polarized i.e., the terminals are designated + and − (positive and negative). This is characteristic of most electrolytic capacitors. The polarity must be observed in parallel as indicated in Fig. 4.22. Furthermore, when electronic capacitors are connected in circuits with dc potentials, the + terminal of the capacitor must be connected to the positive terminal of the circuit. Otherwise, the electrolytic capacitor might be damaged by the dc potential.

Connecting capacitors in series

The total capacitance of two or more capacitors connected in series is always less than the value of the smallest capacitor. One way to remember how to compute the value of series capacitors is that it is the same method used to compute resistors in parallel.

Three equations are available for computing the net capacitance of capacitors connected in series. Figure 4.23 shows typical series capacitor circuits.

Two capacitors in series (Fig. 4.23A):

$$C_T = \frac{C_1 \times C_2}{C_1 + C_2} \text{ farads}$$ **(Eq. 4.12)**

Three or more capacitors in series (Fig. 4.23B):

$$C_T = \frac{1}{\dfrac{1}{C_1} + \dfrac{1}{C_2} + \dfrac{1}{C_3} + \dots} \text{ farads}$$ **(Eq. 4.13)**

Any number n capacitors in series where the value of each capacitor is equal to the value C (Fig. 4.23C):

$$C_T = \frac{C}{n} \text{ farads}$$ **(Eq. 4.14)**

Capacitive reactance

The alternating charge and discharge action of a capacitor allows an ac current to flow in a circuit. However, a capacitor, like a resistor, offers opposition to the flow of the ac current. This opposition, or capacitive reactance, depends upon the value of capacitance and the frequency of

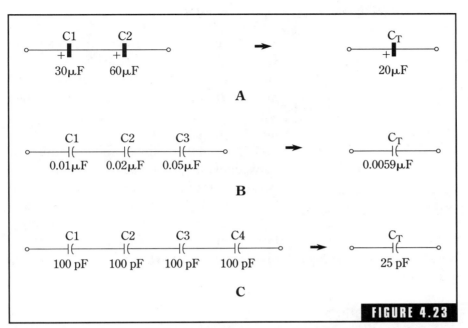

FIGURE 4.23

Series capacitor circuits. (A) Two capacitors in series. (B) Three capacitors in series. (C) Four capacitors of equal value in series.

the ac voltage being impressed on the capacitor. This relationship is expressed as follows:

$$X_c = \frac{1}{2\pi fC} \qquad \textbf{(Eq. 4.15)}$$

where

X_c is the capacitive reactance in ohms
π is the constant pi, which is equal to 3.1416
f is the frequency in hertz
C is the capacitance in farads

For convenience in computations, this equation may be reduced to

$$X_c = \frac{0.159}{fC} \text{ ohms}$$

The units of frequency and capacitance must be expressed in hertz and farads, respectively, when using these above equations. Note that when several capacitors are connected in series or parallel, the resulting capacitance from such combinations can be used to find the total capacitive reactance involved.

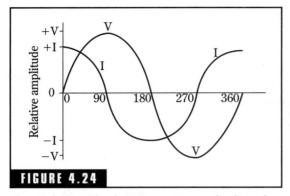

FIGURE 4.24

Phase relationships between current and voltage in a capacitor.

Phase angle of capacitors

The current flow in a capacitor leads the voltage across the capacitor by 90 electrical degrees. This means that current flow must start before any voltage is developed across the capacitor. This action was demonstrated in Fig. 4.21.

Figure 4.24 shows the relationship between voltage and current when an ac voltage is impressed across the capacitor. At 0 and 180 degrees, the current flow is at a maximum and the voltage across the capacitor is zero volts. Conversely, at 90 and 270 degrees, the voltage levels are at a maximum while the current flow is zero amperes.

Testing capacitors

Any capacitor can be quickly tested with a multimeter (set to measure resistance) to determine if it is shorted or leaky. With the exception of electrolytic capacitors, the resistance of virtually all other types should be infinite in value. For example, capacitors with air, mica, glass, ceramic, and similar dielectric materials should indicate an open circuit (infinite resistance) when tested with an ohmmeter.

CAUTION: *Don't try to test the capacitor when it is soldered in a "live" or powered-up circuit. Turn off all power to the circuit being tested by unplugging the ac power cord! Otherwise, you might experience a severe or fatal electrical shock. In many cases, you might also damage the multimeter or the circuit being tested. Damage to the multimeter can range from virtual destruction to requiring replacement of an internal fuse within the instrument.*

When you try to test a capacitor that is soldered in a circuit, adjacent resistors or other components connected in this circuit might show resistance on your multimeter, falsely indicating a defective capacitor. Unsolder one end of the capacitor before you test for a short or leaky condition. Any observable resistance reading for these types of capacitors indicates a defective component.

Unfortunately, the above continuity test will not indicate a defective capacitor with an open circuit. Further tests with a capacitor tester are required to ensure that the capacitor has the required capacitance. Some digital multimeters, including the portable handheld types, also have a capacitance test capability. Before buying any multimeter for your ham shack, you might want to consider one with this expanded test capability.

Many electrolytic capacitors, ranging from about 1 microfarad to many thousands of microfarads, will show some leakage resistance and still perform satisfactorily in a circuit. With a little experience, you will be able to check almost any type of capacitor with a multimeter. In most cases, the capacitor must be removed from the circuit for the test to be meaningful.

Inductors and Inductance

A simple inductor can be described as a coil with one or more turns of wire, wound around a material known as the *core.* Some inductors employ air as the core material. Like capacitors, inductors are used primarily in ac circuits. Inductors perform many important functions in electronic circuits such as transferring ac energy from one circuit to another and as frequency selective elements in RF circuits.

Although many types and sizes of inductors are available from commercial sources, they are not used as often as resistors and capacitors. Most inductors are manufactured for specific applications involving a given frequency or band of frequencies. In some instances, you might have to build a coil required for a specific application. However, coil forms and wire are available for most of these "home-brew" coils. Figure 4.25 shows a collection of inductors found in ham radio and other electronic equipment. The coils shown on the left are RF chokes. At the top center are two torroid inductors, one molded in plastic and the other a "home-brew" torroid for a project. The coils shown at the bottom are adjustable inductors with powdered-iron cores. An adjustable IF transformer, shown in the upper right, is used in 455-kHz IF amplifier circuits.

Inductance

Inductance is the ability of a conductor to develop an induced voltage when a varying or alternating current flows through the conductor.

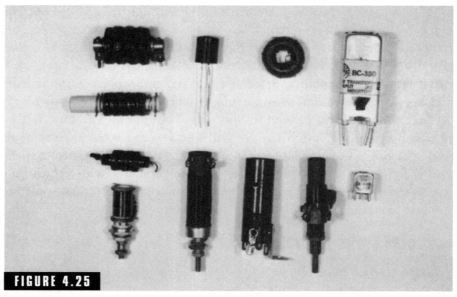

FIGURE 4.25

Typical inductors used in amateur radio and other electronic equipment.

This induced voltage is produced by the expanding and contracting magnetic field associated with the varying current.

When the conductor is wound in the form of a coil, the inductance is increased many times depending on the number of turns. The coiled conductor concentrates the magnetic field in a smaller volume and the expanding and contracting field cuts across each turn of wire, simultaneously generating a higher voltage.

The inductance of a coil can be increased many times by changing the core material. For example, a powdered-iron or ferrite core will increase the inductance of an air-core coil many times. These materials are used in radio-frequency coils to achieve higher efficiency of operation.

The unit of inductance is the henry (H). One henry is defined as the amount of inductance that allows one volt to be induced across a coil when the current through the coil changes at a rate of one ampere per second. Most coils used in audio- and radio-frequency circuits possess values of inductance in the millihenry and microhenry range.

1 millihenry = 1 mH = 0.001 H = 1×10^{-3} H

1 microhenry = 1 H = 0.000001 H = 1×10^{-6} H

A unique characteristic of inductances involves the action of the transformer. If the magnetic field created by one coil cuts across another coil, a voltage is induced in the second coil. In this manner, we can transfer ac energy from one circuit or another without any direct physical connection. Transformers are covered in more detail at the end of this section.

Series inductors

When two or more inductors are connected in series, as in Fig. 4.26, the total inductance is equal to the sum of the individual values of inductances, or:

$$L_T = L_1 + L_2 + L_3 + \ldots \qquad \textbf{(Eq. 4.16)}$$

where

L_T is the total inductance of the series combination
L_1, L_2, L_3, etc., are the values of the individual inductances

All values of inductances must be expressed in the same units of henrys.

This equation is not valid if the inductors are placed close together, resulting in a "transformer action." *Mutual inductance* is developed when the magnetic field from one coil cuts across another coil. In a series-aiding connection, additional inductance will be produced by this mutual inductance. On the other hand, if the coils are connected in a series-opposing manner, the mutual inductance will have a canceling effect and the total inductance of the two coils will be less than the sum of the individual values of inductance.

Parallel inductors

The equations for series and parallel inductors are similar to those for series and parallel resistor combinations.

Parallel combinations of inductors are illustrated in Fig. 4.27. The equations used to compute the total inductances are given here. Remember, these equations are not valid if mutual induction exists between any of the coils.

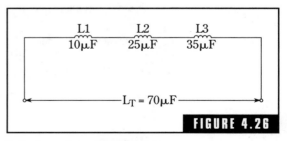

FIGURE 4.26

Inductors connected in series.

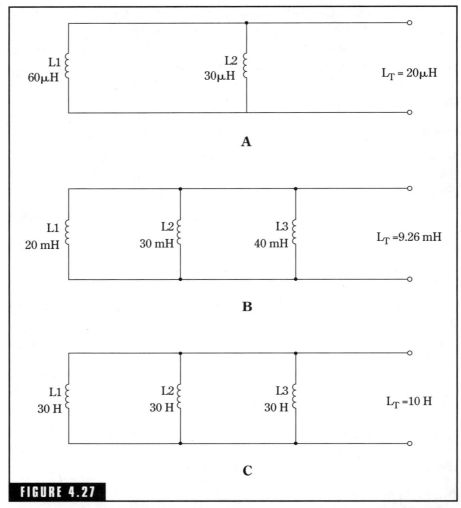

FIGURE 4.27

Inductors connected in parallel. (A) Two inductors connected in parallel. (B) Three inductors of unequal value in parallel. (C) Three inductors of equal value in parallel.

Two inductors, L_1 and L_2, connected in parallel (Fig. 4.27A):

$$L_T = \frac{L_1 \times L_2}{L_1 + L_1} \text{ henrys} \qquad \textbf{(Eq. 4.17)}$$

Two or more inductors connected in parallel (Fig. 4.27B):

$$L_T = \cfrac{1}{\cfrac{1}{L_1} + \cfrac{1}{L_2} + \cfrac{1}{L_3} + \ldots} \text{ henrys} \qquad \textbf{(Eq. 4.18)}$$

Any number n inductors in parallel where the value of each inductor is equal to the value L (Fig. 4.27C):

$$L_T = \frac{L}{n} \text{ henrys} \qquad \textbf{(Eq. 4.19)}$$

Inductive reactance, X_L

Inductive reactance is the opposition to the flow of ac current developed by an inductance. It is due to the opposing induced voltage created by the expanding and contracting magnetic field. This opposition, which does not include the dc resistance of the inductor, is measured in ohms.

Inductive reactance varies directly with both inductance and the operating frequency. An increase in either factor results in a corresponding increase in inductive reactance. This relationship is expressed as:

$$X_L = 2\pi fL \text{ or} \qquad \textbf{(Eq. 4.20)}$$

$$X_L = 6.28fL \qquad \textbf{(Eq. 4.21)}$$

where

X_L is the inductive reactance in ohms
π is the constant pi (3.1416)
f is the frequency in Hz
L is the inductance in hertz

Phase angle of inductors

A varying or alternating current flow produces an induced voltage across an inductor. In an ideal inductor (one with no dc resistance), the current lags the induced voltage by 90 electrical degrees. This means that the phase angle between the voltage and current is 90 degrees, as shown in Fig. 4.28.

This 90-degree phase angle is due to the fact that the induced voltage depends on the rate of change in the current flowing through the inductor. For example, at 0, 180, and 360 degrees, the current is

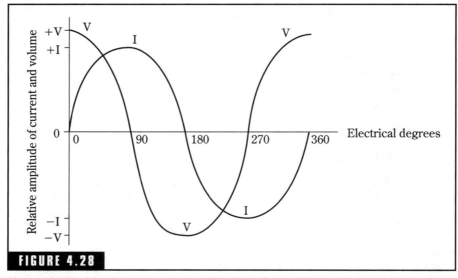

FIGURE 4.28

Phase relationships between current and voltage for an inductor in an ac circuit.

changing at maximum rate while at 90 and 270 degrees, the rate of current change approaches zero. We are not referring to actual values of current, but to how fast the current is changing in value. The voltage levels at 0, 180, and 360 degrees are at a maximum because the rate of current change is highest at these angles. However, the induced voltage drops to zero at 90 and 270 degrees because the current is not changing in value at these angles.

Transformers

The transformer is an electromagnetic device for transforming or changing alternating voltage or current from one level to another. In its simplest form, a transformer consists of two inductors placed close together so that a magnetic field generated in one inductor will cut across the other inductor. This induces a voltage in the second inductor.

Transformers are used in radio and electronic equipment for many applications. The ac power transformer shown in Fig. 4.29 can be used in the power supply circuit of an amateur receiver to transform the 120-Vac commercial power to 24 Vac. The power transformer also performs another important function, that of isolating the 120-Vac power

FIGURE 4.29

A typical 120-Vac transformer used in amateur radio and other electronic equipment.

line from the common ground connections of the receiver. This helps to eliminate potential shock hazard to the amateur operator.

The same receiver might contain an audio output transformer to drive the low-impedance speaker, or RF and IF transformers to couple RF and IF signals between stages.

Basic transformer concepts

Transformers can be classified as one-to-one, step-up, or step-down, depending on the ratio of the number of turns in the primary winding to the number of turns in the secondary winding. A one-to-one transformer produces a voltage in the secondary winding equal to that being applied to the primary winding. The step-up transformer produces a higher output voltage, when compared to the input voltage applied to the primary winding.

The turns ratio determines the general operating characteristics of a transformer: one-to-one, step-up, or step-down. The voltage ratio in terms of turns ratio is:

$$\frac{V_P}{V_S} = \frac{N_P}{N_S}$$

(Eq. 4.22)

where

V_P is the voltage across the primary winding in volts
V_S is the voltage across the secondary winding in volts
N_P is the number of turns in the primary winding
N_S is the number of turns in the secondary winding

Equation 4-22 can be rearranged to solve for the secondary voltage when the primary voltage and the turns ratio are known:

$$V_s = V_P \frac{N_s}{N_P}$$

(Eq. 4.23)

Step-up and step-down transformers are illustrated in Fig. 4.30 along with calculations required to determine the secondary voltages for each example.

The ratio of current in the primary winding to current in the secondary winding is the inverse of the turns and voltage ratios. This can be expressed as:

$$\frac{I_S}{I_P} = \frac{N_P}{N_S} = \frac{V_P}{V_S}$$

(Eq. 4.24)

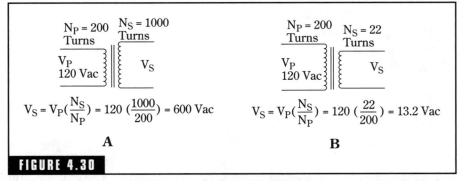

$$V_S = V_P(\frac{N_S}{N_P}) = 120\ (\frac{1000}{200}) = 600\ \text{Vac} \qquad V_S = V_P(\frac{N_S}{N_P}) = 120\ (\frac{22}{200}) = 13.2\ \text{Vac}$$

A   B

FIGURE 4.30

Examples of iron core transformers showing the effect of the turns ratio on the secondary voltage. (A) Step-up transformer provides increased secondary voltage. (B) Step-down transformer provides reduced secondary voltage.

where

I_s is the current in the secondary winding in amperes

I_P is the current in the primary winding in amperes

Equation 4-24 can be rearranged to solve for the secondary current, I_s, as follows:

$$I_S = I_P \frac{V_P}{V_S} = I_P \frac{N_P}{N_S} \qquad \textbf{(Eq. 4.25)}$$

One interesting aspect of step-up or step-down transformers is that the step-up transformer reduces the output current available while the step-down transformer provides for an increase in current in the secondary winding.

Some transformers have more than one secondary winding. The equations just given for finding output voltage and current levels must be used for each secondary winding in relationship to the primary winding.

Testing inductors and transformers with a multimeter

The multimeter can be used to check inductors and transformers for open windings and, in some instances, shorted windings. For example, if the manufacturer's specifications regarding the resistance of windings are available, the multimeter, particularly the digital multimeter with precise readout of resistance, can be used to check these resistance values. If the resistance readings indicated by the multimeter are lower than those specified by the manufacturer, one or more turns of the winding are probably shorted.

CAUTION: *Don't try to test the inductor or transformer when it is connected in a "live" or powered-up circuit. Turn off all power to the circuit being tested by unplugging the ac power cord! Otherwise, you might experience a severe or fatal electrical shock. In many cases, you can also damage the multimeter or the circuit being tested. Damage to the multimeter can range from virtual destruction to requiring replacement of an internal fuse within the instrument.*

Large audio inductors and power transformers will exhibit a resistance of a few ohms to 20 ohms or more. Small inductors used in RF and IF applications will normally show a resistance of one ohm or less. The actual resistance will depend on the size of the wire and the number of turns.

Any inductor or transformer that exhibits an abnormal resistance reading or an open winding should be replaced with a new component of equal specifications.

ac Circuit Analysis

Resistors, capacitors, and inductors all offer opposition to the flow of current in an ac circuit. Although this opposition is always measured in units of ohms, we find that each component behaves differently in the circuit. The resistor merely impedes the flow of ac current and dissipates energy in the form of heat. The capacitor and the ideal inductor offer opposition to ac current flow but do not consume energy. The energy temporarily stored in each component—an electric field in the capacitor and a magnetic field in the inductor—is returned to the circuit during a portion of each cycle.

A complete study of ac circuit analysis is beyond the scope of this book and the requirements for the Technician Class license. However, we will review a few concepts that will help to explain the operation of ac circuits in amateur radio transmitters and receivers.

Impedance and phase angles

Impedance is a measure of the total opposition that a component or circuit presents to an ac current (power, audio signal, or RF signal). Impedance can contain resistance, capacitive reactance, and inductive reactance in series and/or parallel combinations.

Impedance must be considered when connecting ac circuits, components, or equipment together. For example, the impedance of an HF antenna for 40 meters, the RF transmission line, and the RF output of an amateur transmitter are all related. If the impedances of these devices are not matched, RF power will be lost and the efficiency of radio transmissions will be reduced. Impedances of other devices such as microphones, audio amplifiers, and speakers must be considered when they are connected together as an audio system. In fact, almost every circuit within a transmitter or receiver relies on the proper matching of impedances for the required operational characteristics.

Any analysis of impedance must take into account the phase angle of the component or circuit involved. Voltages (or currents) in phase are additive in nature while those out of phase tend to cancel each other.

Resonance and tuned circuits

The concept of resonance is one of the most important factors in radio communications. Without resonant or tuned circuits, we would be unable to separate the desired signal from the many signals intercepted by the antenna. The selectivity of the receiver's RF and IF amplifier stages, and the frequency-determining networks of the local and beat-frequency oscillators, are all related to the phenomenon of resonance.

The radio transmitter employs tuned circuits for generating, amplifying, and modulating RF signals. Tuned circuits allow the RF signals to be coupled to the transmission line while attenuating harmonics or other undesired signals generated within the transmitter. Finally, the antenna acts as a tuned circuit, resonant only at the frequency of the signal being transmitted.

The series LC network in Fig. 4.31A acts as a short circuit at the resonant frequency. This is due to the fact that the capacitive reactance is equal but opposite to the inductive reactance. As a result, the reactances

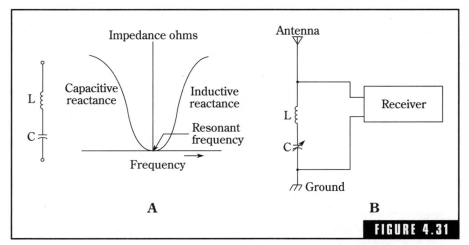

FIGURE 4.31

The resonant series LC network. (A) Characteristics of a resonant series LC network. (B) The resonant series LC network used as a wave trap.

cancel and the impedance of the network is zero ohms. However, above and below the resonant frequency, the LC network exhibits inductive and capacitive reactance, respectively. Series resonant networks are sometimes used as wave traps in antenna circuits to bypass interfering signals (of a different frequency) to ground. Figure 4.31B illustrates this type of application.

Parallel resonant LC networks, such as shown in Fig. 4.32, exhibit maximum impedance at the resonant frequency. Above or below the resonant frequency, the network appears as a capacitive or inductive reactance, respectively. When this type of network is connected across the input of a tube or transistor amplifier, only the resonant-frequency signals are allowed to be amplified. Signals above or below the resonant frequency are shunted to ground.

The resonant frequency for either the series or parallel LC network can be computed by the following equation:

$$f_r = \frac{1}{2\pi\sqrt{LC}}$$ (Eq. 4.26)

where

f_r is the resonant frequency in hertz
π is the constant pi (3.1416)
L is the inductance in henrys
C is the capacitance in farads

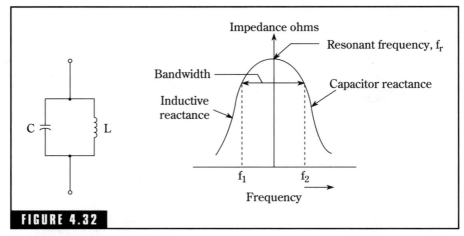

FIGURE 4.32

The parallel resonant LC circuit.

The Q of resonant circuits

The Q of a series or parallel resonant circuit is defined as the figure of merit or quality factor for that particular circuit. In general, the Q of a circuit describes the selectivity or sharpness of the response curve for a resonant circuit. A high Q means that the circuit has a narrow selectivity with a corresponding low bandwidth factor. The bandwidth of the parallel LC network in Fig. 4.32 is indicated by the frequencies f_1 and f_2. Actually, the bandwidth for this circuit is equal to $f_2 - f_1$. Frequencies f_1 and f_2 are the points where the response of the parallel LC network drops to 0.707 of the maximum response at the resonant frequency, f_r.

The Q of an inductor is equal to the ratio of the inductive reactance to the effective resistance of the coil. For example, a coil wound with a large size of wire would exhibit very little dc resistance. Accordingly, the Q of this coil would be much higher than a similar coil constructed with smaller size wire. A smaller dc resistance in the coil winding means that the coil will dissipate less energy as heat.

The capacitor also can be described in terms of its Q. This is essentially a ratio of the energy stored in the capacitor to the energy dissipated as heat due to the effective resistance of the capacitor dielectric losses. In most capacitors, the effective resistance values are in the hundreds to thousands of megohms. Therefore, the capacitor will exhibit an extremely high Q.

Power Relationships

The power associated with an electrical circuit, either ac or dc, is defined as the rate at which work is being performed. For example, a current flowing in a resistor performs work in the sense that electrical current is being converted to heat. A radio transmitter performs work by generating RF energy that can be fed into an antenna, thereby creating electromagnetic waves.

The basic unit of electrical power is the watt. In radio communications, power is often expressed in terms of:

1 microwatt = 1 μW = 0.000001 watt = 1×10^{-6} watt

1 milliwatt = 1 mW = 0.001 watt = 1×10^{-3} watt

1 kilowatt = 1 kW = 1000 watts = 1×10^3 watts

1 megawatt = 1 MW = 1,000,000 watts = 1×10^6 watts

Electrical power can be converted to mechanical power. An electric motor, for example, converts electricity into a rotating mechanical motion. Mechanical power is expressed in units of horsepower. The relationship between horsepower and the watt is:

1 horsepower = 746 watts

Power in a dc circuit

The power delivered to a resistive load by a direct-current source is determined by the voltage across the load times the current through the resistive load, or:

$$P = VI \qquad \text{(Eq. 4.27)}$$

where

P is the power in watts
V is the voltage in volts across the load
I is the current in amperes through the load

Ohm's law can be applied to develop two additional power equations. These equations are sometimes very useful when computing the power in a dc circuit.

$$P = \frac{V^2}{R} \text{ watts} \qquad \text{(Eq. 4.28)}$$

$$P = I^2R \text{ watts} \qquad \text{(Eq. 4.29)}$$

where

R is the resistance of the load in ohms

Figure 4.33 illustrates three examples for computing dc power.

Power in an ac circuit

Calculations involving power delivered to an ac circuit are more complex than dc power calculations. The resistive elements in an ac circuit dissipate power in the form of heat. This is known as the *real power* delivered to the ac circuit. This power is usually computed using Equation 4-29, above.

The energy absorbed by the capacitive and inductive components

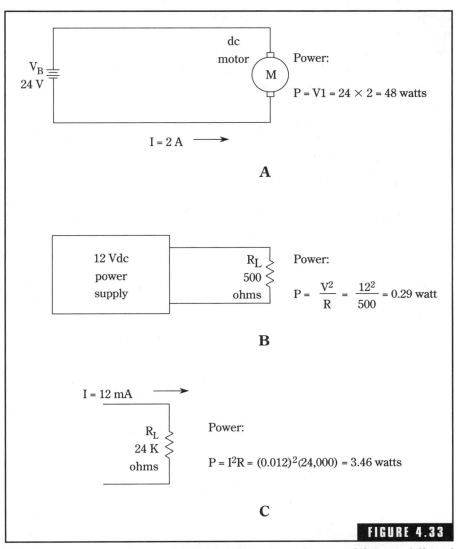

Typical dc power computations. (A) Power delivered to a dc motor. (B) Power delivered to a resistor. (C) Current through a resistor dissipates power.

of an ac circuit is temporarily stored and is returned to the circuit during a later portion of each cycle. This is known as *reactive power*.

The total power delivered to an ac circuit, referred to as *apparent power*, consists of the real power consumed by the resistive elements and the reactive power absorbed (but not dissipated) by the reactive elements. Apparent power, which is defined as the product of the current through the circuit and the voltage impressed across the circuit, is expressed in volt-amperes instead of watts.

$$P_a = VI \qquad \text{(Eq. 4.30)}$$

where

P_a is the apparent power in volt-amperes

Maximum power transfer

Maximum power is transferred to a load when the internal resistance (or impedance) of the power is equal to the load resistance (or impedance). This condition is valid for both ac and dc circuits.

Tubes and Semiconductors

Tubes and semiconductor devices form the basis for all electronic circuits involving rectification, amplification, oscillation, digital, and other related functions. Tubes (such as triodes and pentodes) and transistors are called active devices because they are amplifying devices capable of boosting very weak signals to high-level amplitudes. For example, audio amplifiers can be used to amplify low-level audio signals from microphones or phonograph pick-up devices, to high signal levels capable of driving loudspeakers. Radio frequency amplifiers (RF amplifiers) are used in radio receivers to amplify extremely weak RF signals (in the microvolt range) from an antenna for subsequent signal processing and ultimate conversion to audio signals for driving speakers or headphones.

The vacuum tube, once the mighty workhorse in radio communications, has been relegated to the back seat by the smaller, more efficient transistor. Most all new radio equipment on today's market employs solid-state technology. Only a few specialized applications, such as high-power final amplifiers in the range of 1000 watts and above still continue to use vacuum tubes.

The triode vacuum tube was invented by Lee De Forest in 1906. One of the most important inventions of the twentieth century, the vacuum tube spurred the development of radio communications, radar, television, space-age electronics, computers, and the list goes on.

The present investment in vacuum tube electronic equipments, particularly amateur radios, is enormous. For example, vacuum tube HF transceivers will continue to give excellent service for years to come. You can usually find some of these old, reliable ham radios at local swap-fests and amateur radio conventions. Also, many major electronic companies continue to manufacture quality HF "1 kilowatt final power amplifiers" using rugged power tubes. The FCC requires that all amateur radio operators develop a working knowledge of vacuum tubes and includes questions about this subject on all amateur examinations. Some introductory material on vacuum tubes is provided here to help you pass the FCC examinations as well as to set up and operate vacuum tube ham radios. Also, from time to time, you will find some interesting construction articles involving home-built vacuum tube transmitters and receivers in amateur radio magazines such as *CQ*, *QST*, and *73*.

Like the vacuum tube, the invention of the transistor in the late 1940s and the integrated circuit in the late 1950s produced a major impact on all fields of electronics. Although the first semiconductor devices (crystal detectors using silicon or carborundum) were invented about 1906, scientists at that time could not explain the theory behind the crystal rectifier. The major breakthrough came in 1947 when scientists at the Bell Telephone Laboratories developed the present-day semiconductor theory. This immediately led to the fabrication of the transistor, the first solid-state amplifying device; junction diodes; and thyristors. The disadvantages of the vacuum tube—fragility, bulkiness, requiring high-voltage power supplies, and use of filaments that require warm-up time and excessive amounts of power—are eliminated by the rugged and compact transistor.

Let's examine the tube and semiconductor devices in more detail. You won't need all this theory and practical information for the Technician Class examinations; only one question on vacuum tubes is included in these examinations. However, a good background on tubes and semiconductors will help you understand the operation of your amateur radio equipment as well as prepare you for upgrade to higher class amateur licenses.

Definitions

Semiconductor A class of materials that are neither good conductors nor good insulators of electricity. These materials, such as germanium and silicon, possess relatively few free electrons for conducting electrical current. In their pure form, semiconductors have little use in electronic applications.

PN junction diode The pn junction diode, a two-terminal semiconductor device, is the simplest form of semiconductor used in electronics. This device consists of two tiny blocks of semiconductor material referred to as *n-type* and *p-type*. These two blocks are, in effect, joined together as one continuous block with leads connected to each end. The pn junction diode allows current to flow easily in one direction but literally blocks current flow in the opposite direction. The two leads or terminals of the pn junction diode are referred to as the anode and cathode. A pn junction diode can be constructed from either germanium or silicon. PN junction diodes are used as rectifiers in power supplies, detectors in radio receivers, and in other applications where rectification action is required.

Zener diode The zener diode, a two-terminal device, uses the reverse-voltage characteristics of a pn junction to cause a voltage breakdown that results in a high current through the diode. This action produces a constant voltage drop across the diode and exhibits voltage-regulating characteristics. Zener diodes are extremely useful as voltage regulators in power supplies.

Bipolar transistor The bipolar transistor consists of three layers of n-type and p-type semiconductor material arranged either in a npn or pnp configuration. The three leads from the npn or pnp transistor are called the *emitter*, *base*, and *collector*. The bipolar transistor is a current-amplifying device and is used in amplifier, oscillator, and other types of electronic circuits.

Field-effect transistor Field-effect transistors, or FETs, are constructed from various layers of n-type and p-type semiconductor materials. FETs are available in two general types: the junction, field-effect transistor (JFET), and the insulated-gate field-effect transistor (IGFET). Sometimes the IGFET is referred to as a metal-oxide semiconductor field-effect transistor, or MOSFET. The FET transistor is a voltage-amplifying device, and as such possesses a very high input resistance.

FETs are useful in amplifier, oscillator, and other types of electronic circuits where an active device is required.

Vacuum tube The vacuum or electron tube is a device that consists of a number of electrodes mounted in an evacuated enclosure of glass or metal. The vacuum tube employs a controlled flow of electron current to perform functions such as rectification or amplification.

A vacuum tube can contain one or more of the following electrodes or elements:

- *Cathode* The cathode produces a stream of electrons that are directed towards the anode electrode. In most tubes, the cathode is heated indirectly, resulting in thermionic emission of electrons.

- *Anode* The anode, or plate, is the electrode to which the stream of electrons flow.

- *Grid* A grid is an electrode normally placed between the cathode and plate to control the flow of electrons. The grid has one or more openings to permit passage of the electrons. A vacuum tube can use one or more grids for controlling the flow of electrons. The grids are called control, screen, suppressor or space-charge, depending on the function to be performed.

Diode The diode is the simplest form of a vacuum tube and consists of two electrodes, a cathode and plate. The word diode is derived from *di*, meaning two, and *ode*, for electrode. A positive potential, or voltage applied to the plate electrode, will attract the electrons emitted from the cathode. This results in one-way flow of current through the diode tube. Diodes are used primarily to rectify ac signals.

Triode A major step in vacuum tube development, the triode provides a capability to amplify weak audio and radio-frequency signals. The triode is a three-electrode device having a control grid positioned between the cathode and plate. The grid controls the flow of electrons to the plate. In addition to amplification, we can use the triode for other important functions in electronic circuits, such as frequency generation, conversion, modulation, and demodulation.

Pentode The pentode, a five-element vacuum tube, can be described as the ideal amplifier in vacuum tube development. Scientists had invented the tetrode, a four-element vacuum tube to overcome the limitations of the triode. Although an improvement over the triode, the tetrode still possessed limitations. The pentode, possessing a cathode, control grid, screen grid, suppressor grid, and plate provides higher signal amplification and more stable operating characteristics.

Vacuum Tubes

Some typical vacuum tubes used in amateur radio and other electronic equipment are illustrated in Fig. 5.1. The tube on the left is a heavy-duty full-wave rectifier used in power supplies. The next two tubes are transmitting power tubes and are rated at 35 and 15 watts, respectively. The three tubes on the right are used in receiving or low-power applications and are referred to as 9-pin miniature, subminiature, and nuvistor types, respectively. (The nuvistor is the small metal tube below the miniature tube.)

Thermionic Emission

Thermionic emission is produced when energy such as heat is applied to a metallic material. For example, if we cause a sufficiently high electrical current to flow through a wire, the heat produced in the wire will cause a cloud of free electrons to be emitted from the surface of the wire.

Thermionic emission is caused by the acceleration of electrons within a material due to increased energy. When the energy imparted to the electrons reaches a given level, the electrons will break away

FIGURE 5.1

Typical receiving and transmitting vacuum tubes.

from the surface of the material, forming a cloud of free electrons. Unless influenced by an external force, such as an electric field, the free electrons will be attracted back to the surface of the material while other electrons are being emitted into the cloud. For a given material and a given temperature, the number of free electrons contained in the cloud will remain relatively constant.

As shown in Fig. 5.2A, the cathode within a vacuum tube produces a cloud of free electrons by either direct or indirect heating methods. Once produced, the cloud of free electrons can be made to flow in a specified direction to other electrodes within the tube, performing functions such as rectification and amplification. Each particular type of vacuum tube requires a specific heater voltage and current. Heater voltage values range from about 1.4 volts for battery-operated equipment and 6.3 volts to 117 volts for ac-operated equipments. Current requirements range from tens to hundreds of milliamperes for receiving tubes to several amperes for high-power transmitting tubes. A vacuum tube handbook or *The ARRL Handbook* will provide specific information on the voltage and current requirements for each tube's heater, as well as other tube operating characteristics. Either ac or dc voltage can be used to power the

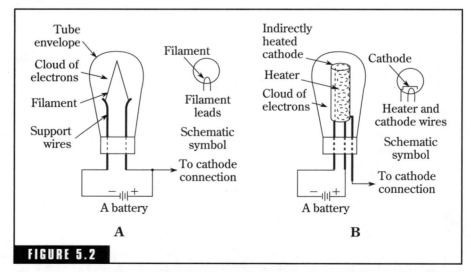

FIGURE 5.2

Pictorial representation and schematic symbols for directly and indirectly heated cathodes. (A) Directly heated cathode. (B) Indirectly heated cathode.

heater in most vacuum tubes—the indirectly heated cathode using mostly ac power, and directly heated cathode using either ac or dc power.

General Types of Vacuum Tubes

Many types of vacuum tubes have been developed over the past sixty years. Specific types used in amateur radio equipment include the diode, triode, tetrode, pentode, and cathode ray tube or CRT. Table 5.1 provides a summary of these tubes, their applications and commonly used symbols.

Diodes

The diode is a simple two-electrode tube containing a cathode and plate. The heater cathode emits a cloud of free electrons that possess negative charges. If the plate voltage of the diode is made positive with respect to the cathode, the free electrons are attracted to the plate, resulting in a flow of electrical current. Figure 5.3A provides a test circuit that illustrates the action of a diode in a dc circuit. An analysis of this circuit, as well as the following ac rectification circuit, will provide you with an insight into the theory of the diode tube as well as troubleshooting techniques.

The 6AL5 used in this test circuit is a twin diode installed in a 7-pin miniature glass envelope. Only one of the diode sections (Pins 1 and 7) is used in this circuit. Figure 5.3B shows the basing diagram for the electrical connections within the tube. Basing diagrams are usually referenced from the bottom view of the tube.

Each diode section in the 6AL5 can be used independently of the other; there are no electrical connections between the two sections. Many multipurpose tubes contain two or more independent sections such as diode-triode, triode-triode, or triode-pentode. The 6AL5 has a common heater for the two diodes, which requires 6.3 volts (ac or dc) at 0.3 ampere. Note that the 6AL5 in Fig. 5.3A is connected to a 6-volt battery. There is no electrical connection between the heater and cathodes.

If we set switch S1 to position A, the potential on the plate electrode is zero volts. As a result, none of the electrons being emitted by the cathode will be attracted to the plate. To verify that no current is

Table 5.1 Types of Vacuum Tubes Used in Amateur Radio Equipment

Tube type	No. of elements	Major applications	Symbol
Diode	Two: Cathode Plate	1. RF Detector and demodulator 2. ac Rectifier 3. Peak detector and pulse shaping circuit	Plate Cathode — Filament (Heater)
Triode	Three: Cathode Control grid Plate	1. Amplifier 2. Oscillator 3. Modulator 4. Mixer	Plate Control grid Cathode Filament
Tetrode	Four: Cathode Control grid Screen grid Plate	1. Improved amplifier (More stable) 2. Oscillator 3. Modulator 4. Mixer	Plate Control grid — Screen grid Cathode Filament
Pentode	Five: Cathode Control grid Screen grid Suppressor grid Plate	1. Improved amplifier (More stable and higher gain) 2. Oscillator 3. Modulator 4. Mixer	Plate — Suppressor grid Control grid — Screen grid Cathode Filament

flowing in the plate circuit, you can measure the voltage across the load resistor R1 with a dc voltmeter. A zero voltage indicates no voltage drop; thus, no current through the resistor.

When we move switch S1 to position B, a positive voltage potential is placed on the plate and free electrons will flow to the plate. The quantity of electrons contained in this flow to the plate depends on several factors, including the number of electrons being emitted from the cathode and the amplitude of the positive potential placed on the

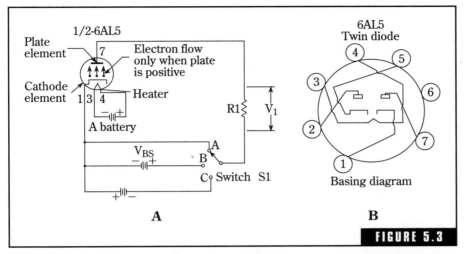

FIGURE 5.3

Diode demonstration circuit using the 6AL5 twin diode vacuum tube. (A) Electron flow in a diode. (B) 6AL5 base connections.

plate. Again, we can use a dc voltmeter to check for the presence of current and also the magnitude of this current. Connect the voltmeter across resistor R1 and determine the voltage value. Using Ohm's law, we can find the actual plate current (amperes) by dividing the plate voltage (volts) by the resistance (ohms) of R1. An alternate method for finding the plate current is to break the circuit at the plate terminal and connect a dc ammeter. However, you have to physically break the conductor to use the ammeter method.

Moving switch S1 to position C results in making the plate negative with respect to the cathode. This negative potential repels the negatively charged electrons, prohibiting any current flow within the diode.

The diode as an ac rectifier

The great utility of the diode is its ability to conduct current in only one direction. Thus we can use a diode to change ac current into dc current. This process is known as *rectification*. Figure 5.4 illustrates how a diode rectifies an ac current. The waveform for the ac input signal is shown at the top of Fig. 5.4B. This signal varies from +5 volts peak to −5 volts peak amplitude. Because the diode conducts in only one direction, only the positive portion of the ac waveform is conducted by the diode. As a

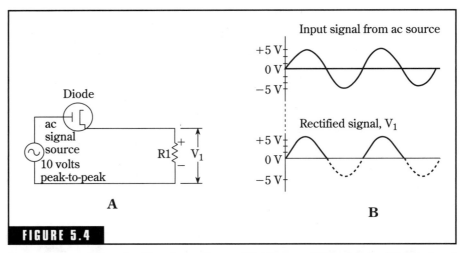

FIGURE 5.4

Basic diode rectifier circuit and waveforms. (A) Diode rectifier circuit. (B) Input and output voltage waveforms.

result, a positive voltage is developed across load resistor R1. The waveform of the rectified signal is shown at the bottom of Fig. 5.4B. In practical circuits, the frequency of the ac signal can vary from power-line frequencies of 60 Hz to radio frequencies in the hundreds of MHz.

In some circuits, diodes are used to demodulate signals at RF or IF frequencies. Demodulation is a form of rectification where the modulating signal, such as voice or tones, is separated from the carrier frequency.

Triodes

The triode, a three-electrode tube, has a control grid positioned between the cathode and the plate. The purpose of this control grid is to control the flow of electrons from the cathode to the plate. The addition of the control grid to the vacuum tube greatly enhanced the capabilities of the tube and revolutionized the electronics field. For the first time, signals, such as audio and RF could be amplified. Electronic oscillators, modulators, and other specialized circuits made possible the design and fabrication of sensitive radio receivers, stable high-power transmitters, and a host of other equipment.

The control grid, sometimes simply referred to as the grid, usually consists of a fine wire that is wound on insulated supports and surrounds the cathode. This wire can be wound in the form of a helix or a ladder type of construction. The spacing between the wire is large compared to the size of the wire. This allows most of the electrons from the cathode to flow unobstructed to the plate.

Dimensions, shapes, and spacing of the three electrodes used in triodes vary in relation to the intended applications. In some instances, two or three triodes are installed inside of one envelope with independent operation of each triode section. Figure 5.5A illustrates the basic construction used in triodes. The physical supports for the electrodes are not shown. Triodes, like diodes, can have either directly or indirectly heated cathodes. Figure 5.5B shows the symbols for typical triode tubes.

How triodes work

The operation of a triode is based primarily on the polarity and magnitude of the voltage impressed on the control grid. A positive voltage potential on the grid attracts the negatively charged free electrons, causing more of them to be accelerated towards the plate electrode.

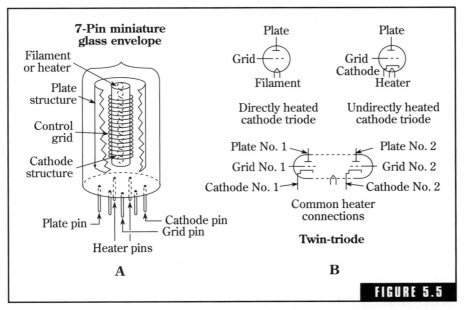

The triode vacuum tube. (A) Cutaway view of triode tube. (B) Symbols used for triode tubes.

The result is an increased current flow through the tube's plate circuit. Alternately, a negative voltage potential on the grid repels the free electrons, causing many of them to be returned to the cathode's surface. Thus the resulting current flow from cathode to plate is reduced or even cut off entirely.

This grid voltage can be either ac or dc. dc voltages are used for our description of the operation of this tube. Figure 5.6 illustrates three conditions for a triode. You can use a dc voltmeter to verify these operating conditions. dc voltage readings showing grid-to-ground and plate-to-ground voltage differences can help you troubleshoot defective tube circuits in your receiver or transmitter. Again, be extremely careful in making such measurements; you can experience severe or fatal electrical shock. Also, you might destroy your dc voltmeter if it is not switched to the proper function and voltage range.

With no voltage applied to the grid, the triode acts like a diode. The current flow between the cathode and plate depends on the polarity

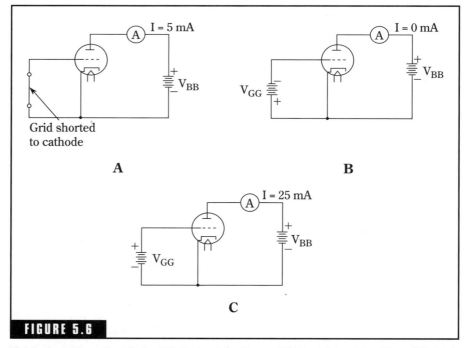

FIGURE 5.6

The basic action of a triode. (A) Zero grid voltage. (B) Negative grid voltage. (C) Positive grid voltage.

and amplitude of the plate voltage. Because the grid is now at a zero voltage potential, it has virtually no effect on the negatively charged electrons being emitted by the cathode.

When a negative voltage is applied to the grid, a negative electrostatic field is created between the grid and cathode. This negative field opposes or repels the electrons being emitted by the cathode. As a result, fewer electrons are able to pass through the grid and reach the plate.

As the negative potential of the grid is increased, the plate current is eventually reduced to zero. This condition is known as *cutoff*, and the amplitude of the grid voltage necessary to cause this state is referred to as the *cutoff voltage*.

A positive grid voltage causes an increase in plate current. This is due to the positive electrostatic field around the grid that attracts the electrons and accelerates them toward the plate. If the positive voltage on the grid is increased beyond a given amplitude, all electrons being emitted by the cathode will flow to the plate. No further increase in plate current is possible. This condition is known as *saturation*. In normal receiving-tube applications, saturation does not exist because the grid voltage is not driven to excessive positive levels. Most triode tube circuits use a *bias* or fixed dc grid voltage to establish a plate current level between the tube's cut-off and saturation levels.

The circuits in Fig. 5.6 do not show the filament or heater power supplies that heat the cathode. All subsequent symbols for vacuum tubes will assume that the filaments or heaters are connected to a power source unless it is pertinent to show these connections. You will find most diagrams of radio and electronic equipment do not show the filament or heater connections with the tube symbols. All of the filament or heater connections are normally given in a separate section of the schematic diagram that is associated with the power supply.

Triode amplification action

The great utility of the triode is that it provides an amplification action. A small change of grid voltage, representing an input signal, produces a much larger change of voltage in the output plate circuit. Thus, the triode serves primarily as a voltage amplifier. Figure 5.7 shows a simplified triode circuit that can be used as an audio amplifier for speech or music applications. This circuit uses a fixed bias supply of −4

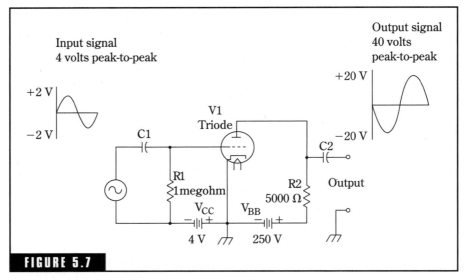

FIGURE 5.7

The triode amplifier circuit.

volts and a plate supply voltage of +250 volts. Capacitors C1 and C2 isolate the signal input and output connections from the tube's dc voltages. Using an oscilloscope to measure the input and output peak-to-peak signal levels, we can determine the voltage gain of this or other similar amplifiers. The voltage gain or amplification for this circuit is 40/4, or 10. Called a *common-cathode configuration*, this amplifier inverts the output signal. That is, the output is 180 degrees out of phase with the input signal.

Limitations of triodes

The main limitation of the triode tube is that of interelectrode capacitance, especially the capacitance between the control grid and the plate. When we use the triode as an amplifier in circuits where the frequency of operation is high, the grid plate capacitance provides a feedback path. This feedback effect results in unstable operation and may cause the amplifier to oscillate. At audio frequencies, the grid-plate capacitance is negligible and, in most cases, this limitation can be ignored. However, at intermediate frequencies (IF) and radio frequencies (RF), the grid-plate capacitance of a triode must be considered in any design of amplifier circuits. For example, neutralizing capacitors are employed with high-power transmitting triodes to balance out the effects of the interelectrode capacitances. Triodes are seldom used as IF or RF amplifiers.

Another major limitation of the triode is that it does not satisfy all the amplifying or gain requirements needed in receivers, transmitters, and other electronic equipment. For example, signal levels in the high-frequency spectrum might be on the order of a few microvolts at the antenna terminals. Triode amplifiers typically provide a gain of 5 to 20. Thus, many triodes would be required to amplify these signals to sufficient level for driving headphones or loudspeakers.

Tetrodes and Pentodes

Limitations of triodes led to the development of tetrodes, pentodes, and related types of vacuum tubes. These tubes have very little grid-plate interelectrode capacitance and they possess high gain characteristics. Tetrode and pentode symbols are given in Fig. 5.8.

The tetrode

The grid-plate interelectrode capacitance of the triode can be reduced to an insignificant value by adding a fourth electrode, called the *screen grid*, between the grid and plate. Having four electrodes, this tube is designated as a tetrode. The screen grid, like the control grid, does not materially interfere with the flow of electrons from the cathode to the

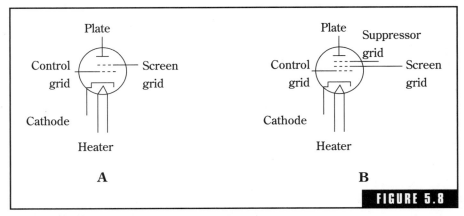

FIGURE 5.8

Symbols for the tetrode and pentode. (A) Tetrode. (B) Pentode.

plate because of the wide spacing of the screen-grid wires. Figure 5.9 shows a basic tetrode amplifier circuit.

Unlike the control grid, the screen grid is operated at a high positive potential. This screen-grid voltage, normally lower than the plate voltage, produces another desirable advantage over the triode. The plate current is independent of plate voltage because the screen grid supplies a positive potential that pulls the electrons from the cathode to the plate. As long as the plate voltage is higher than the screen voltage, the plate current in the tube depends primarily on the screen-grid voltage and very little on the plate voltage. This action of the tetrode results in a higher amplification factor and increases plate resistance as compared to the triode.

Limitations of the tetrode

Although the tetrode is a considerable improvement over the triode, secondary emission limits the performance of this tube. Secondary emission of electrons occurs when high-velocity electrons from the cathode strike the metallic plate surface. During this process, several secondary electrons might be knocked free from the plate by each high-velocity electron striking the plate. As long as the plate is more positive than the screen grid, these secondary electrons will be attracted back to the plate. However, if the plate voltage drops below the screen voltage due to a high input signal, the secondary electrons will be attracted to the screen grid. The net result is that the amplifi-

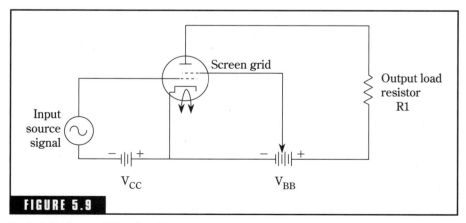

FIGURE 5.9

Basic tetrode amplifier circuit.

cation of the tube is decreased and the output signal is distorted.

Tetrodes are used primarily as power transmitting tubes in many amateur transmitters. The 6146 and the 2E26 are representative of medium-power tetrode transmitting tubes.

The pentode

The pentode, a five-electrode tube, has a suppressor grid placed between the screen grid and plate electrode. In most pentodes, the suppressor grid is internally connected to the cathode. Thus, the suppressor grid is essentially at a ground or slightly positive voltage potential in most pentode circuits. Figure 5.10 shows a basic pentode amplifier circuit.

The purpose of the suppressor grid is to eliminate the effects of secondary emission encountered in the tetrode. Because the suppressor grid is at a negative potential with respect to the plate, it repels the secondary electrons and forces them back to the plate. The addition of the suppressor grid further reduces the grid-plate interelectrode capacitance. Therefore, the pentode can provide higher amplification with reduced feedback instability problems when compared to the tetrode. Voltage gain or amplification for pentode amplifiers can range up to 200 or more for receiving pentodes, such as the 6AG5, 6AU6, and the 6CB6. Pentodes are used almost exclusively in tube-type amateur receivers and transceivers for RF and IF amplifiers.

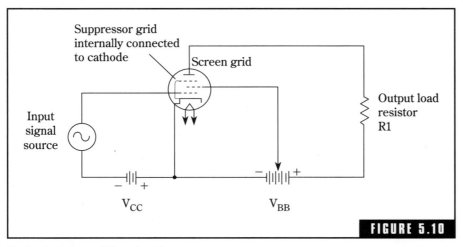

FIGURE 5.10

Basic pentodeamplifier circuit.

The pentode as an amplifier

Figure 5.11 shows a 6AU6 pentode connected as a simple audio amplifier. This amplifier has a frequency response of approximately 100 Hz to 10 kHz, which is more than adequate for the voice frequency spectrum. Resistance-capacitance (RC) coupling is used in the input grid and output plate circuits for coupling the low-level signal into the amplifier and amplified output signal to the next stage. RC coupling isolates the low-grid and high-plate dc voltage levels from the input and output circuitry.

Another important aspect of this circuit is that of *self-bias*. This eliminates the need for a separate grid voltage power supply. The self-bias grid voltage is developed by the cathode resistor Rk, which develops a negative grid bias voltage. The voltage drop across the cathode resistor due to the cathode current raises the cathode to a positive potential with respect to the control grid. The cathode bypass capacitor Ck across the cathode resistor bypasses the cathode AF signals to ground, leaving a dc voltage drop across the cathode resistor. Thus, the dc voltage on the grid electrode with respect to the cathode electrode has a negative voltage potential, providing a grid voltage for the tube

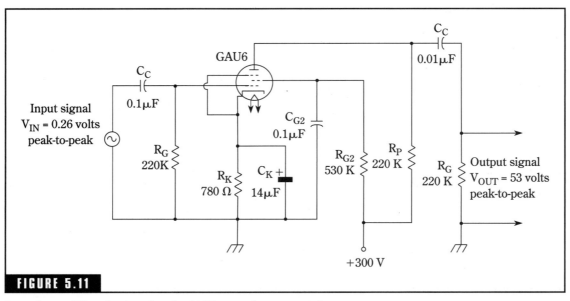

FIGURE 5.11

Pentode amplifier circuit using the 6AU6 pentode vacuum tube.

circuit. This eliminates the need to use a separate negative power supply for grid biasing.

The 6AU6 pentode amplifier develops a voltage gain of approximately 200. This means that a 50-millivolt input signal to the control grid will be amplified to 200 times its original level. The output signal across the plate load resistor will be 200 times 50 millivolts (200 × 0.05), or 10 volts in amplitude.

Specific vacuum tube operating specifications and design data for triodes, tetrodes, and pentodes are available in many vacuum-tube manuals and electronic handbooks such as *The ARRL Handbook*. This type of information is useful in servicing tube-type amateur equipment.

Beam-power tubes

The beam-power tube combines the advantages of the tetrode and pentode to provide for an increased power handling capability. This type of tube has a cathode, control grid, screen grid, plate, and beam-forming electrodes that concentrate beams of electrons to the plate. Although the exterior appearance of beam-power tubes are identical to other tubes, the tube electrodes are larger to accommodate the additional power developed by these tubes. Typical beam-power tubes used in present-day amateur receivers and transceivers include the 6AQ5, 6DS5, and the 6V6.

Gas-filled tubes

The thyratron, a gas-filled triode tube, and the cold-cathode voltage-regulator tube are representative of special-purpose tubes that employ a gaseous atmosphere within the envelope of the tube. These tubes conduct current due to the ionization of the gaseous atmosphere inside of the tube.

The thyratron initially conducts current only when the control grid is made positive. However, after the tube begins to conduct current, the grid no longer has an effect on the flow of current. The plate voltage must be reduced below the ionization potential of the gas in order to cut off the current flow. The thyratron is used primarily to control ac current.

Cold-cathode voltage-regulator tubes are used to maintain constant voltage drops for dc power supply applications. Sometimes referred to as VR tubes, these tubes are available in a number of voltage ratings, such as 75, 90, 105, and 150 volts.

Cathode-ray tubes

Cathode-ray tubes are used to provide a visual display of electrical signals such as a sine-wave display on an oscilloscope or a video signal on a television receiver. The *electron gun* within a cathode-ray tube produces a thin beam of high-speed electrons that is directed toward a fluorescent screen. When the electrons strike the fluorescent screen, visible light is produced. This stream of electrons can be deflected by electrostatic plates or by a magnetic field to provide the required display of information. Figure 5.12 shows the two types of cathode-ray tubes currently in use. High voltages on the order of 1000 to 15,000 volts are required to produce the stream of high-speed electrons.

Diodes, Transistors, and other Semiconductor devices

Semiconductor diodes and transistors have replaced virtually all types of vacuum tubes used in modern amateur radio equipment. These semiconductor devices operate on unique principles not encountered in conventional conductors of electrical current, such as copper wire or vacuum tubes. In contrast, pure semiconductor materials—primarily germanium and silicon—possess relatively few free electrons available for conducting electrical current. The semiconductor material has a crystaline atomic structure where virtually all of the electrons are locked together in the outer orbit of each atom. Energy in the form of heat liberates some electrons to support current flow. However, the maximum current flow in a small block of semiconductor material is too small to be of any practical use in electronic circuits.

The addition or *doping* of impurities within semiconductor materials produces either excess numbers of free electrons or mobile *holes*, depending on the type of material used as the impurity. Figure 5.13 shows a pure semiconductor material and semiconductor materials that have been doped with certain impurities. Both germanium and silicon semiconductor materials exhibit these characteristics. Both types of modified semiconductor materials, referred to as n-type and p-type semiconductors, can be made to produce usable amounts of current flow. Figure 5.13 shows a relative comparison of pure and modified semiconductor material.

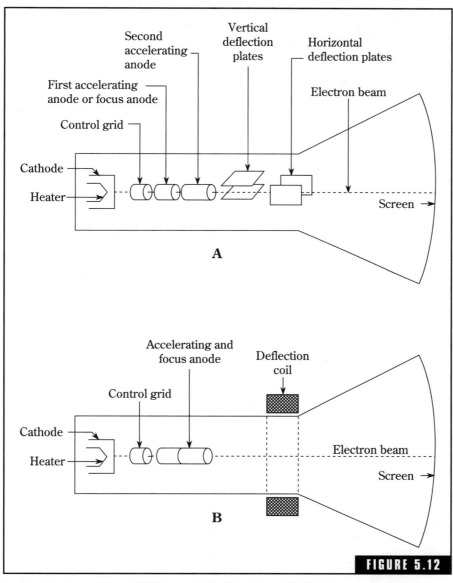

FIGURE 5.12

Basic cathode-ray tubes. (A) Electrostatic deflection. (B) Electromagnetic deflection.

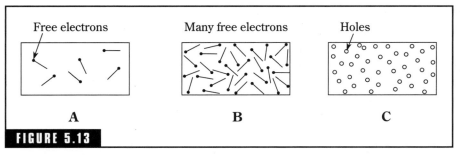

Pure and modified semiconduction material. (A) Pure semiconductor material. (B) n-type material. (C) p-type material.

n-type semiconductor

If a tiny amount of doping material such as arsenic, antimony, or phosphorus is combined with the semiconductor material, the number of free electrons is increased by a substantial amount. The n-type semiconductor material will support practical and usable amounts of electrical current from microamperes to amperes. By itself, n-type silicon is used to manufacture resistors; the amount of resistance is a function of the level of concentration of impurities and the geometric shape of the section of the n-type material. Tiny n-type silicon paths of resistance are deposited in ICs to provide the required resistance for IC circuit operation.

p-type semiconductor

P-type semiconductor materials represent a relatively new concept in electronics—"hole" current flow. When a semiconductor material such as silicon is doped with aluminum, boron, or gallium, a substantial number of mobile or free holes are created in the atomic crystalline structure. Actually, the holes represent the absence of an electron in an outer electron orbit in many semiconductor atoms. These holes, capable of carrying positive charges through the p-type silicon, can support a practical current from microamperes to amperes.

How semiconductor devices are made

Semiconductor diodes and transistors are formed from alternate layers of n-type and p-type semiconductor materials. For example, pn junction diodes consist of a layer of n-type and p-type material. pnp or npn transistors can be fabricated by forming three tiny blocks of p-n-p or n-p-n semiconductor materials, respectively. Figure 5.14 illustrates the pn junction diode and pnp and npn transistor constructions, typical schematic symbols, and representative packaging or cases. The type of case is not always indicative of the device contained within the case.

The pn junction

The pn junction acts as a diode to provide rectification or one-way current flow, similar to the diode vacuum tube. Two pn junctions are used

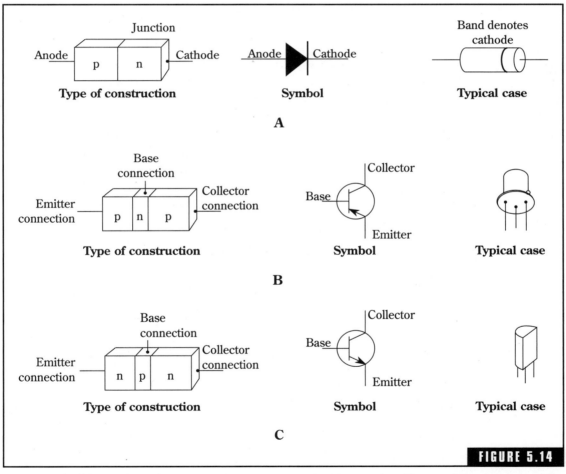

FIGURE 5.14

Construction, symbol, and typical case for a pn junction diode, a pnp and an npn transistor. (A) pn junction diode. (B) pnp transistor. (C) npn transistor.

in "bipolar" pnp or npn transistors to provide amplification of small input signals to large output signals similar to the triode vacuum tube. This section concentrates on how the pn junction diode works.

A pn junction is formed when one half of a block of pure semiconductor material is doped with a trivalent impurity and the other half is doped with a pentavalent impurity. The resulting pn junction creates an electrochemical force due to the excess holes in the p-type material and the excess free electrons in the n-type material. This electrochemical force is responsible for the operation of the pn junction diode and the transistor.

When the pn junction is formed, as shown in Fig. 5.15A, free electrons from the n-type material drift across the junction and immediately combine with the numerous holes present in the p-type material. Thus some atoms in the p-type material gain additional electrons and exhibit a net negative charge. As a group, these atoms, or negative ions, form a negative electric field adjacent to the junction.

The loss of free electrons by the n-type material in the vicinity of the junction leaves some atoms with a net positive charge. These atoms, or positive ions, form a positive electric field next to the junction. Each pair of positive and negative ions are referred to as a *dipole* (not to be associated with the dipole antenna, covered in Chapter 10).

The formation of negative and positive electric fields at the junction are shown in Fig. 5.15B. This region is referred to as the *depletion zone*. The depletion zone shown in Fig. 5.15B is not drawn to scale. Actually, the depletion zone is very thin, virtually a few atoms thick. The resulting electric field, or barrier potential, varies with the type of semiconductor material being used. For example, the barrier potential at room temperature is approximately 0.3 volt for a germanium pn junction and approximately 0.7 volt for a silicon pn junction. These values of barrier potential are the most useful in describing the action of pn junction diodes and transistors. Note that the p-type material in the pn junction diode is referred to as the *anode* and the n-type material is referred to as the *cathode*. These designators represent a carryover from the vacuum tube era.

If the temperature of the pn junction is increased, the barrier potential is decreased by a slight amount due to the increased number of electron-hole pairs, or dipoles. Temperature effects in semiconductor materials must be taken into account when designing solid-state electronic circuits.

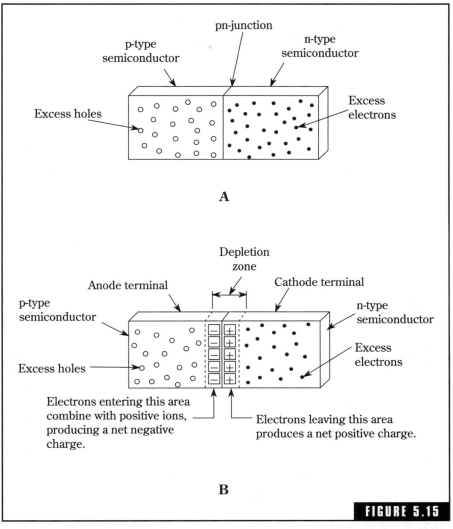

The pn junction diode. (A) Initial condition when a pn junction diode is formed. (B) Final condition of a pn junction diode after electric field is stabilized.

The pn junction described in the previous section is used as pn junction diodes. Figure 5.16 shows the schematic symbol and several case outlines used to house these diodes. These case outlines represent a general cross section of the numerous types of diodes available on the commercial market. Semiconductor handbooks should be consulted for specific technical information on particular diodes. The

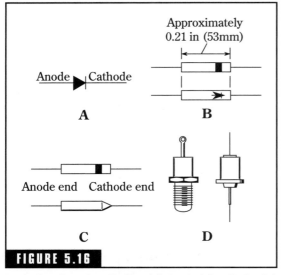

Approximately
0.21 in (53mm)

Anode ▶ Cathode

A

B

Anode end   Cathode end

C

D

FIGURE 5.16

Typical pn junction diodes. (A) Symbol for diode. (B) Low-power or signal diodes. (C) Miniature rectifier or power diodes. (D) High-power rectifier diodes.

diodes in Fig. 5.16B are low-power or signal diodes. These types of diodes are used in RF detector, demodulator, and switching circuits. The miniature rectifier or power diodes shown in Fig. 5.16C are commonly used in low-power applications involving current up to about 1 ampere. Such diodes can be used in ac power supplies involving receivers and low-power transmitters. The high-power rectifier diodes shown in Fig. 5.16D will conduct current levels up to 10 amperes or more with peak inverse voltage ratings of 50 to 600 volts depending on the type. These high-power rectifiers are used in the power supplies of high-power amateur radio transmitters.

The pn junction diodes are available in many sizes and case types. The basic pn junction diode is used as a rectifier, either in high-current, low-frequency rectification circuits or in low-current, high-frequency, and high-speed switching circuits. Power rectifier diodes are generally larger, containing a larger junction area with increased capacitance. Due to this larger capacitance, these diodes are not suitable for high-frequency or switching circuits. High-frequency signal diodes and switching diodes used in digital circuits are physically smaller and have less internal capacitance. Consequently, their current conducting ability is less than the power diodes.

How diodes work

The operation of pn junction diodes can be explained in terms of forward and reverse bias conditions. Forward biasing a diode simply means that the anode terminal is connected to a positive voltage potential, and the cathode terminal is connected to a negative voltage potential. Figure 5.17A shows a battery connected across a diode in a forward bias arrangement. (In practical circuits, this direct connection will burn out the diode if the battery voltage is too high. A forward bias voltage of several volts would destroy most diodes.)

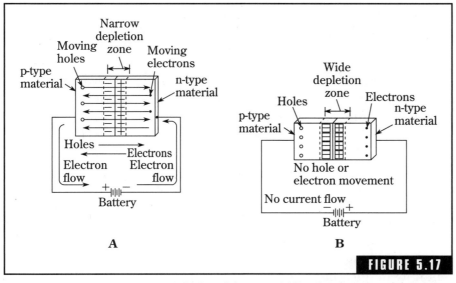

Forward and reverse biasing a pn junction. (A) Forward-biased pn junction. (B) Reverse-biased pn junction.

When the voltage across the diode is greater than the barrier potential of the pn junction, a forward current flows in the diode. This current flow is due to the electric field produced by the voltage across the junction, which forces both free electrons and holes toward the junction. The barrier potential for a germanium pn junction is approximately 0.3 volt. When the voltage across the junction exceeds 0.3 volt, this higher potential pushes free electrons and holes across the junction. Thus free electrons in the n-type material cross the junction and drift toward the anode. At the same time, holes from the p-type material cross the junction, drifting toward the cathode end. These actions result in a current flow that is proportional to the applied voltage.

The movement of holes and free electrons toward the junction reduces the size of the depletion zone. As the external battery voltage is increased, the depletion zone becomes progressively smaller and is eventually eliminated for all practical purposes.

Reverse bias

Reverse biasing a pn junction diode simply means making the cathode more positive than the anode. Figure 5.17B shows a battery connected across the diode in a reverse bias arrangement. The reverse current in this circuit is very small, practically zero.

The action of the reverse voltage places a positive potential at the cathode terminal or n-type material of the diode. This potential attracts the free electrons and causes them to drift toward the cathode terminal. At the same time, holes in the p-type material drift toward the anode terminal because they are attracted to the negative potential. This increases the size of the depletion zone because the departing free electrons and holes create more positive ions in the n-type material and more negative ions in the p-type material, respectively. The depletion zone expands to result in a barrier potential equal to the applied reverse voltage. Thus, the current flow in the diode is virtually zero due to equal and opposing voltages. Only a small leakage current is present in the reverse biased diode. The pn junction acts as a one-way switch, allowing current to flow in one direction, but restricting current flow in the other direction.

Diode circuits

The pn junction diode serves as a rectifier, converting alternating current to unidirectional or one-way current flow. The simplest diode circuit is the half-wave rectifier, shown in Fig. 5.18. Input and output voltage ratings of power transformers are normally given in rms or effective voltage levels. However, the diode rectifies the peak value of the voltage waveform and provides a pulsating dc output

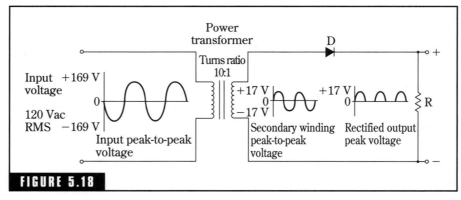

FIGURE 5.18

Half-wave rectifier circuit.

voltage level. The peak ac voltage is obtained by multiplying the rms value by 1.414.

Two diodes can be connected to form a full-wave rectifier circuit when the secondary winding of the transformer has a center tap. Figure 5.19 illustrates this type of full-wave rectifier circuit. The diodes conduct on alternating half cycles of the ac waveform, each referenced from the center tap of the secondary winding. Thus, only one-half of the total peak-to-peak output voltage is used for each diode. Diode D_1 conducts only when the ac voltage is positive, and diode D_2 conducts when the ac voltage is negative. The ground return is connected to the center tap of the secondary winding. The current from either diode flows through load resistor R in the same direction. Thus, the output voltage is a varying dc voltage ranging from 0 to +8.5 volts. The primary disadvantage of the full-wave rectifier circuit is that the rectified voltage peak is one half or less of the rated secondary voltage of the transformer. The pulsating dc output of a full-wave rectifier has a frequency that is twice the ac input frequency.

The pn junction diodes can be used as clipper diodes, eliminating either the positive or negative half of the input ac signal. Figure 5.20 shows a positive clipper circuit. The diode is forward biased during the positive half of the ac signal and conducts this part of the signal to ground. The negative half of the ac signal is unaffected because the diode is reverse biased and does not conduct. A negative clipper circuit is shown in Fig. 5.21.

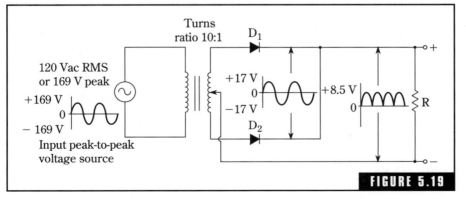

FIGURE 5.19

Full-wave rectifier circuit.

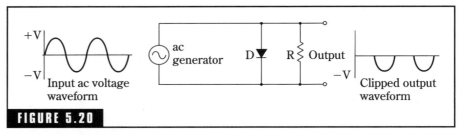

Positive clipper circuit.

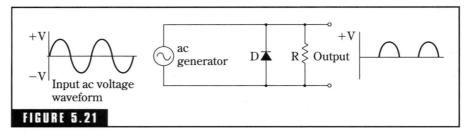

Negative clipper circuit.

Special-Purpose Diodes

The unique properties of the pn junction make possible the construction of special-purpose diodes that exhibit useful characteristics. By varying the doping process and manufacturing techniques, special-purpose diodes can be made to perform as voltage regulators, voltage-controlled capacitors, and negative-resistance devices.

The zener diode

As noted earlier, ordinary pn junction diodes can be destroyed if they are subjected to the reverse bias breakdown voltage. The zener diode is a unique silicon pn junction diode designed to operate in the reverse bias breakdown mode. By controlling the doping levels, zener diodes can be manufactured for specific breakdown voltage levels from about 2 to 200 volts. The constant breakdown-voltage-level characteristic exhibited by the zener diode provides an extremely useful voltage-regulating device.

The zener diode is heavily doped, which results in a very narrow depletion zone. When a reverse bias voltage is applied to the diode, the intense electric field developed across the junction is sufficient to dislodge valence or outer-orbit electrons in the material. These electrons are free to flow across the junction, resulting in current flow. This current flow must be kept within manufacturers' limits by connecting current-limiting resistors in series with the external voltage source.

Symbols for the zener diode are given in Fig. 5.22. A typical application of the zener diode as a voltage regulator is shown in Fig. 5.23. In this example, a 9-volt zener with a power dissipation of 3 watts is used to provide a stable 9-volt source from a varying 11- to 14-volt

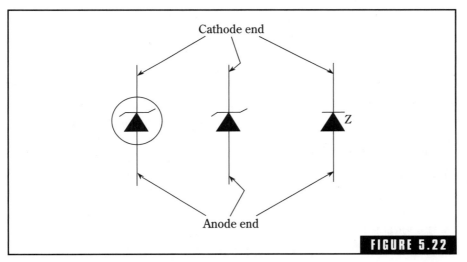

FIGURE 5.22

Common symbols used for zener diodes.

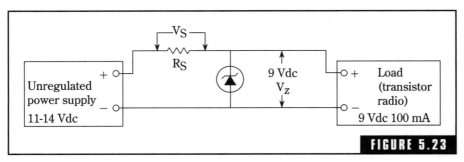

FIGURE 5.23

Typical voltage regulator circuit using a zener diode.

power supply. This type of voltage regulator can be used to power a transistor radio from the electrical system of an automobile. With the engine running, the electrical system of the automobile can supply up to 14 volts from the generator or alternator. When the engine is shut down, the voltage drops to about 12 volts if the battery is charged. Regardless of the varying voltage supplied by the system, the zener regulating circuit maintains a constant 9-volt supply to the transistor.

The varactor diode

The varactor diode, or simply *varactor*, is a voltage-controlled, variable-capacitance pn junction diode used for tuning radio-frequency circuits to a desired resonant frequency. The varactor, a rugged, compact semiconductor device, represents an important advance in electronic technology. It is replacing the bulky mechanical variable capacitors used in radio communications equipments for the past some 75 years. Unlike the variable capacitor, the varactor is not affected by dust or moisture and can be remotely controlled. Symbols for the varactor are given in Fig. 5.24.

The capacitance of the pn junction diode can be increased by special doping processes. This increased capacitance, on the order of 10 to 60 picofarads, allows the varactor to be operated in the reverse biased mode with negligible current flow. Figure 5.25 shows a typical circuit application using the varactor to tune an LC tank circuit. Coil L1 and capacitor C1 form a resonant tank circuit, which is tuned to frequencies by the varactor diode. Capacitor C2 isolates the tank circuit from the dc power supply. Radio frequency choke RFC prevents any RF coupling between the tank circuit and the power supply. The varactor diode is reverse biased and draws essentially no current from the power supply. The varactor diode can provide a wide variation in capacitance, allowing for a considerable change in the resonant frequency of the tank circuit.

Transistors

Transistors, like triode, tetrode, and pentode tubes, have the ability to amplify

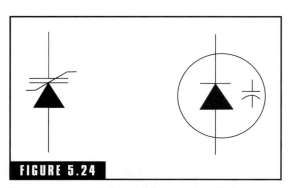

FIGURE 5.24

Two common symbols used for varactor diodes.

weak signals. Also, like tubes, transistors are designed for varying power applications. Small-signal transistors are used for amplifying weak or low amplitude signals to power levels of about one watt or less. Power transistors are designed to handle power levels in excess of one watt. Figure 5.26 shows a group of typical transistors used in amateur radio and other electronic equipment. Transistors at the top of the

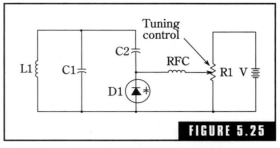

FIGURE 5.25

Typical varactor tuning diode tuning circuit.

photograph are the small signal types and employ either metal or plastic cases. Power transistors are shown at the bottom. For specific information on transistor types and case information, transistor handbooks are available at most electronic supply houses, such as Radio Shack.

Transistors fall into two general categories: bipolar (or junction) transistors and field-effect transistors. Each class possesses unique characteristics that are useful in various applications.

The bipolar transistor consists of three doped regions either in a pnp or an npn arrangement. This creates two independent pn junctions that govern the operational characteristics of these devices. Either type of bipolar transistor can be constructed from germanium or silicon semiconductor material.

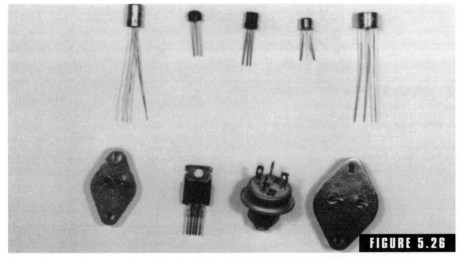

FIGURE 5.26

Typical transistors used in amateur radio and other electronic equipment.

Bipolar transistors are current-operated devices, producing a large change in output current for a small change in input current. Therefore, the bipolar transistor has a low input resistance when compared to the almost infinite input resistance of vacuum-tube circuits. Basic amplifier circuits for the pnp and npn transistors are compared with that of the triode vacuum tube in Fig. 5.27. The npn transistor circuit is similar to the triode circuit in terms of power supply polarities. The triode is connected as a common-cathode amplifier and the transistors are connected as common-emitter amplifiers. The bias circuits, V_{CC} for the triode and V_{BB} for the transistors, are shown in terms of separate batteries to illustrate the potentials on each control element. However, most tube and transistor circuits operate from a single power source and the dc bias voltages are established by resistive networks.

The bipolar transistor is capable of providing high gain characteristics. Current and voltage gain ranges between 50 to 200 for most small-signal transistors. Some transistors will exhibit current gain as high as 1000. In terms of gain characteristics, the bipolar transistor is superior to the triode and is comparable to the pentode.

Field-effect transistors, or FETs, are different from the bipolar transistor in that an electric field is used to control the flow of current within the device. Thus, the FET is a voltage-operated device with an input resistance that is virtually infinite. The characteristic operating curves for the FET are similar to those of the pentode tube. However,

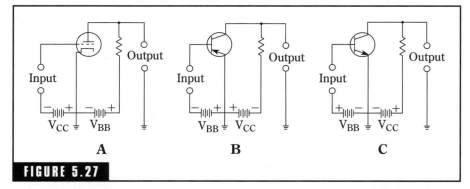

FIGURE 5.27

A comparison between a triode vacuum tube amplifier circuit and amplifier circuits utilizing pnp and npn transistors. (A) A triode vacuum tube amplifier. (B) A pnp transistor amplifier. (C) An npn transistor amplifier.

FETs do not possess the high gain characteristics exhibited by bipolar transistors or pentodes.

Field-effect transistors are generally divided into two types: the junction FET (or JFET), and the insulated-gate FET (or IGFET). Sometimes the IGFET is referred to as a metal-oxide semiconductor FET or MOSFET. Symbols for the JFET and IGFET are given in Fig. 5.28.

Bipolar transistor construction

At first glance, the bipolar transistor might appear to be simply two pn junction diodes connected back to back. However, each of the elements (emitter, base, and collector) is carefully manufactured with different specifications to obtain different characteristics. Figure 5.29 shows the basic layout for an npn transistor. The n-type emitter is heavily doped, producing many free electrons that can be emitted toward the base region. The base is a very narrow layer of lightly doped p-type material containing a small number of holes. Finally, the n-type collector, moderately doped to produce a limited number of free electrons, is designed to collect free electrons that drift across the base material from the emitter. The collector is the largest element because it must dissipate more heat than either the emitter or base elements.

npn transistor operation

Figure 5.30 shows the required biasing for the operation of npn transistors. The base-collector pn junction is reverse biased. Without the

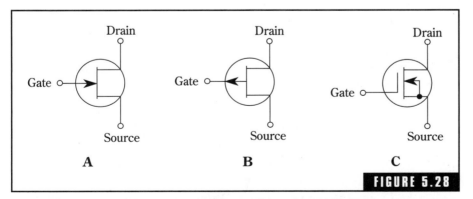

FIGURE 5.28

Three common types of field-effect transistors. (A) n-channel JFET. (B) p-channel JFET. (C) n-channel MOSFET.

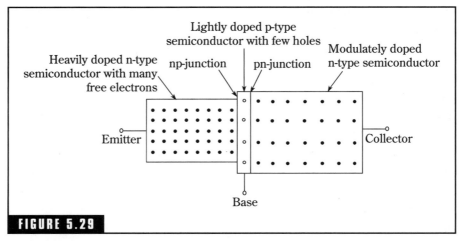

FIGURE 5.29

Basic construction of an npn bipolar transistor.

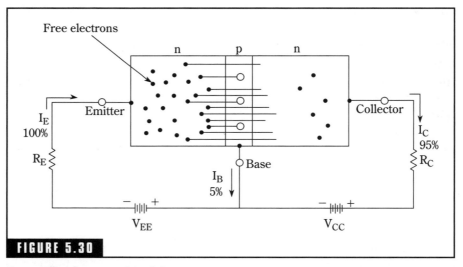

FIGURE 5.30

Current flow in an npn transistor.

emitter circuit, the base-collector pn junction prevents any flow of current. However, the emitter-base np junction is forward biased and this causes a flow of free electrons toward the base region.

Most of the free electrons entering the base drift across the narrow region and pass into the collector. Only a few of these free electrons combine with the limited number of holes in the base region. This results in a small current of about 5 percent or less of the total emitter current. Thus, some 95 percent of the free electrons pass through the base to the positive collector. This action is responsible for the amplifying ability of the transistor.

pnp transistor operation

A pnp transistor operates in the same basic manner as an npn transistor except that holes are emitted by the emitter instead of electrons. Note in Fig. 5.31 that the polarities of the batteries are reversed when compared to those for the npn transistor. The emitter-base pn junction is forward biased, causing holes in the emitter to drift toward the base. Most of these holes drift across the narrow region of the base and into the collector. Some 5 percent or less of the holes crossing into the base material combine with free electrons and produce a small base current. Thus, 95 percent or more of the current produced by the emitter-base pn junction flows through the base to the negative collector.

Common-base amplifying circuits

The npn and pnp transistors in Figs. 5.30 and 5.31 are connected in the common-base configuration. This is one of the three basic circuits applicable to transistors. The common-emitter and the common-collector configurations are discussed later.

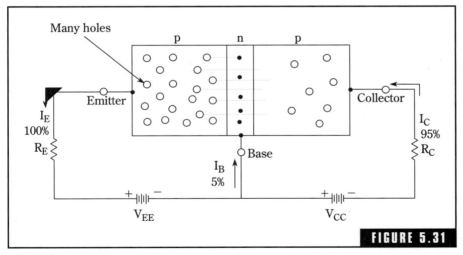

FIGURE 5.31

Current flow in a pnp transistor.

The relationship between the emitter, base, and collector currents is given by:

$$I_E = I_B + I_C \qquad \text{(Eq. 5.1)}$$

where

I_E is the emitter current
I_B is the base current
I_C is the collector current

An important parameter or characteristic of the common-base configuration is the current gain. This is referred to as alpha (α) and represents the ratio of collector current to emitter current, or:

$$\alpha = \frac{I_C}{I_E} \qquad \text{(Eq. 5.2)}$$

Because I_C can never exceed I_E the α of a transistor is always less than one. The value of α will range from about 0.95 to 0.99, depending on the gain characteristics of the transistor.

Figure 5.32 shows a basic pnp common-base amplifier circuit. The signal from the ac generator is coupled to the emitter via a dc-blocking capacitor, C1. The output signal is taken from the collector through the dc-blocking capacitor, C2, and can be connected to a following stage for additional amplification if required. The emitter-base pn junction is forward biased by battery V_{BB}. Emitter resistor R_E is a current-limiting resistor. Thus, battery V_{BB} and resistor R_E establish a dc operating current before an ac input signal is applied to the amplifier.

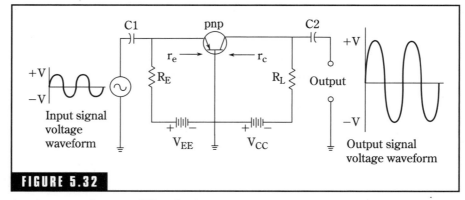

FIGURE 5.32

A pnp common-base amplifier circuit.

The amplifying action of the common-base configuration is based on the ac input signal varying the forward bias across the emitter-base pn junction. This varying voltage produces a change in the current flowing from the emitter toward the base region. As stated earlier, most of the current flows through the narrow base region into the collector. Thus, the current flowing into the collector circuit varies essentially the same as the emitter current and produces an ac voltage drop across the load resistor, R_L.

The ac output voltage developed across R_L is large compared to the ac input signal voltage. This is due to R_L being many times the value of the input resistance of the emitter-base circuit. The input resistance of the common-base configuration, referred to as the emitter resistance R_e, is typically less than 100 ohms and can be as low as 10 ohms or less. This is due to the fact that the emitter-base junction is forward biased, presenting a low resistance. Conversely, the collector output resistance is extremely high, on the order of 300 k to 500 k.

To provide a useful output from a common-base amplifier, such as driving a subsequent stage of amplification, load resistor R_L is typically on the order of 2 k to 10 k. The voltage gain of the common-base configuration can be approximated by:

$$A_V = \frac{\alpha R_L}{R_e} \qquad \textbf{(Eq. 5.3)}$$

where

A_V is voltage gain

α is alpha or current gain

R_L is the collector load resistance

R_e is the emitter resistance

Example 1   Find the voltage gain of the circuit in Fig. 5.32 if the alpha of the transistor is 0.98. The input emitter resistance is 25 ohms and the collector load resistor is 2000 ohms.

Solution

Compute the voltage gain, A_V, by substituting the given values in Equation 5.3:

$$A_V = \frac{\alpha R_L}{R_e}$$

$$= \frac{0.98 \times 2000}{25}$$

$$= 78.4 \text{ (Answer)}$$

The common-base configuration does not provide current gain because the collector output current is always slightly less than the input current in the emitter circuit. Due to this limitation and the extremely low input resistance, the common-base configuration is not widely used in solid-state circuits.

Common-emitter amplifier circuits

The common-emitter configuration is the most useful of the three basic transistor circuits. Both npn and pnp common-emitter circuits are illustrated in Fig. 5.33.

A dc base current, on the order of a few microamperes for small-signal transistors, flows in the forward-biased base-emitter pn junction.

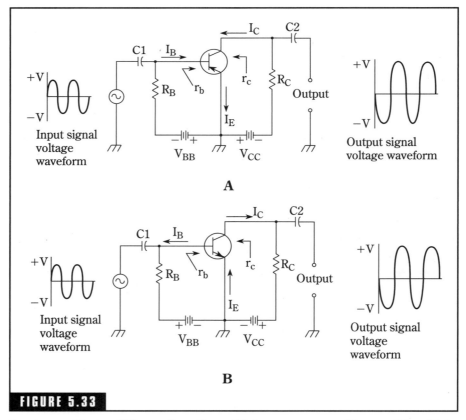

FIGURE 5.33

Common-emitter amplifier configurations. (A) A pnp common-emitter amplifier. (B) An npn common-emitter amplifier.

The collector current, on the order of a few milliamperes, flows in the collector-emitter circuit.

The ac signal input to this amplifier is coupled to the base via a dc blocking capacitor, C1. Battery V_{BB} and base-current-limiting resistor R_B provide a dc operating base current. The ac input signal voltage variations produce a corresponding variation in the dc base-emitter current. This action in turn varies the collector current, resulting in current amplification.

Current gain in a common-emitter configuration is determined by the ratio of change in collector current to change in base current. This quantity is expressed as beta (β). In mathematical terms, beta is given as:

$$\beta = \frac{I_C}{I_B} \qquad \textbf{(Eq. 5.4)}$$

where

β is beta or current gain

I_C is the change in collector current caused by the varying input signal

I_B is the change in base current caused by the varying input signal

The beta of most small-signal transistors will range from about 40 to 200, with some high-gain transistors having a beta of up to 1000. For example, if a 10-microampere change in base current produces a 500-microampere change in collector current, the resulting beta is 500/10, or 50.

The base input resistance of common-emitter transistor circuits is normally on the order of 500 to 5000 ohms. The collector output resistance is about 50,000 ohms. These moderate input and output resistances, along with high voltage and current gain, make the common-emitter configuration a valuable circuit for solid-state design.

The voltage gain of the common-emitter configuration is comparable to that of the common-base configuration—normally on the order of 100 to 1000. This voltage gain along with high current gain produces a power gain of 10,000 or more for the common-emitter configuration. The phase of the amplified output signal produced by the common-emitter amplifier is reversed when compared to the input signal, i.e., as the ac input signal increases from zero to a positive level, the ac output signal drops from zero to a negative level. This is due to the increasing positive voltage on the base that causes the collector current to increase, thus decreasing the collector voltage. Conversely, an increasing negative voltage on the base reduces the collector current, causing the collector voltage to rise.

Common-emitter design considerations

Solid-state design is a complex and wide-ranging aspect of electronics that includes many types of semiconductor devices and circuits. Many excellent texts are available that cover practical and theoretical details of this subject. Because space is not available in this handbook to cover all aspects of solid-state design, a procedure for designing a simple common-emitter amplifier circuit is given here.

Most transistor circuits operate from a single common power source. The simple npn amplifier in Fig. 5.34 uses a single 9-volt battery for both base-biasing and collector-supply requirements. The current flowing in base-biasing resistor R_B establishes a dc operating current for a no-signal input condition.

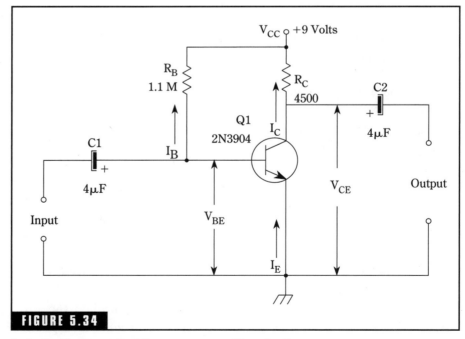

FIGURE 5.34

A simple biasing method for a common-emitter circuit.

A 2N3904 npn silicon transistor is used in this circuit. The 2N3904, similar to the Sylvania ECG123A or the Radio Shack RS-2030, is designed for general-purpose amplifier and switching applications. Applicable ratings for the 2N3904 transistor, available in most transistor handbooks, are as follows:

Maximum Voltage and Current Ratings	
Collector-base voltage, V_{CB}	60 V
Collector-emitter voltage, V_{CE}	40 V
Base-emitter voltage, V_{BE}	6 V
Collector current, I_C	200 mA
Junction operating temperature	135° C
Small-Signal Characteristics	
For $I_C =1$ mA, $V_{CE} = 10$ V, and Signal Frequency $=1$ kHz:	
Current gain (Beta)	100 to 400, Average = 125
Input impedance	1 k to 4 k, Average = 3.5 k

The 2N3904 is capable of handling currents of up to 200 mA. However, for small-signal applications an I_C of about 1 mA is recommended. This smaller static or dc current will result in a higher input impedance (or resistance) for the base-emitter circuit and ensure linear amplification of small ac signals. Also, the small collector current results in very little power dissipation within the transistor and minimizes power supply drain. For battery-powered applications, this latter advantage is all important.

The dc collector-emitter voltage, V_{CE}, is normally established as one-half of the supply voltage, V_{CC}. In this example, V_{CE} will be equal to one-half of 9 V, or 4.5 V. The value of R_C can be calculated from the following equation:

$$R_C = \frac{V_{CC} - V_{CE}}{I_C} \text{ ohms} \qquad \textbf{(Eq. 5.5)}$$

Substituting the required values, R_C is:

$$R_C = \frac{V_{CC} - V_{CE}}{I_C} = \frac{9 - 4.5}{0.001} = 4500 \text{ ohms}$$

The base current, I_B, is related to the collector current, I_C, and the beta of the transistor. Thus, I_B can be calculated as follows. Because:

$$\beta = \frac{I_C}{I_B}, \text{ then } I_B = \frac{I_C}{\beta} \qquad \textbf{(Eq. 5.6)}$$

Substituting $I_C = 0.001$A and $\beta = 125$ in Equation 5.6,

$$I_B = \frac{0.001}{125} = 0.000008 \text{ A or } 8 \ \mu A$$

The base resistor, R_B, can now be calculated based on $I_B = 8 \ \mu$A. The base-emitter pn junction is forward biased, resulting in a voltage drop of approximately 0.7 volt for silicon transistors. (If a germanium transistor is used in this type of circuit, the base-emitter voltage, V_{BE}, is approximately 0.3 volt.) Therefore, if V_{CC} is 9 volts and V_{BE} is 0.7 volt, the voltage drop across base resistor R_B is $9 - 0.7$ or 8.3 volts. The equation for determining R_B is:

$$R_B = \frac{V_{CC} - V_{BE}}{I_B} \text{ ohms} \qquad \textbf{(Eq. 5.7)}$$

R_B can now be calculated as follows:

$$R_B = \frac{9 - 0.7}{8 \ \mu A} = \frac{8.3}{0.000008} = 1,030,000 \text{ ohms} \cong 1.0 \text{ megohm}$$

The values of dc blocking capacitors C1 and C2 are selected to provide a low-impedance path for the input and output signals. For example, if the amplifier is designed for an audio frequency response from 200 Hz to 3 kHz, the capacitive reactance, X_C, of each capacitor should not exceed a given value over this frequency range. Because X_C varies inversely with frequency (remember, $X_C = \frac{1}{2} \pi fc$), its value will be greatest at the lower frequencies. A 4-μF capacitor will present an X_C of about 200 ohms at 200 Hz. Considering the input resistance (about 3500 ohms) and the output resistance (about 4500 ohms) for the transistor amplifier, the 200-ohm capacitive reactance of the blocking capacitors will offer little opposition to the input and output signals.

The preliminary design for the small-signal amplifier using the 2N3904 transistor is now complete and can be breadboarded for initial testing. The current gain is approximately 125, based on the average beta for this transistor. However, what happens if the 2N3904 transistor obtained for this circuit has a different beta, say 300? This poses a problem that is inherent in this design procedure.

Variations in transistor characteristics

Unlike vacuum tubes, the characteristics of transistors will vary during manufacturing processes. For example, the common-emitter current gain, or beta, for the 2N3904 transistor might range from about 100 to 400. If the previous circuit design was based on a beta of 125, a 2N3904 with a beta of 400 will not perform as expected in this circuit. The increased current gain will result in a higher collector current, lowering the desired collector-emitter voltage.

An experimental approach may be used to determine the optimum value of the base resistor, R_B. The value of R_B is critical in controlling the collector current. Thus, R_B can be determined experimentally by substituting a variable resistor in the circuit and varying the resistance for the required collector-emitter voltage of 4.5 volts. A recommended procedure for determining the proper value of base resistor R_B is:

- With V_{CC} (9 volts in this circuit) applied to the amplifier, measure the value of V_{CE}. If the measured value of V_{CE} is within about 10 percent of the required value, the original calculations are adequate and no changes are required. If the measured value of V_{CE} is not within 10 percent of the required value, that is, below 4 volts or above 5 volts, the value of R_B should be changed to obtain the correct V_{CE}.

- Figure 5.35 illustrates how a variable resistor network can be temporarily installed in the base-biasing circuit (between points A and B). Note that a fixed resistor is used along with the potentiometer in order to prevent excessive base current. The total range of this resistive network can be varied from 0.5 to 3.5 megohms.

- With 9 volts applied to the circuit, adjust the variable resistor to obtain a V_{CE} of 4.5 volts. Remove power from the circuit and

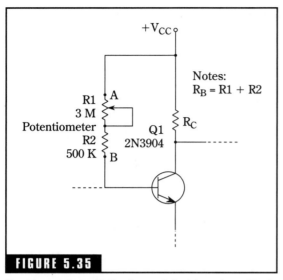

FIGURE 5.35

Experimental circuit for determining the value of Rg in a simple biased common-emitter transistor amplifier.

measure the resistance of the temporary network with an ohmmeter. A fixed resistor of this approximate value can now be connected permanently in the circuit in place of the temporary network. This completes the design and the amplifier can be used for small-signal amplification.

Voltage gain of common-emitter amplifier circuits

Beta, the current gain for the circuit in Fig. 5.34, was determined from manufacturers' specifications. The voltage gain of this amplifier is defined as the ratio of the output signal voltage to the input signal voltage, or:

$$A_v = \frac{V_{\text{out}}}{V_{\text{in}}} \qquad \textbf{(Eq. 5.8)}$$

where

A_v is the voltage gain

V_{out} is the amplitude of the output signal

V_{in} is the amplitude of the input signal

The approximate voltage gain can be determined by the following equations:

$$A_v \cong \frac{i_c \times R_c}{i_b \times r_b}, \text{ or} \qquad \textbf{(Eq. 5.9)}$$

$$A_v \cong \beta \, \frac{R_c}{r_b} \qquad \textbf{(Eq. 5.10)}$$

where

i_c is the change in collector current caused by the ac input signal

i_b is the change in base current caused by the ac input signal

R_C is the collector load resistor

β is the current gain

r_b is the input base-emitter resistance of the transistor

Note that the current ratio, i_c/i_b, is the beta of the transistor. Sometimes the beta for a specific dc operating level, such as a specified collector current of 1 mA, is termed the *ac current gain* or A_i.

The approximate voltage gain, A_v, for the circuit in Fig. 5.34 is calculated using a beta of 125, an input base-emitter resistance of 3.5 k, and a 4.5-k collector load resistor, R_c.

$$A_v = \beta \, \frac{R_c}{r_b} = 125 \times \frac{4500}{3500} = 161$$

If a 10-millivolt, peak-to-peak signal is applied to the input of this amplifier, the amplifier output signal will be 161 times 10 millivolts, or 1.6 volts, peak-to-peak.

If the manufacturers' specifications do not include the value of r_b, the base-emitter resistance, it can be approximated by the following equation:

$$r_b = \frac{25\beta}{I_E \,(\text{mA})} \ \text{ohms} \qquad \textbf{(Eq. 5.11)}$$

The value of r_b for the 2N3904 transistor with a collector current, I_c, of 1 mA is calculated as follows:

$$r_b = \frac{25\beta}{1_E(\text{mA})} = \frac{25 \times 125}{1} = 3125 \text{ ohms}$$

This compares favorably with the 3500 ohms specified by the manufacturer.

This completes the design calculations for the small-signal amplifier using the 2N3904 transistor. In general, this design is valid for audio and low-frequency applications. When radio-frequency amplifier circuit design is required, many factors must be taken into consideration. For example, the frequency response of the transistor will affect circuit design.

Frequency limitations of transistors

Many factors affect the frequency response of a transistor. These factors include physical construction characteristics, electron transit time between the emitter and collector, interelement capacitance, and circuit configuration. Some transistors provide usable gain in the hundreds of megahertz while others will not operate at frequencies over 50 kHz.

Manufacturers' specifications or transistor handbooks usually provide one or more parameters that describe the performance of transistors as a function of frequency. The three most common frequency parameters are as follows:

- Common-emitter or beta cutoff frequency, f_b. This is the frequency at which the ac gain for the common-emitter configuration drops to 70 percent of its value at low frequencies. Sometimes this parameter is referred to as f_{ae} or f_{hfe}. For example the f_b for the 2N2342 npn silicon transistor is rated at 550 kHz. The ac beta or ac current gain for this transistor at low frequencies will range from 10 to 40 with an average of about 25. If the frequency of operation is increased to 550 kHz, the average ac beta of 25 will drop by 70 percent, or to about 17.5. For frequencies above 550 kHz, the ac beta drops sharply.

- Common-base or alpha cutoff frequency, f_T. Sometimes referred to as f_{ab} or f_{hfb}, this parameter describes the frequency at which the ac alpha gain drops to 70 percent of its value at low frequencies.

- Common-emitter gain-bandwidth product frequency, f_T. The f_T is defined as the frequency for which beta or ac current gain has dropped to unity or 1.

The common-collector configuration

The third basic transistor configuration is the common-collector or emitter follower, illustrated in Fig. 5.36. Either npn or pnp transistors can be used in these circuits. The only difference is that the power supply polarity is reversed.

The three useful characteristics of the common-collector configuration are:

- Extremely high input resistance, ranging up to 500 k.

- Very low output resistance, from about 50 to 1000 ohms.

- Current gain approximately equal to the beta rating of the transistor.

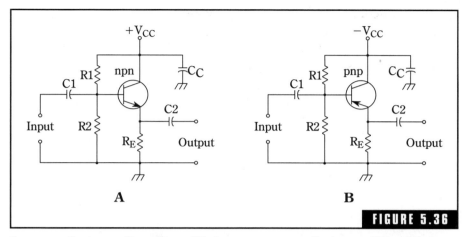

Common-collector configurations. (A) An npn common-collector amplifier. (B) A pnp common-collector amplifier.

The primary advantage of the common-collector configuration is its ability to act as an impedance-matching device, transforming a high-impedance, low-current signal to a low-impedance, high-current signal. In this sense, the common-collector configuration acts as a step-down transformer. Unlike the transformer with a limited frequency response and no power gain, the common-collector configuration operates over a wide frequency range, approximately equal to the beta cutoff frequency, f_b, rating of the transistor. The current gain of the common-collector circuit is approximately equal to the beta of the transistor.

The common-collector circuit is similar to the common-emitter circuit in terms of signal input and biasing arrangements. However, the common-collector configuration does not provide voltage gain. Note that the output signal is taken from the emitter terminal that is always at a lower potential than the base terminal. This difference of potential is approximately 0.7 volt for silicon transistors and 0.3 volt for germanium transistors. The collector terminal in the common-collector configuration is connected directly to the power source, V_{CC}. Capacitor C_c allows the signal developed at the collector to be bypassed to ground. This places the collector terminal at an ac ground potential and the output signal is developed across the emitter resistor, R_E.

The common-collector circuits in Fig. 5.36 employ voltage-divider biasing, also used for the common-emitter configuration. This circuit compensates for the wide variations in current gain (beta), temperature characteristics, and aging encountered in transistors. However, the simple resistor bias network or a separate bias supply can be used if desired.

Summary of transistor characteristics

Technical information and specifications for particular transistors are available in manufacturers' specification sheets and transistor handbooks. Most handbooks provide a cross-reference listing for replacement purposes. Transistor characteristic curves, available in many of these publications, are useful in specific design problems. These curves usually show collector current-and-voltage relationships for common-base and common-emitter configurations.

Gain and resistance characteristics for the three basic transistor configurations are given in Table 5.2. This data is typical of the voltage gain, current gain, input resistance, and output resistance for small-signal transistors.

Transistor testing

In-circuit testing of equipment using transistors is best accomplished with the aid of a schematic diagram that provides key voltage and resistance test data. Defective transistors or other components often can be quickly located by analyzing voltage and resistance readings.

Table 5.2 Bipolar Transistor Characteristics

Characteristic of transistor	Common-base configuration	Common-emitter configuration	Common-collector configuration
Input resistance (ohms)	30 to 150	500 to 1500	5 k to 500 k
Output resistance (ohms)	100 k to 500 k	30 k to 50 k	50 to 1000
Voltage gain	500 to 1500	300 to 1000	Less than 1
Current gain	Less than 1	25 to 100	25 to 100
Power gain	500 to 1500	Up to 10,000	Up to 100

A close visual inspection for loose connections, short circuits, or poor solder joints is a prerequisite for any troubleshooting of transistor circuits. Such preliminary efforts can save many hours of investigating erroneous indications.

Voltage measurements should be made with voltmeters possessing a sensitivity of 20,000 ohms per volt or greater. Otherwise, the internal resistance of the voltmeter might change the transistor biasing circuit and show incorrect voltage readings. Vacuum tube voltmeters (VTVM) or transistorized voltmeters (TVM) with input impedances on the order of 11 megohms are preferred for these types of measurements. Digital voltmeters are especially suited to voltage measurements in transistor circuits. Key voltage measurements in any transistor circuit should include base-emitter bias voltage and collector and emitter voltages.

Commercial in-circuit transistor testers are available for checking individual transistors wired into the circuit. However, some knowledge of the particular circuit might be required to ensure a thorough test of the transistor. For example, coils, transformers, resistors, and forward-biased diodes can produce low-resistance shunts across the transistor elements.

Transistor testing with ohmmeters

Front-to-back resistance measurements involving both pn junctions is a simple but effective way to determine the general condition of a transistor. However, this test does not give the alpha or beta current-gain values or other performance characteristics expected from the transistor. The ohmmeter test is best accomplished with the transistor removed from the circuit. Only a detailed examination of the particular circuit involved will indicate if the transistor can be tested while wired in the circuit.

Before attempting to use the ohmmeter for testing transistors, determine the polarity of the ohmmeter test leads due to the internal battery. Normally, the positive terminal of the battery will be connected to the red (positive) test lead. An external voltmeter can be used to verify this polarity. This information is helpful in identifying npn or pnp transistors.

Avoid using any ohmmeter range that can cause excessive current to flow within the transistor. For example, the maximum base-emitter voltage for many transistors will range between 3 to 6 volts. Ohmmeters employing internal battery voltages exceeding 3 volts can damage or destroy a good transistor. Also, the R × 1 scale in most ohmmeters will allow excessive current to flow in the pn junctions of the transistors, causing permanent damage. Never use the R × 1 scale; the R × 10 or R × 100 scale is generally safe for testing transistors and will provide adequate resistance readings.

The test procedure is based on considering the transistor as two individual pn junction diodes. Figure 5.37 shows the pnp and npn transistors in the initial stage of testing forward biasing. Each internal pn junction is represented as a diode. The solid-line connections are used for checking the forward resistance of the collector-base pn junction of either transistor. When this reading is obtained, the lead connected to the collector is moved to the emitter terminal. This provides the forward resistance reading for the emitter-base pn junction. To obtain the reverse resistance readings for each junction, simply reverse the test leads and repeat the procedure just described.

In either type of transistor, pnp or npn, the ohmmeter is used to check the forward-biased resistance of the collector-base and emitter-base junctions. Record these readings and reverse the ohmmeter test

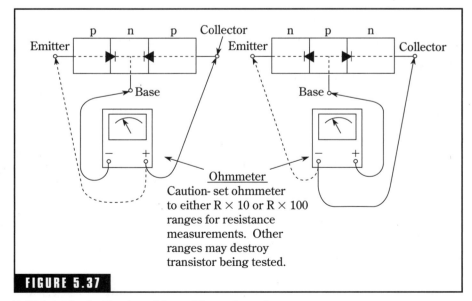

FIGURE 5.37

Initial set for testing transistors with an ohmmeter.

leads and repeat the resistance reading. This provides the reverse-biased resistance readings for both junctions. In general, a high front-to-back ratio of resistance (about 100 to 1) should be obtained for each junction of germanium transistors. The front-to-back resistance ratio for silicon transistor junctions can exceed 1,000,000 to 1. Table 5.3 shows typical values of resistance measurements for both small-signal and power transistors.

Field-Effect Transistors

Field-effect transistors, like vacuum tubes, use an electric field to control the flow of current within the device. This permits the control element, called a *gate*, to exhibit an extremely high input resistance in the range of hundreds to thousands of megohms. This desirable characteristic, along with an ability to provide amplification, makes the field-effect transistor, or FET, a valuable component for solid-state circuit design.

There are basically two types of field-effect transistors—the junction field-effect transistor, or JFET, and the insulated-gate field-effect

Table 5.3 Resistance Measurements for Transistors

Condition for resistance measurement	Low-power transistors		Medium- to high-power transistors	
	Germanium	Silicon	Germanium	Silicon
Forward collector-to-base resistance	100 ohms or less	200 ohms or less	50 ohms or less	100 ohms or less
Forward emitter-to-base resistance	100 ohms or less	200 ohms or less	50 ohms or less	100 ohms or less
Reverse collector-to-base resistance	200,000 ohms or less	Almost infinite	1000 ohms or less	Almost infinite
Reverse emitter-to-base resistance	200,000 ohms or less	Almost infinite	1000 ohms or less	Almost infinite

transistor, or IGFET. The IGFET is also referred to as a metal-oxide semiconductor FET, or MOSFET.

The junction field-effect transistor

The JFET is basically a tiny bar of doped silicon that acts as a resistor. This bar, referred to as a *channel*, can be constructed from either n-type or p-type silicon material. Figure 5.38 shows a cross-sectional view of a simple n-channel JFET. The n-channel material contains many free electrons capable of sustaining a usable current through the device. The terminals connecting each end of the n-channel to an output circuit are designated as *source* and *drain*.

The *gate* consists of doped silicon material opposite to that of the channel material. The gate is normally embedded in the channel material by a diffusion process. In Fig. 5.38, the gate is made of p-type silicon material.

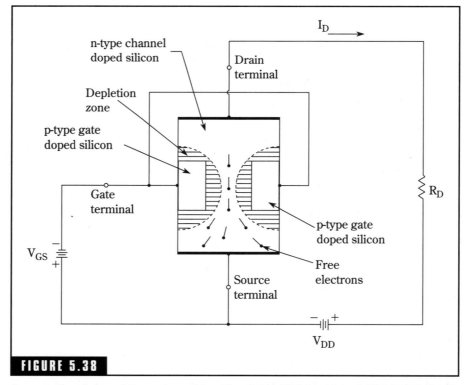

FIGURE 5.38

Cross-sectional view of the n-channel junction field-effect transistor with biasing circuit.

The pn junctions formed within the device produce depletion zones, the widths of which can be controlled by placing a voltage potential between the gate and channel material. Reverse biasing these pn junctions results in expanding the depletion zones and restricting the flow of free electrons within the n-channel. Thus, current flow depends on the size of the depletion zone, which in turn is controlled by the voltage on the gate terminal. If the depletion region extends across the width of the n-channel, current flow is cut off. The value of the gate voltage, V_{GS}, which reduces the current flow to zero is called the *pinch-off* voltage.

Maximum current flow in the channel material occurs when the gate voltage is zero. This is equivalent to the gate being shorted to the source terminal. Most JFET specification sheets provide the typical value of the shorted-gate drain current. This current is designated as I_{DSS}. The subscript DDS is defined as drain to source with shorted gate. The value of I_{DSS} is used in designing JFET circuits.

Because the pn junctions in JFETs are operated in the reverse-biased mode, negligible current flows in the junction. This accounts for the extremely high input resistance at the gate terminal.

The p-channel JFETs are constructed with silicon p-type channel material and n-type gates. The p-channel contains many holes or positive charge carriers, which are capable of conducting a usable current through the device. As anticipated, the power supply polarities and direction of current flow are reversed for p-channel JFETs when compared to n-channel. Figures 5.39 and 5.40 show both n- and p-channel JFETs with different biasing circuits. The circuits in Fig. 5.39 employ separate power sources for gate biasing. Figure 5.40 shows n- and p-channel JFET amplifiers, each with single power sources and voltage-divider biasing networks. The dc gate bias voltage, V_{GS}, is developed by voltage-divider network R1 R2 and the voltage drop across source resistor R_s. Capacitor C_s bypasses any ac signals to ground while maintaining a dc voltage level between the source terminal and ground.

Characteristics of junction field-effect transistors

One of the major advantages of the JFET is its extremely high input resistance, which approaches infinity for most circuits. However, this high input resistance results in less control over output drain current, or in effect, less voltage amplification. The effective voltage gain of a

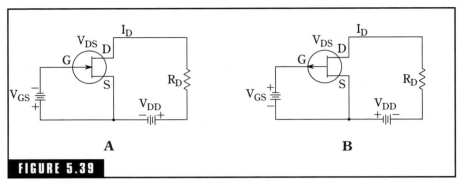

FIGURE 5.39

Biasing circuits for n- and p-channel JFETs using separate power sources for gate and drain requirements. (A) n-channel JFET. (B) p-channel JFET.

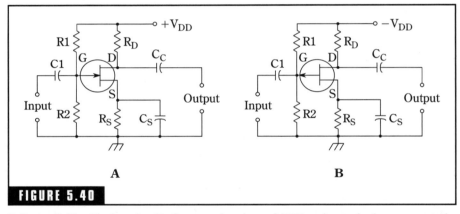

FIGURE 5.40

Voltage-divider biasing circuits for n- and p-channel JFETs using a single power supply. (A) n-channel JFET. (B) p-channel JFET.

JFET is substantially less than that of the bipolar transistor. The voltage gain of JFET amplifiers is typically on the order of 5 to 50 as compared to 100 to 200 for bipolar transistors. The current gain characteristics of JFETs are much superior to that of bipolar transistors. In fact, these gain values approach infinity because virtually no current flows in the JFET input circuit.

JFET circuit configurations

JFETs can be employed in three basic circuit configurations: common source, common gate, and common drain. Similar in performance to the three basic bipolar transistor configurations, these circuits are also valid for the insulated-gate field-effect transistors, or IGFET devices.

Figure 5.41 and Table 5.4 show these three basic circuits and related characteristics. The transistors used in these circuits are n-channel devices. For p-channel devices, the power supply polarities and directions of current flow are reversed.

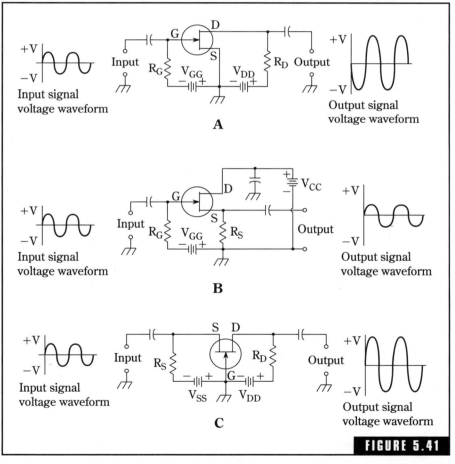

The three basic JFET circuit configurations. (A) Common-source n-channel JFET amplifier. (B) Common-drain n-channel JFET amplifier. (C) Common-gate n-channel JFET amplifier.

Table 5.4 Characteristics of Three Basic JFET Amplifier Configurations

Characteristics	Common-source configuration	Common-drain configuration	Common-gate configuration
Input resistance	Extremely high	Extremely high	Very low
Output resistance	Moderate	Low	Moderate
Voltage gain	High	Less than 1	Moderate
Current gain	Extremely high	Extremely high	Less than 1
Power gain	Extremely high	Extremely high	Moderate
Phase input/output relationship	180°	0°	0°
Primary use	Voltage amplifier	High-to-low impedance matching	Stable High-frequency amplifier

Insulated-gate field-effect transistors

IGFET transistors, sometimes referred to as metal-oxide semiconductor field-effect transistors, or MOSFETs, are similar in performance to JFETs. Gate biasing methods used for JFETs are, in general, applicable to IGFETs.

There are two basic types of insulated-gate field-effect transistors: *depletion-mode* and *enhancement-mode* IGFETs. The depletion-mode IGFET operates in a manner similar to that of the junction field-effect transistor. However, the depletion-mode device does not use a pn junction to produce the depletion zone. As illustrated in Fig. 5.42, the gate is insulated from the channel material by a thin layer of metal oxide. The n-type material, heavily doped to provide many free electrons, is embedded in a base of lightly doped p-type material. This base material is referred to as the *substrate*.

Depletion-mode IGFET operation

Either n- or p-channel semiconductor material can be used to construct IGFETs. The depletion-mode IGFET in Fig. 5.42 shows an n-channel type with one gate. This gate and the narrow n-channel form a capacitor with the metal-oxide insulator acting as a dielectric.

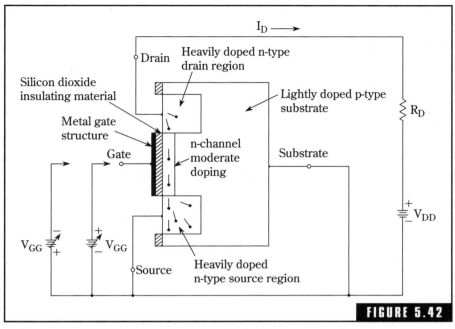

Depletion-mode n-channel insulated-gate field-effect transistor (IGFET).

When the gate potential is zero volts, current flows through the n-channel due to the abundance of free electrons or negative-charge carriers. The positive potential on the drain terminal attracts the free electrons within the n-channel, causing current flow. The depletion-mode IGFET is considered a *normally on* device—current flows in the channel material when the gate is at zero potential.

If a positive voltage is connected to the gate terminal, the gate side of the capacitor assumes a positive potential and the n-channel material assumes a negative potential. This causes more free electrons to be induced into the n channel and, hence, more current flow within the device. Note in contrast that the gate on an n-channel JFET is never allowed to become positive. This is one major difference between the operation of IGFETs and JFETs.

When the IGFET gate is connected to a negative potential, the capacitance of the gate and the n-channel produces a positive charge in the n-channel material. This reduces the number of free electrons within the n-channel material, in effect creating a depletion zone. Subsequently, the current flow within the device is reduced. As the negative gate voltage is increased, the drain current is finally reduced to zero.

Depletion-mode IGFETs can be constructed with p-channel semiconductor material embedded in an n-type substrate. When using these p-channel devices, power supply polarities must be reversed from that used for n-channel devices. Note that the direction of current flow will also be reversed.

Dual-gate depletion-mode IGFETs contain two independent gate structures, each capable of controlling the drain current. These devices are useful as mixers or gain-controlled amplifiers in mixer, RF, and IF stages of receivers.

Some IGFETs are available with zener-diode input networks to protect the gates from accidental exposure to high voltage such as static electrical discharges. The extremely thin metal-oxide insulating material can easily be punctured and permanently damaged by excessive high voltage between the gate and other terminals of the device. The zener diodes are connected back-to-back across each gate to provide a discharge path for excessive voltages appearing across these terminals. During normal operation, the zener diodes do not conduct and, thus do not load the input circuit.

Symbols for the depletion-mode IGFETs are given in Fig. 5.43.

Enhancement-mode insulated-gate field-effect transistors

A major difference in the operation of enhancement-mode IGFETs as compared to depletion-mode IGFETs is that they act as *normally off* devices. Drain current flows only when a potential, equal in polarity to that of the drain supply, is applied to the gate terminal.

Enhancement-mode IGFETs are designated as either p-channel or n-channel devices. Figure 5.44 illustrates a p-channel depletion-mode IGFET. This device consists of a tiny block of lightly doped n-type material with separate drain and source p-type regions embedded or diffused in the block. The metal gate, separated from the block by a thin layer of metal-oxide insulating material, overlaps the drain and source regions. As in the case of the depletion-mode IGFET, the gate and semiconductor material form two plates of a capacitor, with the insulating material serving as a dielectric.

When there is no potential on the gate terminal, no conduction path exists for source-to-drain current flow. When a negative poten-

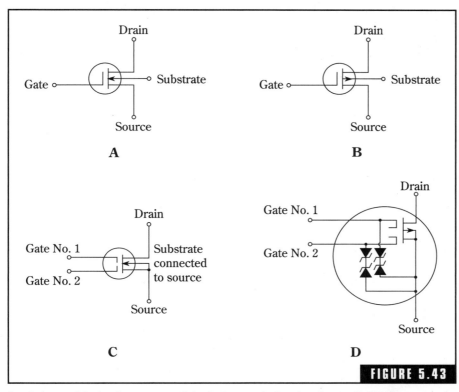

Drain

Gate ○── ─○ Substrate

Source

A

Drain

Gate ○── ─○ Substrate

Source

B

Drain

Gate No. 1 ○──
                    Substrate
                    connected
Gate No. 2 ○──   to source

Source

C

Drain

Gate No. 1 ○──

Gate No. 2 ○──

Source

D

FIGURE 5.43

Symbols used for various types of depletion-mode insulated-gate field-effect transistors. (A) Single-gate n-channel. (B) Single-gate p-channel. (C) Dual-gate n-channel. (D) Dual-gate diode-protected p-channel.

tial is applied to the gate, holes or positive-charge carriers are induced into the region forming the capacitor plate. Because the drain is at a negative potential, these holes flow into the drain region. Therefore, a positive charge flows within the device. Increasing the negative potential on the gate terminal produces a corresponding increase in drain current.

The n-channel enhancement-mode IGFETs are similar in construction to the p-channel device. However, the substrate is formed from lightly doped p-type material, and drain and source regions are embedded or diffused n-type material. As anticipated, the power source polarity, gate polarity, and direction of current flow are reversed when compared to the n-channel device. A positive gate potential is required to "turn on" the drain current in the n-channel

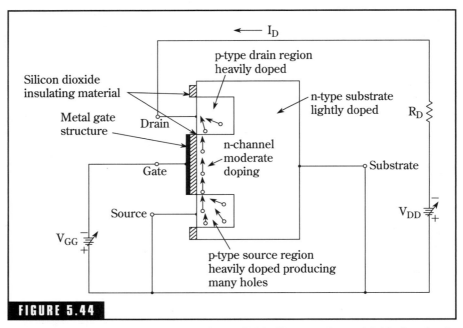

P-channel enhancement-mode insulated-gate field-effect transistor with biasing circuit.

enhancement-mode IGFET. Symbols for the enhancement-mode IGFETs are given in Fig. 5.45.

Care and handling of insulated-gate semiconductor devices

Insulated-gate field-effect transistors are extremely susceptible to permanent damage when subjected to excessive high voltages and currents. For example, normal static electricity that exists on the human body and test probes can puncture the gate structure on IGFETs that do not have zener-diode protective networks. Maximum voltage and current ratings as specified in manufacturers' data sheets or transistor handbooks should never be exceeded.

Extreme care must be exercised in handling, testing, and installing insulated-gate devices. The device leads should be kept shorted together until the device is installed and soldered in the required circuit. Most IGFETs are shipped by the manufacturer with the leads shorted together in some fashion such as by a shorting wire or conductive foam packing.

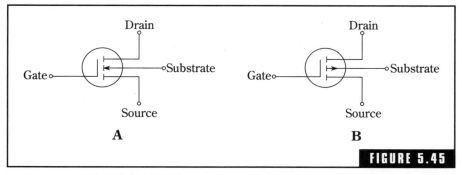

FIGURE 5.45

Symbols for enhancement-mode insulated-gate field-effect transistors. (A) n-channel.
(B) p-channel.

As IGFETs cannot be safely tested with ohmmeters, two alternate means of testing are recommended if a commercial IGFET tester is not available:

- Check the device installed in the original circuit using the manufacturer's recommended test procedures. This might include specific resistance measurements with the power off and voltage and signal waveform analysis with the power on. This type of test procedure usually leads to the identification of the defective component.

- Replace the suspect IGFET with a substitute IGFET that is known to be good. If no direct replacement is available, a similar device can be found using cross-reference listings in transistor handbooks. The device leads should be left shorted together until it is installed and soldered in the circuit. Soldering iron tips and tools should be shorted to the circuit ground during installation. Finally, no test signals should be applied to the IGFET input circuit with the drain power source turned off.

Optoelectronic Devices

Semiconductor technology has made possible the development of many specialized semiconductor devices for electronic functions such as light detection and display. The optoelectronics family includes the light emitting display (LED), liquid crystal display (LCD), photodiode,

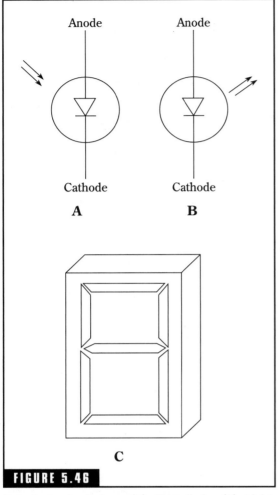

FIGURE 5.46

Optoelectronic devices. (A) Photodiode. (B) Light-emitting diode. (C) Seven-segment display, LCD or LED.

phototransistor, and optocoupler. The electronic display devices are used to display frequency and other information in amateur radios, and alphanumeric displays for computers. Photodiodes and phototransistors are used to detect the presence and intensity of light energy. Many modern amateur radios use the above devices to control and display operating information such as frequency, modulation modes, signal power levels, and memories.

Photodiodes

Photodiodes are light-sensitive devices made with semiconductor materials. There are two basic types of photodiodes, the photoconductive and photovoltaic diodes. Each type detects the presence of light.

The photoconductive diode requires a reverse bias voltage for proper operation. Light waves striking an exposed pn junction cause a sizable increase in current flowing through the device. This increase in current can be used to switch on a transistor which, in turn, can be used to activate a relay or control a larger current.

A photovoltaic diode converts the energy contained in light waves striking the device into electrical energy. The typical photovoltaic diode has a large exposed pn junction upon which the light waves are directed. The schematic symbol for the photodiode is shown in Fig. 5.46A.

As you might suspect, exposed pn junctions of bipolar npn or pnp transistors can be used as sensitive light detectors. Such phototransistors are used in fiber-optic and other applications.

Light-emitting diodes

The light-emitting diode, or LED, represents a major improvement in visual display capability. Virtually all new amateur radios employ LEDs to provide frequency display. With its rugged construction, small compact size, and almost unlimited life expectancy, the LED can be switched on and off at a very high rate.

LEDs use a forward-biased pn junction that radiates energy in the form of light waves. This radiation is due to free electrons crossing the pn junction into the p-type region and combining with holes in this region. Usually a high-efficiency semiconductor material, such as gallium or phosphorus, is used in manufacturing LEDs. These materials allow construction of LEDs with radiation wavelengths in the red, green, or yellow regions of a visual spectrum.

Most LEDs require a forward bias voltage of about 1.2 volts and a current of about 15 to 20 milliamperes. Segmented displays for letters or numbers consist of many LEDs connected in series. Digital frequency displays for receivers and the numeric displays of pocket calculators are typical examples of LED display devices.

Figure 5.46B and 5.46C shows the schematic symbols for LEDs and a typical seven-segment LED display.

LCD displays

The LCD, or liquid crystal display, is a popular display device used in amateur radios, pocket calculators, miniature TV sets, and portable or "laptop" computers to display alphanumeric and graphic information. Most LCDs are configured as seven-segment displays to permit display of numeric and limited alpha characters. The operation of an LCD is different from the LED in that the LCD does not generate a light; it suppresses reflected or incident light on an LCD segment. This produces a dark area on the LCD segment. LCD displays cannot be used in the dark or low light levels. In many applications, an external light source is provided to permit reading the displayed information.

The operation of LCDs is based on polarization characteristics of light. The nematic liquid crystal material in the nonactivated LCD segments reflects incident rays of light in a normal manner; this causes the LCD segment to appear invisible. Applying a voltage potential to the LCD segment and backplane activates the LCD segment, producing a random scattering of incident light. This effect makes the LCD segment appear dark to an observer.

Unlike the LEDs and other optical displays, LCDs require very low operating voltages and currents. Such an advantage is important in the use of battery operated equipments. Another advantage is that LCDs are readable in bright sunlight; LEDs tend to "fade out" under these conditions. Thus, LCD indicators are very popular in mobile amateur radios. Figure 5-46C shows a schematic symbol for an LCD seven-segment display device.

Integrated Circuits

The development of the integrated circuit, or IC, in the late 1950s has revolutionized the electronics industry. The ability to implement a complete electronic circuit consisting of thousands of transistors, diodes, resistors, and capacitors on a single tiny chip simply could not be envisioned in the days of the vacuum tube. This fantastic reduction in size, along with mass production techniques and resultant cost reductions, is one of the major accomplishments of this century. The cost of a typical IC is less than that of a conventional vacuum tube.

Types of integrated circuits

There are four basic types of integrated circuits: monolithic, thin film, thick film, and hybrid. Each type involves different methods of construction with varying advantages, performance characteristics, and costs.

The monolithic IC uses a single chip of semiconductor material on which are formed the many individual components and interconnecting circuitry. This base, or substrate, can be either n- or p-type material. Diffusion or growth techniques are used to assemble transistors, diodes, resistors, and capacitors on this tiny chip. Inductors are seldom incorporated within ICs due to the large physical size of practical coils. When required, external connections are provided for separate inductors.

Integrated circuits are produced in batches of about 25 to several hundred chips on a single wafer. When all individual components and interconnecting circuitry have been formed on each chip, the wafer is cut into separate chips. These chips are tested with automatic test machines which check the quality and performance of the mass-produced chips. Figure 5.47 shows an enlargement of a microprocessor or a *computer-on-a-chip*.

The thin-film IC is similar to the monolithic IC except that the base material is composed of an insulating material such as glass or

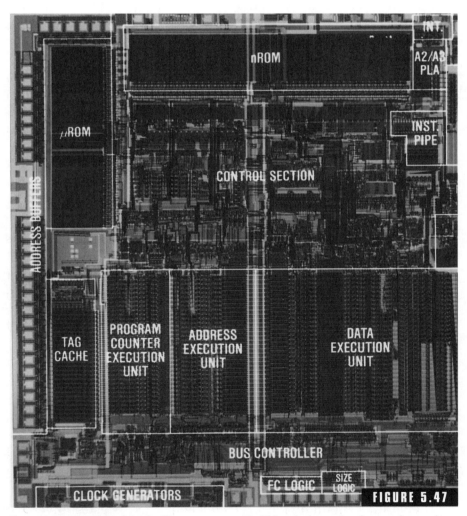

Motorola's MC68020 32-bit Microprocessor IC. This photograph of a modern microprocessor or "computer-on-a-chip" shows the complexity of IC manufacturing technology. With an on-board RAM and ROM memory, this microprocessor can act as a stand-alone computer for many applications, such as radio transceivers and modems.

ceramic. This type of construction is used when a high degree of insulation between components is required.

Thick-film technology uses deposited resistors and capacitors on a common substrate. Individual transistors and diode chips are added as discrete components. Hybrid ICs, similar to thick-film ICs, combine monolithic and thin-film components on a ceramic base. Thick-film and hybrid ICs are usually produced in limited quantities for specialized applications.

A typical IC—the 555 timer

Figure 5.48 shows the technical specifications for the Signetics 555 timer. This IC was used in the code-practice oscillator described in Chap. 2. As indicated in Fig. 5.48, the 555 timer is available in either an 8-pin or a 14-pin DIP (dual in-line package). The 555 timer is a versatile linear IC. It has been used in numerous applications. Many technical articles in radio and electronics magazines have been devoted to projects involving the 555 timer.

Packaging of integrated circuits

Normally IC clips are mounted in plastic cases with connecting leads suitable for either installing in sockets or soldering in printed-circuit boards. Figure 5.49 shows a typical 14-pin DIP or TO-116 case. Other ICs are available in many configurations including the standard TO-5 transistor case, 10-pin ceramic flat packs, and 16- to 40-pin DIP cases.

Linear and digital integrated circuits

In general, ICs can be classified as either linear or digital. Linear ICs can contain many individual transistors, diodes, or complete amplifier circuits designed for specific applications. These linear functions include amplification of ac signals ranging from audio to radio frequencies, modulators, demodulators, and oscillators. Figure 5.50 shows an IC containing all the semiconductors required for an AM radio.

Digital ICs contain digital logic elements, such as gates, inverters, flip-flops, counters, and microprocessors—virtually a complete computer on a single chip. Two expressions are sometimes used to

Description

The 555 monolithic timing circuit is a highly stable controller capable of producing accurate time delays, or oscillation. In the time delay mode of operation, the time is precisely controlled by one external resistor and capacitor. For a stable operation as an oscillator, the free running frequency and the duty cycle are both accurately controlled with two external resistors and one capacitor. The circuit may be triggered and reset on falling waveforms, and the output structure can source or sink up to 200 mA.

Features

- Turn off time less than $2 \mu s$
- Maximum operating frequency greater than 500 kHz
- Timing from microseconds to hours
- Operates in both astable and monostable modes
- High output current
- Adjustable duty cycle
- TTL compatible
- Temperature stability of 0.005% per °C

Applications

- Precision timing
- Pulse generation
- Sequential timing
- Time delay generation

Pin configurations

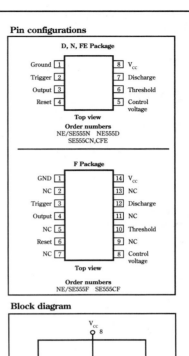

Absolute maximum ratings

Parameter	Rating	Unit
Supply voltage		
SE555	+18	V
NE555, SE555C	+16	V
Power dissipation	600	mW
Operating temperature range		
NE555	0 to +70	°C
SE555, SE555C	−55 to +125	°C
Storage temperature range	−65 to +150	°C
Lead temperature (soldering, 60 sec.)	300	°C

Block diagram

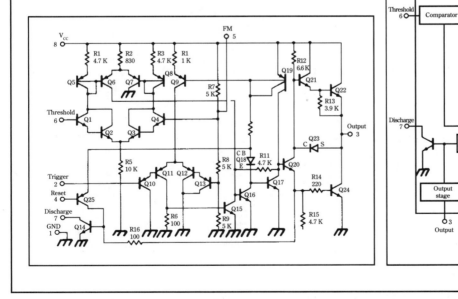

FIGURE 5.48

The 555 timer integrated circuit. *Signetics Corporation, 811 East Arques Avenue, P.O. Box 3409, Sunnyvale, CA 94088-3409, Telephone: (408) 991-2000*

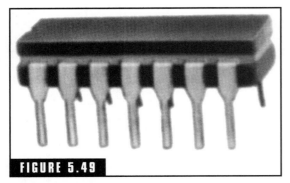

FIGURE 5.49

Typical 14-pin DIP (dual in-line package) case.

describe these ICs: MSI (medium-scale integration) and LSI (large-scale integration). The MSI chip contains up to about 100 individual logic elements such as gates while the LSI chip may contain hundreds to thousands or more of these elements.

Radio communications systems, including some amateur radio equipment, employ a combination of linear and digital ICs to perform the basic functions of signal generation, amplification, conversion, frequency counting, and digital display. The microprocessor and digital memory ICs are being used to perform basic station functions, such as Morse-code keying, encoding, and translation; logging and tuning operations; and special modulation schemes.

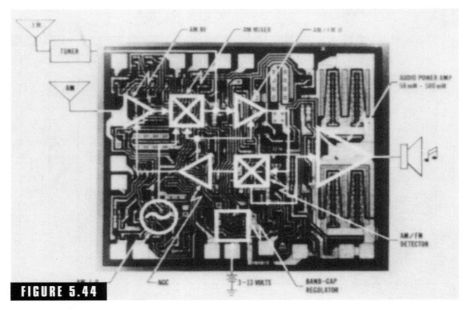

FIGURE 5.44

Integrated circuit for a complete AM radio receiver.

6

Power Supplies

The operation of your amateur radio equipment either solid-state or vacuum-tube type, requires a stable source of dc power. Batteries can be used to provide power for portable, mobile, or low-power "QRP" rigs. However, the ac power line is more appropriate for fixed-station operation involving high-power transmitters. Power supplies are used to convert the ac line power to the required dc operation.

A trend in modern-day amateur equipment, such as HF and VHF solid-state transceivers, is to design for 12- to 13.5-Vdc power requirements. With separate ac-line-operated power supplies, this equipment offers the flexibility of both portable/mobile and fixed-station use. Some transceivers have internal power supplies that are operable from either a 12- to 13.5-Vdc power source or a 120-Vac, 60-Hz power source.

Power-supply design today is focused on two separate types: the more conventional diode rectifer, capacitor filter, and electronic voltage regulator power supply; and the more recently developed electronic "switching" power supply. The first type—covered in this book—is more suitable for home design and construction and will provide stable dc power for amateur equipment with a modest investment of time and money. The electronic switching power supply, a more recent development, is generally much smaller in size and provides improved efficiency. However, the design and construction of the switching power supply is beyond the scope of this book and most beginners in amateur radio.

Definitions

Diode   A diode is a two-element device that allows current flow in one direction but restricts or eliminates current flow in the opposite direction. Diodes are used in power supplies to rectify the ac current. Diodes can be either vacuum tubes or semiconductor devices and are illustrated in Fig. 6.1.

Half-wave rectifier   Normally consisting of a single diode, the half-wave rectifier allows only half of the ac current to flow through the load, producing a pulsating dc current. A basic half-wave circuit is shown in Fig. 6.2.

Full-wave rectifier   The full-wave rectifier, shown in Fig. 6.3, allows both halves of the ac current to flow through the load in the same direction. This produces a pulsating dc with a frequency twice that of the original ac waveform.

Rectification   Rectification is defined as the process of converting an ac current to a dc current. Rectifiers are basically forms of diodes, either vacuum tubes or semiconductor devices.

Filter circuit   The filter circuit in a power supply smooths out the ac ripple in the rectified dc current, producing a relatively constant output voltage across the load. Figure 6.4 shows several types of filter circuits.

Transformers   Transformers are electromagnetic devices consisting of two or more windings that are inductively coupled to transfer electrical

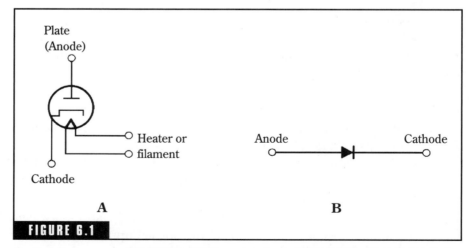

FIGURE 6.1

Typical diodes used as rectifiers. (A) Vacuum tube rectifier. (B) Semiconductor diode.

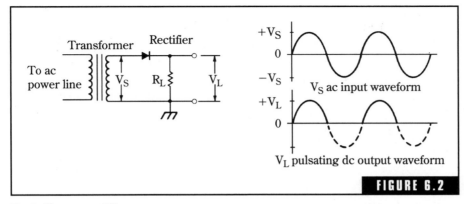

FIGURE 6.2

The half-wave rectifier.

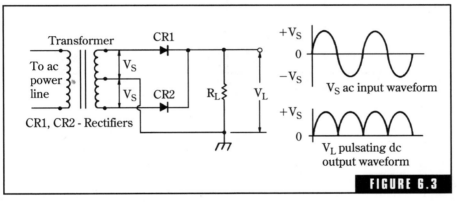

FIGURE 6.3

The full-wave rectifier.

energy from one winding to the other(s). The transformer is used in most ac-operated power supplies to perform two important functions. First, it is used to step-up or step-down the ac line voltage to the required operating levels. The second function of the transformer is that of isolating the power output circuit from the ac power line. This reduces or eliminates the potential shock hazard involved in direct connection to the ac power line. Figures 6.2 and 6.3 illustrate the use of power transformers.

Regulator circuit The purpose of the regulator circuit in a power supply is to maintain a constant output voltage under conditions of varying input ac line voltage or output load resistance. In its simplest form, a regulator circuit might consist of a loading resistor permanently connected

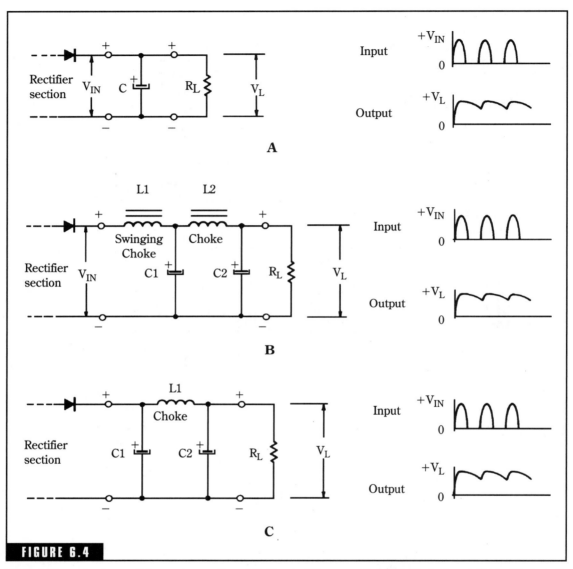

FIGURE 6.4

Typical filter circuits used in power supplies. (A) Simple filter capacitor. (B) Choke input filter. (C) Capacitor-input pi filter.

across the output of the power supply. The resistor acts as a partial load, which reduces the load/no-load voltage variations experienced with simple power supplies. Electronic voltage regulators, using vacuum tubes or semiconductor devices, maintain a constant output voltage within specified limits of variations in the input ac power line voltage and output load conditions. Semiconductor voltage regulators include zener diodes, transistors, or special integrated circuits designed for this purpose.

Ripple voltage Usually associated with the dc output of a rectifier circuit, the ripple voltage is defined as the ac component inherent in the dc output voltage. Filters are used to reduce or eliminate the ac ripple voltage.

Power-Supply Design Considerations

Power-supply design discussed in this section is based on the use of semiconductor devices and relatively low output voltage requirements. Information on vacuum-tube power supplies is available in the many amateur radio publications, which can be obtained at libraries, newsstands, or electronic parts distributors. However, the principles involved in either vacuum-tube or semiconductor power supplies are the same.

The power transformer

Power transformers are normally specified in terms of input and output voltages, and the current rating of the secondary winding. Typical 120-Vac, 60-Hz power transformers have secondary voltages of 6.3 Vac, 12.6 Vac, or 24 Vac. The secondary output voltages are listed in terms of rms or effective ac volts. The rectified peak output voltage is 1.414 times the rms value. Thus a 12.6-Vac rms voltage produces a peak voltage of almost 18 Vdc when rectified. The ac and dc voltage waveforms are illustrated in Fig. 6.5. Transformer secondary or output voltage requirements are based on the following factors: ac power line variations, internal voltage drops of the power supply, and the output load. Internal voltage drops include forward-bias voltages across rectifiers, ac ripple voltages, and the regulator dropout voltage, if applicable. These factors will be considered in the final design criteria.

Secondary-winding current ratings are determined by any internal power-supply current and the maximum load current anticipated for

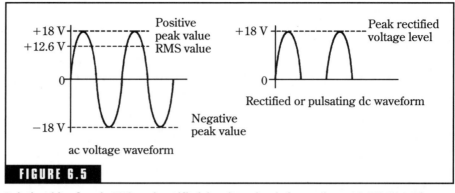

FIGURE 6.5

Relationship of peak, RMS, and rectified dc voltage levels from using a 12.6 Vac transformer.

the power supply. A good rule of thumb is to select a transformer that provides at least 50% more secondary current than is required by the anticipated load requirement. This will prevent overheating problems and increase life expectancy of the power supply.

Rectifier circuits

Figures 6.2 and 6.3 illustrate the half-wave and full-wave rectifier circuits, respectively. The half-wave rectifier is rarely used in power-supply design due to the increased filtering required to reduce the ac ripple voltage to acceptable levels.

THE FULL-WAVE RECTIFIER

The detailed analysis of the full-wave rectifier is given to illustrate the operational characteristics of this circuit. Consider the circuit in Fig. 6.6. A 24-Vac center-tapped transformer, T1, is used to provide 12 Vac to each rectifier diode. Note that 12 Vac rms is approximately equal to a peak ac voltage of about 17 volts. The peak ac values must be considered in any power-supply design.

When the first half of the ac voltage waveform is impressed across the secondary winding, point A becomes positive and allows diode D1 to conduct. Current flows in the circuit as shown by the solid arrows. Because point B is negative during this interval, diode D2 is reversed biased and does not conduct.

During the second half of the ac waveform, point B becomes positive and this causes diode D2 to conduct. Current flow in the circuit

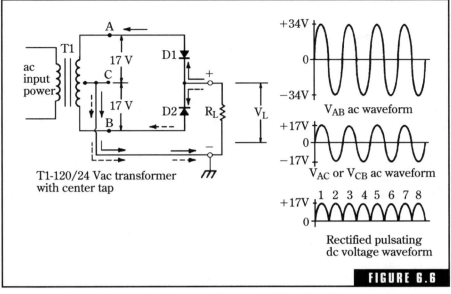

+34V

0

−34V

V_{AB} ac waveform

+17V

0

−17V

V_{AC} or V_{CB} ac waveform

+17V   1 2 3 4 5 6 7 8

0

Rectified pulsating
dc voltage waveform

FIGURE 6.6

T1-120/24 Vac transformer
with center tap

The full-wave rectifier circuit.

due to D2 conducting is illustrated by the dashed arrows. Diode D1 is reversed biased during this portion of the cycle and does not conduct.

The output voltage, V_L, appearing across load resistor R_L is also illustrated in Fig. 6.6. The odd-numbered half cycles are related to current flow in diode D1 while the even-numbered half cycles are due to current flow in diode D2.

The advantage of the full-wave rectifier lies in the fact that both halves of the ac cycle are used, resulting in a ripple frequency of 120 Hz. The main disadvantage of the full-wave rectifier is that a transformer with a center-tapped secondary winding and a higher output voltage is required.

THE FULL-WAVE BRIDGE RECTIFIER

The limitations of the full-wave rectifier can be overcome by the bridge rectifier shown in Fig. 6.7. A description of the operation of the bridge rectifier is given as follows. Diodes D2 and D4 conduct during the positive half of the ac cycle. This occurs when point A is positive and point B is negative. Current flow during this interval is shown by the solid arrows.

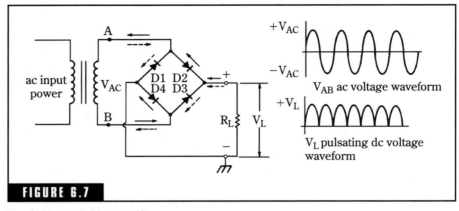

FIGURE 6.7

The full-wave bridge rectifier.

When the ac voltage across points A and B is reversed during the second half of the ac cycle, diodes D1 and D3 conduct. At the same time, diodes D2 and D4 are reversed biased and are cut off. The resulting current flow is shown by the dashed arrows.

The bridge rectifier combines the best features of the half-wave and full-wave rectifiers—the dc output voltage is approximately equal to the secondary voltage and the ripple frequency is 120 Hz.

RECTIFIER VOLTAGE AND CURRENT RATINGS

Three characteristics of rectifiers must be considered in any power-supply design. All are critical to the proper and efficient operation of the power supply.

Peak inverse voltage (PIV)　All rectifiers are rated in terms of peak inverse voltage. This is defined as the maximum reverse voltage that the rectifier can withstand during periods of nonconduction. The PIV ratings of most rectifiers will range from about 50 to 1000 volts. Generally speaking, the PIV rating for a specific rectifier should be at least twice the peak ac voltage output of the power transformer. Many designers add an additional 50% to account for ac line voltage variations and transient voltage spikes.

Average rectified current (I_o)　Most rectifiers are rated in terms of the maximum load current. When individual diodes are used to make up a full-wave or bridge rectifier, the individual current rating of each diode must be considered. In general, each diode in a full-wave or bridge rectifier circuit should have a minimum average current rating of one-half the maximum load current. For maximum circuit reliability, the average current rating of each diode should equal that of the maximum load current.

Encapsulated or packaged rectifier assemblies are normally rated in terms of maximum load current. It is recommended that the current rating of the packaged rectifier assembly be about twice that of the intended maximum load current.

Forward voltage drop (V_F) During conduction, the forward voltage drop across a rectifier diode must be considered as an internal voltage drop within the power supply. Silicon rectifiers will exhibit a forward voltage drop of about 1 to 2 volts for each diode. The forward voltage drop for germanium diodes varies from about 0.5 to 1.0 volt. Because two diodes are always conducting in a bridge rectifier circuit, the total forward voltage drop for the rectifier circuit ranges from about 2 to 3 volts.

Rectifier and power-diode ratings are available in most semiconductor handbooks. In many instances, the ratings are printed on the package or box containing the device. Careful attention should be paid to these ratings when determining the correct specifications for the internal circuit. Otherwise, using an improper rectifier in the power supply will result in marginal performance or destruction of the rectifier. Table 6.1 provides a partial listing of typical rectifiers that are widely available.

Filter circuits

We use filter circuits in power supplies to smooth out the varying dc output from the rectifier. This pulsating dc voltage can cause hum and other poor operational characteristics in electronic circuits such as amplifiers.

The primary element in the filter circuit is the filter capacitor. Figure 6.8 illustrates the action of a filter capacitor during charging and discharging cycles. First, consider the simple half-wave rectifier with no filter capacitor. The rectified or pulsating dc voltage developed across the load resistor, R_L, rises to a maximum value, V_P, during the first half of each ac cycle. Then, this rectified voltage drops to zero and remains at this level during the second half of each ac cycle. This varying dc voltage, shown in Fig. 6.8B, has the same frequency as the ac input voltage.

TABLE 6.1 Ratings for Typical Rectifier Diodes

Type	Peak inverse voltage, PIV	Average rectified forward current, I_0	Forward voltage drop, V_F
1N1199A	50 V	12 A	1.2 V
1N1200A	100 V	12 A	1.2 V
1N1341A	50 V	6 A	1.4 V
1N1342A	100 V	6 A	1.4 V
1N4001	50 V	1 A	1.6 V
1N4002	100 V	1 A	1.6 V
1N4003	200 V	1 A	1.6 V
1N4004	400 V	1 A	1.6 V
1N4005	600 V	1 A	1.6 V
1N4006	800 V	1 A	1.6 V
1N4007	1000 V	1 A	1.6 V
1N4719	50 V	3 A	1.0 V
1N4720	100 V	3 A	1.0 V
1N4721	200 V	3 A	1.0 V
1N4722	400 V	3 A	1.0 V
1N4723	600 V	3 A	1.0 V
1N4724	800 V	3 A	1.0 V
1N4725	1000 V	3 A	1.0 V

Now add the filter capacitor, C_F, to the circuit. For the first half of the ac cycle, the increasing rectified dc voltage charges the capacitor, storing electrical energy within the capacitor. When the rectified dc voltage passes the peak value and starts to fall to zero, the charged capacitor begins to discharge, providing current to the load resistor. The diode prevents the capacitor from discharging current back into the ac source.

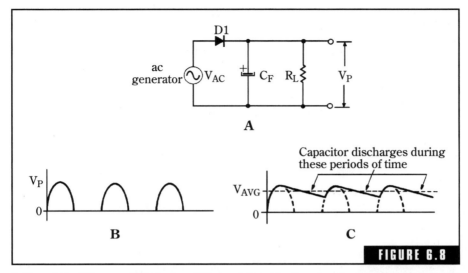

FIGURE 6.8

Action of the filter capacitor in smoothing out the pulsating dc output of the rectifier. (A) Half-wave rectifier with simple filter capacitor. (B) Pulsating dc voltage V_P with no filter capacitor. (C) dc output voltage V_{avg} with filter capacitor in circuit.

If the capacitor has sufficient capacitance, i.e., many microfarads for most filter capacitors, the discharge action will not be completed before the next ac cycle begins. Thus, the next positive half cycle of ac will produce a dc current flow that recharges the capacitor back to its original value. This charge and discharge action, illustrated in Fig. 6.8C, occurs once for every ac cycle.

Although the filter capacitor prevents the dc voltage across R_L from dropping to zero, there remains a varying ac component in the dc voltage. This ac component is known as the *ac ripple voltage*. The frequency of the ac ripple voltage is the same as that of the ac power source, normally 60 Hz for half-wave ac-operated power supplies. Figure 6.9 illustrates the ac ripple voltage contained in the rectified dc voltage. The amplitude of the ac ripple voltage, defined as $V_{max} - V_{min}$, depends primarily on two factors: the capacitance of the filter capacitor, C_F, and the current flowing in the load resistor, R_L. Increasing the capacitance of C_F reduces the ac ripple voltage. Conversely, an increase in load current results in a higher ac ripple voltage. For half-wave rectifiers, the approximate ac ripple voltage contained in a simple, single-capacitor filter circuit is given as:

$$V_R = \frac{16I}{C} \qquad \text{(Eq. 6.1)}$$

where

V_R is the ac ripple voltage in volts

I is the load current in milliamperes

C is the capacitance of the filter capacitor in microfarads

Full-wave and bridge rectifiers produce a rectified dc voltage with an ac ripple frequency twice that of the ac power source. The ripple frequency is normally 120 Hz for ac-operated power supplies using full-wave rectifiers. The approximate amplitude of the ac ripple voltage for single-capacitor filter circuits used with full-wave or bridge rectifiers is determined by:

$$V_R = \frac{8I}{C} \qquad \text{(Eq. 6.2)}$$

As in half-wave rectifier circuits, increasing the value of the filter capacitor in full-wave rectifier circuits reduces the ac ripple voltage. However, for the same value of filter capacitance, the ac ripple voltage in full-wave and bridge rectifiers is approximately one-half that of half-wave rectifiers.

In regulated solid-state power supplies, the filter circuit normally consists of two capacitors. One capacitor, normally ranging from hundreds to thousands of microfarads, is used to reduce the ac ripple volt-

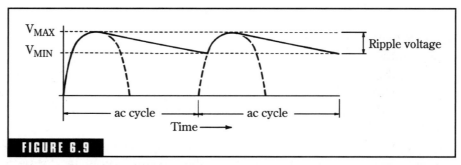

FIGURE 6.9

ac ripple voltage in the rectified output of a typical power supply. Note: The levels are not necessarily to scale because most power supplies have very small ripple voltage characteristics.

age to an acceptable level. The other capacitor, an RF bypass capacitor of about 0.1 to 0.5 µF, is used to short out any transient voltage spikes, which occasionally occur on the ac power line. The second capacitor is required because large-capacitance electrolytic capacitors do not act as capacitors at high frequencies.

The single-capacitor filter reduces the ac ripple voltage in low-voltage regulated power supplies to values from less than 1 volt to about 3 volts. The electronic regulator then takes over and reduces the ac ripple voltage to insignificant values.

Filter circuits for vacuum-tube and solid-state power supplies are illustrated in Fig. 6.10. These circuits are used when electronic regulators are impractical or too expensive. The two-capacitor input circuits in Fig. 6.10A and B are known as *pi filters*, and each component, capacitor, resistor, and inductor contributes to the filtering process. The capacitors smooth out the ac ripple voltage and the resistor or inductor provides opposition to the flow of ac ripple current. The RC filter circuit in Fig. 6.10A also reduces the dc output voltage of the power supply.

The choke input filter in Fig. 6.10C is used in power supplies where the load variation is extensive. For example, the load presented by an HF transceiver varies considerably from receive to transmit conditions. Inductor L_1 is a special choke designed to improve voltage regulation. This inductor is referred to as a *swinging choke*; the inductance varies with the load current. A disadvantage of the choke-input power supply is that the dc output voltage is somewhat less than that of the capacitor-input type.

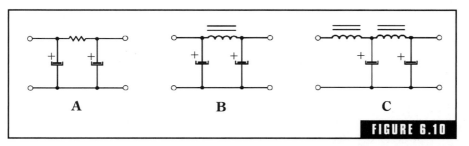

A   **B**   **C**

FIGURE 6.10

Filter circuits used in power supplies. (A) Simple RC filter circuit. (B) Capacitor-input filter circuit. (C) Choke-input LC filter using a swinging choke.

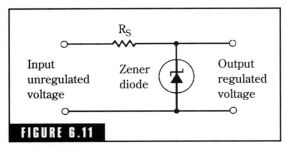

FIGURE 6.11

Zener-diode voltage regulator.

Electronic voltage regulators

Modern solid-state power supplies include some form of electronic regulation to maintain a constant output voltage. The regulator circuit may consist of zener diodes, transistors, integrated-circuit units, or a combination of these devices.

The zener-diode voltage regulator, illustrated in Fig. 6.11, is a simple circuit that provides good regulation for applications requiring low load current.

ZENER-FOLLOWER REGULATOR

The zener-follower regulator, shown in Fig. 6.12, is a series voltage regulator using a power transistor to control the dc output voltage. A zener diode provides a stable reference voltage on the base of the power transistor, thereby maintaining a constant emitter voltage. The power transistor, referred to as a *series-pass transistor*, is connected in a common-collector or emitter-follower configuration.

The collector-emitter voltage, V_{CE}, is equal to the input unregulated voltage, V_{IN}, minus the output regulated voltage, V_{OUT}. Thus the power, P_D, that the pass transistor must be able to dissipate is:

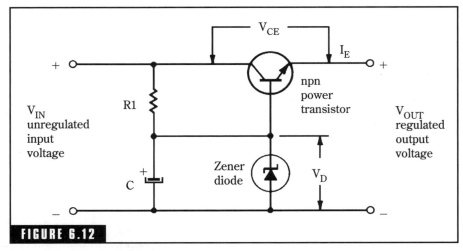

FIGURE 6.12

Zener-follower voltage regulator.

$$P_D = V_{CE} \times I_E = (V_{IN} - V_{OUT}) \times I_E \qquad \textbf{(Eq. 6.3)}$$

The output voltage, V_{OUT}, is determined by the zener diode reference voltage, V_D, less the base-emitter junction voltage, V_{BE}. This junction voltage is approximately 0.3 volt for germanium transistors and 0.7 volt for silicon transistors. For example, if a dc output voltage of about 12 volts is required, a 13-volt zener diode is recommended.

The value of R_1 is based on the power dissipation rating of the zener diode and the required base current of the pass transistor for both full-load and no-load conditions.

A typical 12-Vdc, 1-ampere zener-follower regulated power supply is shown in Fig. 6.13. This power supply is capable of powering amateur equipment with low current requirements such as receivers or low-power QRP rigs. The output voltage can be changed by substituting

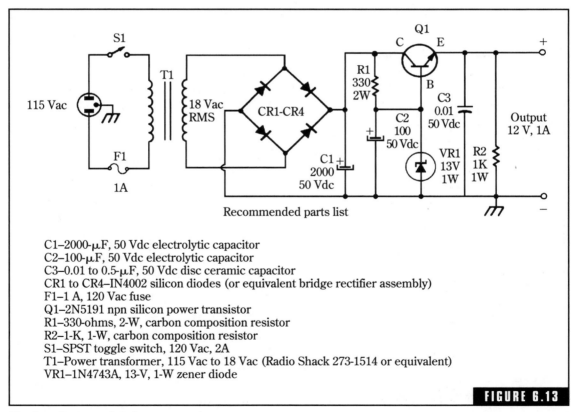

Recommended parts list

C1–2000-μF, 50 Vdc electrolytic capacitor
C2–100-μF, 50 Vdc electrolytic capacitor
C3–0.01 to 0.5-μF, 50 Vdc disc ceramic capacitor
CR1 to CR4–IN4002 silicon diodes (or equivalent bridge rectifier assembly)
F1–1 A, 120 Vac fuse
Q1–2N5191 npn silicon power transistor
R1–330-ohms, 2-W, carbon composition resistor
R2–1-K, 1-W, carbon composition resistor
S1–SPST toggle switch, 120 Vac, 2A
T1–Power transformer, 115 Vac to 18 Vac (Radio Shack 273-1514 or equivalent)
VR1–1N4743A, 13-V, 1-W zener diode

FIGURE 6.13

Simple voltage regulated power supply.

a different zener diode. For example, output voltages of approximately 5, 6, 9, 15, and 18 volts can be obtained with zener diodes rated at 6, 7, 10, 16, and 19 volts, respectively. If the lower voltages of 5 to 9 volts are the maximum that will be required, a power transformer with a secondary rating of 12 Vac at 1.2 to 2 amperes can be substituted for the 18-volt transformer.

The power supply in Fig. 6.13 can be constructed using point-to-point wiring, perforated board, or a "home-brew" printed-circuit board, whichever is most convenient. The only critical item is an appropriate heat sink for the power transistor. The heat sink can be fabricated from a small aluminum bracket or obtained from a local electronics parts supplier.

The zener-follower voltage regulator provides good regulation characteristics with minimum ac ripple voltage for low to medium load current. The disadvantages of this type of regulator are poor performance characteristics with high load current and lack of short-circuit protection. The power transistor can be destroyed if a short is placed across the output terminals. Sometimes a damaged power transistor will have a collector-to-emitter short, causing the entire dc output voltage to appear across the load. This resulting 20 to 25 volts dc can damage ham equipment such as a QRP rig designed to operate from 12 to 14 volts dc.

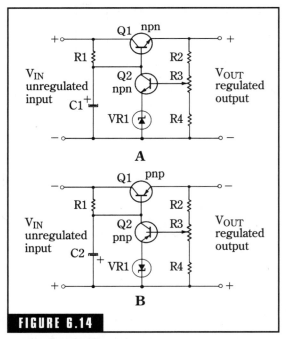

FIGURE 6.14

The basic series-parallel voltage regulator. (A) Positive series-parallel voltage regulator. (B) Negative series-parallelvoltage regulator.

SERIES-PARALLEL VOLTAGE REGULATORS

The series-parallel voltage regulator uses a series-pass transistor controlled by a parallel-connected transistor that senses the output voltage to provide automatic corrective action. Figure 6.14 illustrates this type of regulator circuit. Transistor Q2 is a dc amplifier controlled by a small portion of the output voltage. A voltage-reference or zener diode, VR1, maintains a constant emitter voltage for Q2. Thus, the output voltage sampled at potentiometer R3 is used to control the base current of the series-pass transistor, Q1. If the output voltage drops due to loading effects or power-line variations, transistor Q2 raises the base voltage of series-pass transistor Q1. This causes the voltage drop, VCE, across the series-pass transistor to decrease and, therefore, raises the load voltage back to the required level. Conversely, if the output voltage increases beyond its intended limit, transistor Q2 lowers the base voltage on series-pass transistor Q1. This action increases the collector-emitter voltage across Q1 and thus reduces the output voltage to the specified value.

The regulator output voltage is held constant by the base-emitter voltage of Q2 and the very stable emitter voltage determined by the zener diode, VR1. The zener voltage for VR1 is normally selected at about 6 to 8 volts because this is an optimum value in terms of temperature stability.

Potentiometer R3 allows a wide control of the output voltage level. This output voltage can be adjusted to a precise value, which remains constant despite varying load conditions and power-line variations. Fixed resistors R2 and R4 place safe limitations on the output voltage.

Both positive and negative series-parallel voltage regulators are shown in Fig. 6.14. The primary differences are that pnp transistors are used in the negative voltage regulator and the zener diode is reversed. These voltage regulators provide excellent regulation characteristics and very low ripple-voltage content. However, like the zener-follower regulator, the series-parallel regulator does not provide for short-circuit protection. A fuse in the input circuit of these regulators will help to protect the pass transistor.

INTEGRATED-CIRCUIT VOLTAGE REGULATORS

A significant advance in solid-state power supply technology is the development of complete voltage-regulator circuits on a single chip. These three-terminal devices are available in a variety of voltage and current ratings. Table 6.2 lists some of the typical IC voltage regulators currently available.

In addition to providing excellent voltage-regulation characteristics and extremely low ripple-voltage content, these regulators provide short-circuit protection. The regulator automatically shuts down if a short circuit is placed across the load terminals.

TABLE 6.2 Typical IC Voltage Regulators.

Regulator type	Output regulated voltage (volts)	Maximum load current (amperes)	Internal voltage drop (volts)	Load regulation (percent)	Free air power dissipation (watts)	Maximum input voltage (volts)	Package style
309H	+5	0.2	2.0	2	1.0	35	TO-5
309K	+5	1.0	2.0	2	3.0	35	TO-3
78XXC	+5 to +24						
	(Note 1)	1.5	2.0	2	2.0	35	
						(Note 2)	TO-220
78LXXc	+5 to +24	0.1	2 to 3	2	0.8	30	TO-39 or
	(Note 1)					(Note 3)	TO-92
							TO-220
79XXT	−5 to −24	1.5	1.1	1	(Note 5)	35	
	(Note 4)					(Note 6)	
723C	+2 to +37	0.15	3.0	0.6	0.8	40	TO-5
LM350	+1.2 to +33	3.0	2.0	0.1	(Note 7)	35	TO-3

Notes: 1. Available in specific output voltages of 5, 6, 8, 10, 12, 15, 18, and 24 volts.
2. Maximum input voltage for 7824C (24 V model) is 40 volts.
3. Maximum input voltages for higher voltage models: 78L12C/35 V, 78L15C/35 V, 78L18C/35 V, and 78L24C/40 V.
4. Available in specific output voltages of −5, −5.2, −6, −8, −9, −12, −15, −18, and −24 volts.
5. Estimated to be approximately 1.5 watts.
6. Maximum input voltage for higher voltage model: 7924T/40 V.
7. Internally limited—Maximum Input-Output Voltage Differential is 35 volts for −65 to +150° C.

Figure 6.15 shows a basic power supply circuit for almost any three-terminal IC voltage regulator. Any of the voltage regulators listed in Table 6.2, except the 723C types, can be used in this circuit as long as proper voltage ratings for the power transformer, bridge rectifier, and filter capacitors are observed. Typical values for these components are given in Fig. 6.15 and Table 6.3. Be sure to install the voltage regulator on an appropriate heat sink. Otherwise, extended operation at high load current will cause excessive heat to build up in the regulator case and possibly damage the device.

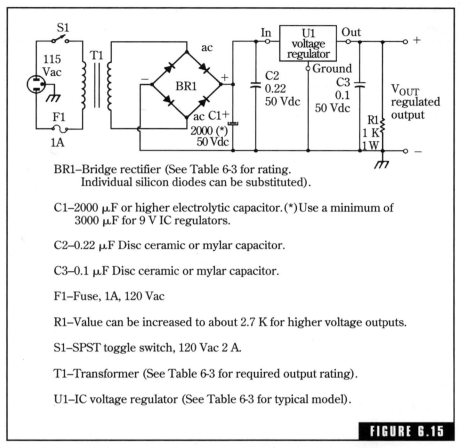

BR1–Bridge rectifier (See Table 6-3 for rating.
Individual silicon diodes can be substituted).

C1–2000 µF or higher electrolytic capacitor. (*) Use a minimum of
3000 µF for 9 V IC regulators.

C2–0.22 µF Disc ceramic or mylar capacitor.

C3–0.1 µF Disc ceramic or mylar capacitor.

F1–Fuse, 1A, 120 Vac

R1–Value can be increased to about 2.7 K for higher voltage outputs.

S1–SPST toggle switch, 120 Vac 2 A.

T1–Transformer (See Table 6-3 for required output rating).

U1–IC voltage regulator (See Table 6-3 for typical model).

FIGURE 6.15

Basic circuit for power supply using an integrated-circuit voltage regulator.

Figure 6.16 shows an internal view of a small 12-Vdc, 1-ampere power supply housed in a 5¼ × 3 × 5⅞-inch metal cabinet. The electronic circuit is wired on a perforated board having 0.1 × 0.1-inch hole centers. The 7812 IC voltage regulator is mounted on a small angle bracket, which serves as a heat sink. The pilot light consists of a small LED and a series 470-ohm resistor connected across the 12-Vdc output.

A word of caution

The construction of ac power supplies requires that the builder exercise extreme caution in assembling and testing circuits that connect to the 120-Vac, 60-Hz power line. All exposed terminals or wires at

TABLE 6.3 Component Values for the Power Supply in Fig. 6.15

Power supply output requirements	T_1 secondary winding ratings	BR_1 bridge rectifier ratings	Typical IC voltage regulator
+5 Vdc, 1A	12 Vac RMS, 2A	50 Vdc, 2A	309K, 7805C
–5 Vdc, 1A	12 Vac RMS, 2A	50 Vdc, 2A	7805T (Note 1)
+9 Vdc, 1A	12 Vac RMS, 2A	50 Vdc, 2A	7808C
+12 Vdc, 1A	18 Vac RMS, 2A	100 Vdc, 2A	7812C
+15 Vdc, 1A	18 Vac RMS, 2A	100 Vdc, 2A	7815C
+24 Vdc, 1A	24 Vac RMS, 2A	100 Vdc, 2A	7824C

Note 1. Negative IC Voltage Regulator

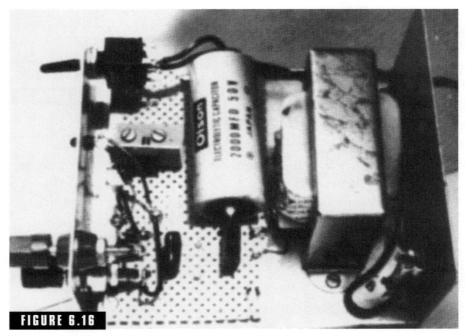

FIGURE 6.16

Internal view of a 12-Vdc, 1-ampere power supply.

120-Vac potential should be taped to prevent accidental contact. The 3-wire-type ac plugs should be used so that the power line ground can be connected to any metal chassis or cabinet used to house the power supply. All primary power wiring should be fused with a fast-blow fuse to prevent damage to the equipment. Remember, the 120-Vac power line can be lethal. Electric shock can also cause burns or other injury to the inexperienced amateur operator. After the photograph in Fig. 6.16 was taken, the 120-Vac switch and the fuse holder were protected with electrical tape to prevent accidental contact with the line voltage at these points.

CHAPTER 7

Electronic Circuits

Semiconductor devices and vacuum tubes are used in electronic circuits to perform a variety of functions related to radio receivers and transmitters. A typical amateur HF transceiver will contain up to ten individual amplifiers and three to four oscillator circuits in addition to modulators, demodulators, and other types of circuits. This chapter covers the amplifier and oscillator, both in terms of theory and practical circuits. The remaining basic receiver and transmitter circuits are covered in Chaps. 8 and 9.

The Technician Class license examination contains only a few questions on electronic circuits. The additional material covered in this chapter will help to increase your knowledge of electronic circuits and help you to upgrade to a higher class of amateur radio license.

Definitions

Amplifier The amplifier is a device or circuit designed to amplify a specified input signal to a required output signal. Amplifiers can contain vacuum tubes, transistors, or integrated circuits as the active amplifying element.

Audio-frequency (AF) amplifier The AF amplifier is a device or circuit designed to amplify audio-frequency signals within the 20-Hz to 20-kHz range of the frequency spectrum. Audio amplifiers are normally operated in Class-A, -B, or -AB modes of operation.

Radio-frequency (RF) amplifier The RF amplifier is a device or circuit designed to amplify radio-frequency signals of a specified frequency or band of frequencies. Radio-frequency amplifiers can be classified as either tuned or untuned, and can be operated in Class-A, -B, -AB, or -C modes of operation.

Class-A amplifier The Class-A amplifier is operated in the linear region of its characteristic curve and produces an output waveform with the same shape as that of the input waveform. Current flows in the output circuit at all times resulting in linear operation.

Class-B amplifier The Class-B amplifier employs two transistors (or vacuum tubes) in a "push-pull" configuration. One transistor (or tube) amplifies the positive one-half cycle of the input signal while the other transistor (or tube) amplifies the negative one-half cycle. The bias for the Class-B amplifier circuit is adjusted to approximately the cut-off bias level; no current flows in the output circuit during the absence of an input signal. With an applied input signal, output current flows for approximately one-half cycle in each transistor (or tube). Push-pull Class-B amplifiers provide an output signal that is a good replica of the input signal waveform as well as good operating efficiency.

Class-AB amplifier This amplifier uses a push-pull circuit configuration, which combines the advantages of Class-A and Class-B amplifiers, namely good linearity and excellent efficiency. The bias for this amplifier is adjusted so that output current flows for slightly more than one-half of the input cycle.

Class-C amplifier The bias for a Class-C amplifier is adjusted so that it exceeds the cutoff bias level. This allows the output current to flow only during a short portion of the input cycle. The Class-C amplifier is essentially nonlinear but provides maximum efficiency of operation.

Preamplifier A preamplifier is an amplifier designed primarily for amplifying low-level or weak signals. The output of the preamplifier is normally fed into a conventional amplifier or a long transmission line. The preamplifier might include specific frequency-bandpass or equalization networks.

Resistance-capacitance (RC) coupled amplifier This describes a method of coupling the output signal from one amplifier to the input

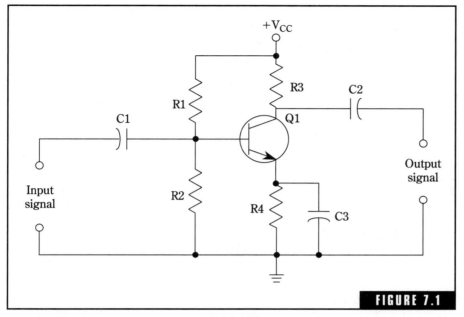

Resistance-capacitance coupled transistor amplifier.

of a following amplifier. Figure 7.1 illustrates an RC-coupled amplifier circuit. The input signal is coupled to the base of Q1 by capacitor C1. This provides dc isolation from the input signal source. Resistors R1 and R2 form a "self-bias" network that biases Q1 for Class-A operation. The output signal voltage is developed across the collector resistor R3. This output signal is coupled to the required load by capacitor C2, which also serves as a dc blocking capacitor. This prevents the dc supply voltage from being impressed on the required load.

Transformer-coupled amplifier Transformers are used to couple the output signal from one amplifier to the input of the next amplifier. Figure 7.2 shows a two-stage transformer-coupled amplifier. Transformer T1 couples the input signal to transistor Q1 for amplification and blocks any dc voltages from interacting between the input signal source and the biasing network for Q1. Transformer T1 also acts as an impedance-matching device, matching the input signal source impedance to the input impedance of Q1. Transformer T2 serves as an interstage-coupling network, matching the output impedance of Q1 to the

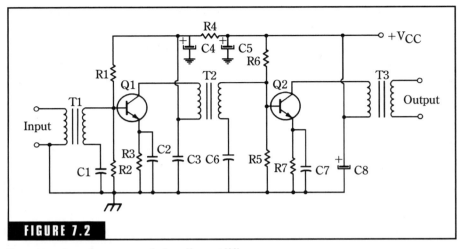

FIGURE 7.2

Two-stage transformer-coupled audio amplifier.

input impedance of Q2. Output transformer T3 matches the output impedance of Q2 into the required load impedance.

Oscillator An oscillator is an electronic device or circuit that generates a sinusoidal or nonsinusoidal ac waveform from a dc power source. The oscillator consists of an amplifier with a positive feedback loop from output to input. Frequency-determining components, such as inductors, capacitors, resistors, or quartz crystals, are included in the feedback network to establish the required frequency of oscillation.

Crystal The crystal used in oscillator circuits consists of a thin wafer of quartz suspended between two metal plates. Acting as the dielectric in a capacitor, the quartz vibrates when subjected to an electric field. This vibration, in turn, produces an alternating voltage across the metal plates. The frequency of this alternating voltage is inversely proportional to the thickness of the crystal wafer. The frequency and amplitude characteristics of the ac voltage developed by the crystal are extremely stable and make possible crystal-controlled oscillators with precise frequency control.

Crystal oscillator This type of oscillator uses a quartz crystal element that determines the frequency of oscillation. Crystal-controlled oscillators are used when an accurate and stable source of oscillation is required. A crystal oscillator having only one crystal element produces a single frequency of oscillation. If multiple frequencies are required, a switching circuit can be added to the oscillator so that the required crystal element can be switched into the feedback network. Crystal oscillators can also be operated at harmonics or multiples of the fundamental frequency of the crystal element. Figure 7.3 shows a typical crystal oscillator circuit.

Colpitts oscillator The Colpitts oscillator employs a series LC tuned circuit with a capacitive voltage divider in the feedback loop. The frequency can be adjusted over a wide range by using variable inductor or variable capacitors. The Colpitts oscillator is illustrated in Fig. 7.4.

Electron-coupled oscillator The electron-coupled oscillator, normally employing either a tetrode or pentrode tube, consists of a tuned control-grid oscillator section coupled into the screen grid. The plate output is isolated from the oscillator section, thereby minimizing the effects of loading on the plate. Figure 7.5 shows an electron-coupled oscillator.

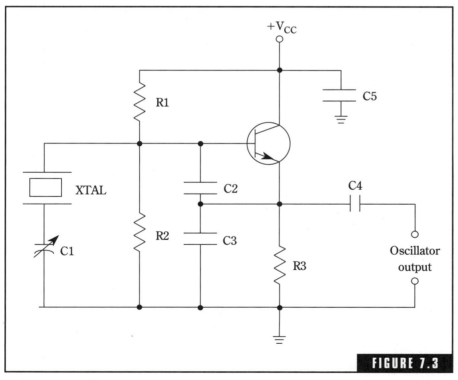

FIGURE 7.3

Basic crystal oscillator.

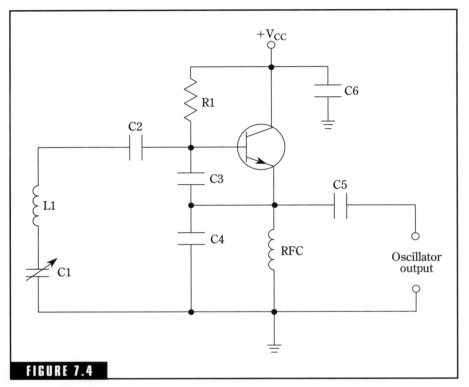

FIGURE 7.4

Colpitts LC oscillator.

Hartley oscillator The Hartley oscillator uses a tuned tank circuit with a tapped inductor in the feedback loop. One end of the tapped inductor is connected to the amplifier input while the other end connects to the amplifier output circuit. The Hartley oscillator is illustrated in Fig. 7.6.

Tuned-plate tuned-grid oscillator The tuned-plate tuned-grid oscillator employs two separate LC tuned circuits, one in the grid circuit of the amplifier tube and one in the plate circuit. Feedback from output to input is accomplished by the internal plate-to-grid capacitance of the tube. The tuned-grid tuned-plate oscillator is illustrated in Fig. 7.7.

Audio- and Radio-Frequency Amplifiers

One of the most important functions of transistors and tubes is that of amplification. Signals on the order of a few microvolts can be amplified to levels of tens to hundreds of volts by the use of amplifiers.

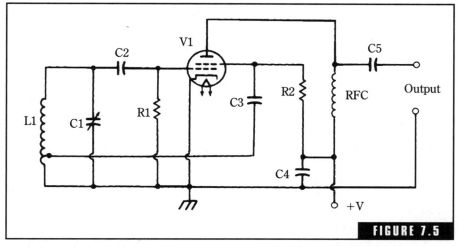

FIGURE 7.5

The electron-coupled oscillator.

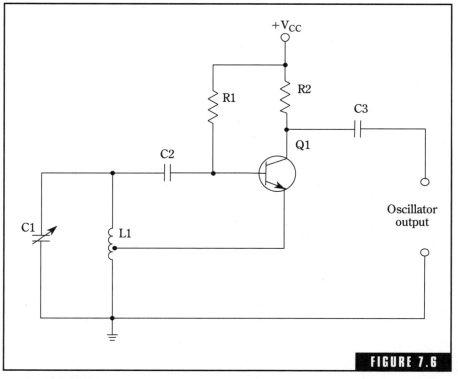

FIGURE 7.6

Hartley LC oscillator.

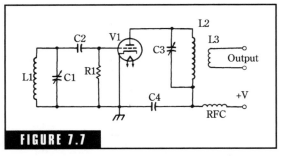

FIGURE 7.7

The tuned-plate tuned-grid oscillator.

Such amplification represents gains on the order of 1,000,000 or more.

Receivers use RF, IF, and AF amplifiers to boost the weak signals collected by the antenna to levels sufficient to drive speakers, headphones, or other output devices, such as radioteletype terminals. An IF amplifier is a fixed-tuned amplifier usually operating at a frequency below that of the incoming RF signal but well above audio frequencies. IF amplifiers are described in the next chapter.

Transmitters use RF amplifiers to boost low-level RF signals to high levels. For example, RF amplifiers in Technician CW transmitters can be used to boost output signal levels up to 250 watts if the maximum authorized power is desired. Voice transmitters have audio amplifiers to amplify the low-level audio signals produced by microphones to levels required for modulation.

Basic Types of Amplifiers

Amplifiers are described in many different ways—voltage, current, power, small-signal, large-signal, radio-frequency, Class-A (-B, -AB, or-C), resistance-capacitance-coupled, transformer-coupled, and other expressions intended to indicate operational characteristics. Amplifiers can use vacuum tubes, bipolar junction transistors, field-effect transistors, or integrated circuits as the active amplifying elements. The term *amplifier* can pertain to a single transistor, or it might describe a complete audio amplifier circuit consisting of many tubes, transistors, or a single integrated circuit.

The ideal amplifier

The ideal amplifier is designed to amplify the input signal in terms of voltage, current, or power. The input signal can be either dc or ac. Figure 7.8 shows two common symbols used for representing amplifiers. Either type of symbol, triangle or rectangle, can be used to indicate a voltage, current, or power amplifier. These symbols are used in block diagrams and sometimes in schematic diagrams when integrated cir-

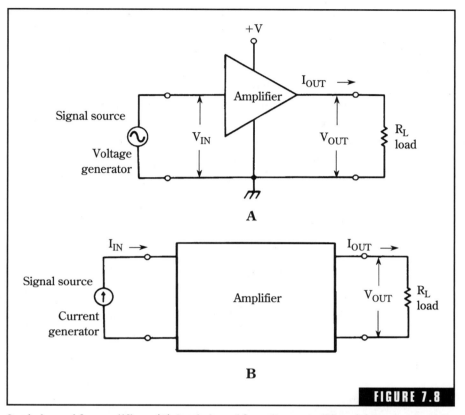

FIGURE 7.8

Symbols used for amplifiers. (A) Symbol used for voltage amplifier. (B) Symbol used for current amplifier.

cuits are involved. In some instances, the biasing networks are not shown. However, in most integrated-circuit devices, the biasing network is included within the chip and no external bias is required.

The voltage amplifier in Fig. 7.8A has an amplification factor, A_v, expressed as:

$$A_v = \frac{V_{out}}{V_{in}}$$

(Eq. 7.1)

where

A_v is the voltage gain

V_{out} is the output signal level in volts

V_{in} is the input signal level in volts

The input current, I_{in}, for the ideal voltage amplifier is zero. For all practical purposes, vacuum tubes and field-effect transistors exhibit this characteristic since the input resistance of these devices is on the order of many megohms.

Figure 7.8B shows a current amplifier, typical of bipolar junction transistors. In the ideal current amplifier, both the input voltage, V_{in}, and the input resistance are zero. The bipolar junction transistor approaches these conditions—the input base-emitter voltage (for the common-emitter configuration) is on the order of 0.7 volt and the input resistance ranges from less than 100 ohms to about 1 k. The gain for the current amplifier is expressed as:

$$A_I = \frac{I_{out}}{I_{in}}$$ **(Eq. 7.2)**

where

A_I is current gain
I_{out} is the output signal current in amperes
I_{in} is the input signal current in amperes

Power amplifiers are designed to deliver output power in appreciable levels, usually one watt or more. The general expression for power amplification is:

$$A_P = \frac{P_{out}}{P_{in}}$$ **(Eq. 7.3)**

where

A_P is the power amplification or gain
P_{out} is the output signal power level in watts
P_{in}, is the input signal power level in watts

The output signal power level can also be expressed as:

$$P_{out} = A_P \times P_{in} = V_{out} \times I_{out}$$ **(Eq. 7.4)**

$$= \frac{(V_{out})^2}{R_L} = (I_{out})^2 \times R_L$$

These equations can help you calculate the output power levels of high-power RF amplifiers used in transmitters.

The Decibel

This is an excellent time to introduce the concept of *decibels*. Many technical articles, books, and equipment specifications refer to the gain (or loss) rating of an amplifier, antenna, or other type of electronic equipment in terms of decibels.

The expression of gain (or loss) in terms of input and output quantities is sometimes awkward or inconvenient. Calculations involving several amplifier stages or other related circuits become complex when using amplification factors. The use of decibels simplifies many of these calculations.

Decibels primarily express a power-change ratio in logarithmic terms. The decibel rating for an electronic device such as an amplifier or antenna indicates the gain (or loss) of electrical energy from input to output. This relationship is:

$$dB = 10 \log \frac{P_{out}}{P_{in}} \qquad \textbf{(Eq. 7.5)}$$

where
  dB is gain (or loss) in decibels
  P_{out} is the signal output power in watts
  P_{in} is the signal input power in watts

Because decibels represent a logarithmic dc ratio of two levels, the gain (or loss) of several circuits connected in cascade is equal to the algebraic sum of their decibel ratings. Figure 7.9 illustrates two amplifiers and a filter network connected in cascade for amplifying a given signal. The bandpass filter has a loss of 3 dB. The total gain of this circuit or related types of circuits is given by:

$$\text{Total Gain}_{dB} = \text{Gain } 1_{dB} + \text{Gain } 2_{dB} + \text{Gain } 3_{dB} + \dots \quad \textbf{(Eq. 7.6)}$$

where
  Total Gain$_{dB}$ is the gain of the total circuit expressed in decibels
  Gain 1_{dB} is the gain of the first component in decibels

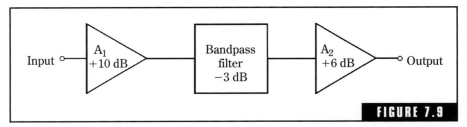

FIGURE 7.9

A typical radio-frequency amplifier consisting of two stages of amplification and a bandpass filter.

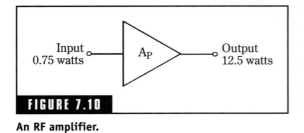

FIGURE 7.10

An RF amplifier.

Gain 2_{dB} is the gain of the second component in decibels,

Gain 3_{dB} is the gain of the third component in decibels.

In this example, the total gain in decibels is:

$$\text{Total Gain}_{dB} = 10 - 3 + 6 = 13 \text{ dB}$$

The following two examples illustrate use of Equations 7.5 and 7.6.

Example 1 The amplifier in Fig. 7.10 produces 12.5 watts of output power when driven by an input signal with a power level of 0.75 watt. Express this power amplification in decibels.

Solution

Step 1. Use Equation 7.5 to set up the initial step for this problem.
Step 2. Table 7.1 discusses the alternate methods that may be used to solve for logarithms. The logarithm of 16.67 is found to be 1.22. Completing the equation in Step 1, the power gain in decibels is calculated:

$$dB = 10 \log 16.67 = 10 \times 1.22 = 12.2 \text{ dB (Answer)}.$$

Example 2 Two RF amplifiers, a transmission line, and an antenna are connected as shown in Fig. 7.11. With the indicated gain (or loss) in decibels for each component, what is the overall gain of this circuit? If 2 watts of RF power is fed into the initial amplifier, what is the RF power radiated by the antenna?

Solution

Step 1. The gain or loss of each section of the circuit can be added algebraically to determine the total gain (or loss) of the total circuit. If quantities are added algebraically, the positive values are all added together and the negative values are subtracted from this total. Therefore,

$$\text{Total Gain}_{dB} = 6 + 10 + 3 - 2 = 17 \text{ dB (Answer)}.$$

Note that the total gain does not have to be equal to a positive number. For example, if the loss in the transmission line was equal to −20 dB, the total circuit loss would be:

$$\text{Total Loss}_{dB} = 6 + 10 + 3 - 20 = -1 \text{ dB}.$$

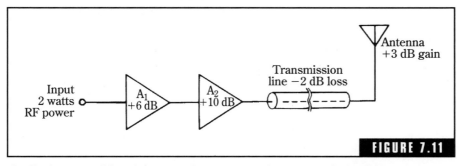

FIGURE 7.11

The final two amplifiers in a transmitter connected to a transmission line and antenna.

A partial short in the transmission line could cause this problem.

Step 2. To find the net output power from the antenna, use Equation 7.5, substituting the known values:

$$dB = 10 \log \frac{P_{out}}{P_{in}}$$

$$17 = 10 \log \frac{P_{out}}{2}$$

$$\log \frac{P_{out}}{2} = \frac{17}{10} = 1.7$$

To find the antilogarithm of 1.7, simply raise 10 to the 1.7th power, or $10^{1.7}$. A scientific pocket calculator or a mechanical slide rule rapidly provides the answer of 50.1. Continuing with the solution to this part of the problem:

$$\text{Antilog} \left(\log \frac{P_{out}}{2} \right) = \text{antilog } 1.7$$

$$\frac{P_{out}}{2} = 10^{1.7} = 50.1$$

Therefore, $P_{out} = 2 \times 50.1 = 100.2$ watts

(Answer).

Voltage and Current Ratios in Decibels

Voltage and current amplification ratios can be expressed in decibels when appropriate. However, these calculations might be meaningless if input and output impedances are not identified. For the special case of $Z_{in} = Z_{out}$ (input impedance equal to output impedance), the actual values of the impedances are not necessary when expressing voltage or current gains in decibels.

The equations for calculating gain in decibels for voltage or current ratios are:

$$dB = 20 \log \frac{V_{out}}{V_{in}}, \text{ and} \qquad \textbf{(Eq. 7.7)}$$

$$dB = 20 \log \frac{I_{out}}{I_{in}} \qquad \textbf{(Eq. 7.8)}$$

where
  dB is decibels
  V_{out} is the output signal level in volts
  V_{in} is the input signal level in volts
  I_{out} is the output signal level in amperes
  I_{in} is the input signal level in amperes

Two examples are given to illustrate the use of these equations.

Example 1 An audio amplifier with input and output impedances equal to 600 ohms is used to amplify voice signals. What is the gain in decibels if a 2-millivolt input signal produces an output of 5 volts?

Solution

Use Equation 7.7 to solve for the voltage gain in decibels.

$$dB = 20 \log \frac{V_{out}}{V_{in}} = 20 \log \frac{5}{0.002} = 20 \log 2500$$

The log of 2500 is found to be 3.40. Continuing with the solution,

$$dB = 20 \times 3.4 = 68 \text{ dB (Answer)}.$$

Example 2 A transistor amplifier produces a 2.5-mA output signal current when driven by a 10-µA signal. Both input and output impedances are specified to be 100 ohms. What is the current gain in decibels for this amplifier?

Solution

Use Equation 7.8 to compute the current gain in decibels.

$$dB = 20 \log \frac{I_{out}}{I_{in}} = 20 \log \frac{0.0025}{0.000010} = 20 \log 250$$

The log of 250 is found to be 2.4. Continuing with the solution,

$$dB = 20 \times 2.4 = 48 \text{ dB (Answer)}.$$

Table 7.1 gives the logarithms for the powers of ten and their relationship to the exponents. A simple two-place logarithmic table is also included. The Technician Class examination will probably not include any questions on decibels or logarithms. However, almost all experienced hams will discuss antenna gain or related measurements in terms of decibels. You will find this concept a handy tool to be used in your amateur radio hobby.

TABLE 7.1. Relationships between exponents and logarithms.

Exponential form	Logarithmic form	Logarithms of basic numbers
$10^4 = 10{,}000$	$\log 10^4 = 4.00$	$\log 1 = 0.00$
$10^3 = 1000$	$\log 10^3 = 3.00$	$\log 2 = 0.30$
$10^2 = 100$	$\log 10^2 = 2.00$	$\log 3 = 0.48$
$10^1 = 10$	$\log 10^1 = 1.00$	$\log 4 = 0.60$
$10^0 = 1$	$\log 10^0 = 0.00$	$\log 5 = 0.70$
$10^{-1} = 0.1$	$\log 10^{-1} = -1.00$	$\log 6 = 0.78$
$10^{-2} = 0.01$	$\log 10^{-2} = -2.00$	$\log 7 = 0.85$
$10^{-3} = 0.001$	$\log 10^{-3} = -3.00$	$\log 8 = 0.90$
$10^{-4} = 0.0001$	$\log 10^{-4} = -4.00$	$\log 9 = 0.95$

The logarithm of any number is expressed as two separate quantities. For example,

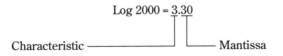

Characteristic —————— Mantissa

Note that the characteristic for numbers above 100 is always equal to the value of the exponent, or one less than the number of decimal digits. An alternate way to express 2000 is 2.0×10^3. The logarithm of this number can be expressed as:

$$\log (2000) = \log (2.0 \times 10^3) = \log 2.0 + \log 10^3$$

From Table 7.1, the log of 2 is 0.3 and the log of 10^3 is 3.0. Adding these two numbers provides the log of 2000 as being 0.3 + 3.0 or 3.30. In this manner, you can solve for the logarithm of any number to about two decimal places. For two-place numbers, such as 15 or 35, you can extrapolate between the given numbers. For example, the log of 15 is approximately halfway between the log of 10 and the log of 20. Solving this example with the aid of Table 7.1 is given as follows:

$$\begin{aligned}
\text{Log } 10 &= \log (1 \times 10^1) = \log 1 + \log 10^1 \\
&= 0.00 + 1 = 1.00 \\
\text{Log } 20 &= \log (2 \times 10^1) = \log 2 + \log 10^1 \\
&= 0.30 + 1 = 1.30
\end{aligned}$$

The log of 15 is approximately one-half way between the values of 1.00 and 1.30, or 1.15 (Answer). Solving this problem with a pocket calculator provides an answer of 1.1761. The error in using the simplified approach is (1.1761 − 1.15)/1.1761, or about 2 percent.

The scientific pocket calculator provides an excellent means for solving problems in decibels and logarithms. An alternate approach is to obtain an inexpensive mechanical slide rule. Either device will help you to sharpen your ability in solving electronics problems. They usually come with detailed instructions that will help you learn some mathematical fundamentals.

Amplifier Power Levels

Amplifiers are normally designed to operate within specified input and output signal levels. These levels are expressed in terms of voltage, current, or power, depending on the type of amplifier circuit and active element (transistor or vacuum tube) used in the circuit. In this manner, the amplifier design can be optimized for maximum desired performance.

The small-signal amplifier is designed to amplify low-level signals in the microvolt to millivolt range. These amplifiers must possess low internal noise so that the low-level signals will not be masked. Design engineers refer to this characteristic in terms of a *usable signal-to-noise ratio*. In most cases, the amplifier linearity and gain stability characteristics are carefully controlled to permit a wide variation of input signal levels. Small-signal transistors or receiving type vacuum tubes, with output load currents of a few milliamperes, are normally used as the active amplifying devices in small-signal amplifiers. The small-signal amplifier is used primarily to boost low-level signals to suitable amplitudes for input to power amplifiers or other related circuits. The small-signal amplifier is rarely designed to produce output power levels over one-half watt.

Power amplifiers are designed to deliver large power levels to the intended load. Audio output amplifiers and transmitter final amplifiers are typical examples of power amplifiers. Generally speaking, any amplifier capable of producing an output power of one watt or more is classified as a power amplifier.

Both small-signal and power amplifiers are used for audio and high-frequency amplification requirements in receivers and transmitters. Most high-frequency amplifiers used for such functions as intermediate- or radio-frequency (IF or RF) stages have frequency-sensitive networks to shape the frequency response of the amplifier in terms of single-frequency or bandpass operation. Sometimes untuned high-frequency amplifiers are used in receivers or transmitters for special-function stages.

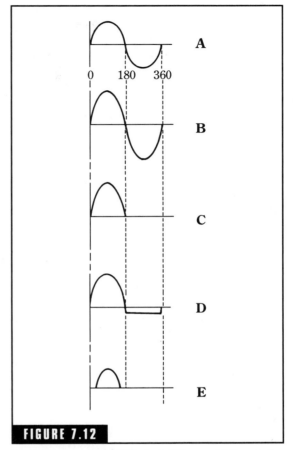

FIGURE 7.12

Basic classes of amplifier operations. (A) Input-signal waveform. (B) Class-A amplifier output-signal waveform. (C) Class-B amplifier output-signal waveform. (D) Class-AB amplifier output-signal waveform. (E) Class-C amplifier output-signal waveform.

Class-A, -B, -AB, and -C Amplifiers

Classification of amplifiers in these terms describes the relationship between the input-signal waveform and the output-signal or load-current flow. This relationship is established by the biasing network and the input-signal amplitude. Figure 7.12 illustrates these four basic classes of amplifier operation along with a summary of the basic characteristics for each class. Any of the previously described amplifiers (audio/high-frequency or small-signal/power) can be operated in Class-A, -B, -AB, or -C modes of operation.

Class-A operation produces a linear output—the output-signal waveform is equal in shape to that of the input signal. The output current flows for the full 360 degrees of each cycle of the ac input waveform. The efficiency of operation ranges from about 15 to 25 percent. In Class-B operation, the output current flows for only 180 degrees due to the biasing network. No current flows during the second half of the ac input-signal cycle. This increases the efficiency of operation up to about 65 percent. The biasing network in a Class-AB amplifier allows a small output current to flow in the absence of an input signal. The output-signal current flows for more than 180 degrees but less than 360 degrees. The efficiency of operation is approximately equal to that of the Class-B amplifier. The Class-C amplifier is biased beyond cutoff and allows the load current to flow for only a portion of the time between 0 and 180 degrees. Efficiency of operation ranges up to 80 percent.

Class-A operation is used primarily for small-signal or power amplifiers using a single amplifier device (tube or transistor). Class-B and -AB amplifiers are normally used in a push-pull configuration for power amplification in audio and high-frequency operation. Class-C amplifiers are used almost exclusively for high-frequency amplification such as in RF amplifiers.

Amplifier Efficiency of Operation

Efficiency of operation for amplifiers is defined as the ratio of useful power developed by the amplifier and the dc power required to operate the amplifier, or:

$$\text{Efficiency of Operation} = \frac{\text{Signal Output Power in Watts}}{\text{dc Input Power in Watts}} \times 100\% \quad \textbf{(Eq. 7.9)}$$

Note that efficiency of operation is always specified in percentage. For vacuum tubes, efficiency of operation is defined as *plate efficiency* and the dc input power is simply plate voltage times plate current. An example is given to illustrate Equation 7.9.

> **Example** A vacuum-tube RF amplifier delivers 50 watts of power when the plate voltage is 400 Vdc and the plate current is 200 mA. What is the plate efficiency of this amplifier?

Solution

Step A. Determine the dc input power

$$dc\ Input\ Power = Plate\ Voltage \times Plate\ Current$$

$$= 400 \times 0.2 = 80\ watts$$

Step B. Use Equation 7.9 to calculate the plate efficiency:

$$Plate\ Efficiency = \frac{Signal\ Output\ Power\ in\ Watts}{dc\ Input\ Power\ in\ Watts} \times 100\%$$

$$= \frac{50\ watts}{80\ watts} \times 100\% = 0.625 \times 100\%$$

$$= 62.5\%$$

Interstage Coupling Techniques

When large amounts of gain are required for specified applications, two or more amplifiers can be connected in a cascade configuration. Figure 7.13 shows a basic cascade amplifier consisting of three individual amplifier stages and two coupling networks.

Four basic types of coupling circuits are used in amateur radio equipment. All of these coupling techniques can be used with both audio and high-frequency amplifiers. Many amateur HF SSB transceivers use all four types of coupling in audio, IF, or RF amplifier circuits.

Direct coupling

As implied, the output signal from one stage is directly coupled to the input of the next stage. Figure 7.14A shows a two-stage, direct-coupled transistor amplifier. This type of coupling provides a wide frequency response and is used in dc amplifiers as well as ac or signal amplifiers. Note that the biasing of Q2 depends on the dc voltage level at the collector Q.

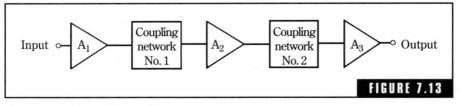

FIGURE 7.13

Three cascaded amplifiers with coupling networks.

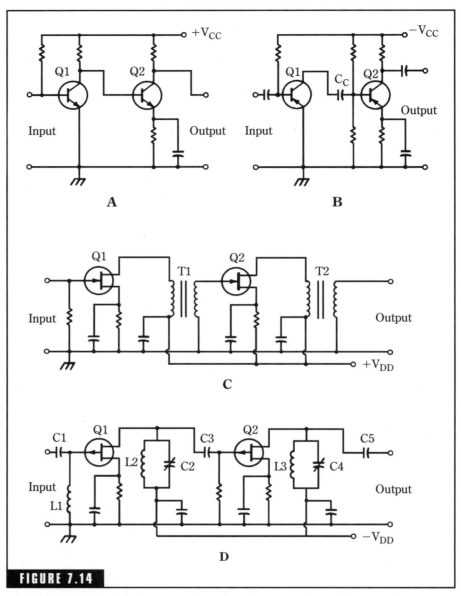

FIGURE 7.14

Basic methods for coupling amplifiers. (A) Direct-coupled amplifier. (B) Resistance-capacitance coupled amplifier. (C) Transformer-coupled amplifier. (D) Impedance-coupled amplifier.

Resistance-capacitance (RC) coupling

Probably the most widely used type of coupling in both vacuum-tube and transistor audio amplifier circuits, this type of coupling isolates the dc supply from each stage. This permits the use of individual bias networks and allows replacement of defective tubes or transistors without having to readjust the bias network for proper operation. Figure 7.14B shows a two-stage transistor audio amplifier using RC coupling. Coupling capacitor C_c allows the ac signal to flow from the output of Q1 to the input of Q2, but blocks the dc supply voltage. The primary disadvantage of this inexpensive type of coupling is an inherent loss of gain when coupling between unequal impedances. For example, coupling the high-impedance output of one common-emitter transistor stage into the low-impedance input of a similar following stage results in a loss of gain.

Transformer coupling

The transformer-coupled amplifier, illustrated in Fig. 7.14C, provides maximum gain because the transformer acts as an impedance-matching device. However, transformer coupling is limited in frequency response and the transformer usually represents a high-cost item. Transformers are usually used in IF and RF amplifier circuits where a specific frequency response is required and gain is important. The transformers also provide dc isolation between stages. In other words, the dc voltage level at the output of one stage will not affect the input of the next stage. Transistors Q1 and Q2 in Fig. 7.14C are n-channel JFETs with source resistor biasing.

Impedance coupling

This type of coupling, shown in Fig. 7.14D, combines the frequency-response characteristics of the transformer and the simplicity of the RC-coupling technique. Impedance coupling is used in IF and RF amplifiers where limited frequency response is required.

Transistor Q1 and Q2 in Fig. 7.14D are n-channel JFETs with source resistor biasing. Capacitor C1 and coil L1 form the input circuit of the signal source. Tank circuit L2C2 is a tuned network, which is normally resonant at the signal frequency. Coupling capacitor C3 provides dc isolation between Q1 and Q2. Tank circuit L3C4 is also normally resonant at the signal frequency. Capacitor C5 is an output coupling capacitor to the intended load. Note that each tank circuit employs a variable capacitor for tuning to the required resonant frequency. Many high-frequency circuits in both receivers and transmitters use this type of coupling between stages.

Push-Pull Operation

Single Class-B and -AB amplifiers, sometimes referred to as *single-ended amplifiers*, cannot be used for linear amplification due to the distorted output characteristics. The push-pull configuration eliminates this distortion by amplifying each half of the input signal with separate dedicated amplifier sections. One section amplifies the positive half and the other section amplifies the negative half. Class-B and -AB push-pull amplifiers provide good linearity and efficiency of operation. Class-C push-pull amplifiers provide even better efficiency of operation but they are nonlinear. Class-B and -AB amplifiers are used as high-level audio modulators in amplitude-modulated (am) transmitters and as linear RF amplifiers in single-sideband (SSB) transmitters. The Class-C amplifier is often used in a push-pull configuration for developing large amounts of radio-frequency power in audio-modulated (am) and continuous-wave (CW) transmitters. Additional information concerning the operation of radio-frequency amplifiers is provided in Chap. 8.

Figure 7.15 shows the basic circuit for a Class-B push-pull audio amplifier. Class-AB operation can be obtained by adding the appropriate biasing network to the base of each transistor. Transformer T1 couples the input signal to the base of each transistor in the proper phase. Note that the positive half of the input signal appears as a positive signal at the base of transistor Q1 and the negative half of the input signal appears as a positive signal at the base of transistor Q2. Thus, the first half of the input signal drives Q1 and the second half of the signal drives Q2 in an alternating or push-pull operation. Transistor Q2 does not conduct during the first half of the input signal while transistor Q1 does not conduct during the second half of the input signal. The ac signal waveforms are indicated for each transistor—the solid curved lines represent the amplified output of each amplifier section and the dotted curved lines show the periods of nonconduction. Transformer T2 combines both halves of the amplified signals from transistors Q1 and Q2, producing a complete,

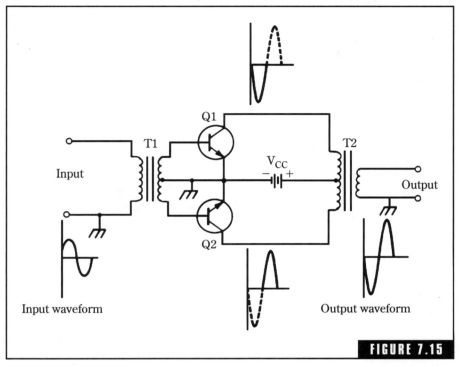

FIGURE 7.15

Class-B push-pull amplifier.

undistorted output signal. Note that the output signal has been inverted 180 degrees with respect to the input signal.

A popular audio push-pull amplifier, made possible by low-impedance transistors, is the complementary-symmetry amplifier illustrated in Fig. 7.16. This configuration eliminates the need for transformers and the low-impedance output can drive loudspeakers or other low-impedance loads on a direct basis. Matched npn and pnp transistors are required for this amplifier circuit. Note that biasing for this type of circuit is critical.

Integrated-Circuit Amplifiers

Integrated-circuit (IC) amplifiers are being used in increasing numbers in all forms of electronic equipment. Specific amplifiers are available for many applications including audio and high-frequency devices.

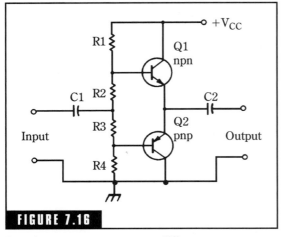

FIGURE 7.16

Complementary symmetry amplifier.

You can use these tiny packages in "home-brew" ham rigs to perform a variety of functions. Probably the most popular IC amplifier is the audio-frequency type, which contains a complete audio circuit within an 8- to 14-pin in-line package. Figure 7.17 illustrates a typical circuit using a 377 2-watt, dual-channel IC amplifier.

The schematic diagram in Fig. 7.17A provides details for a complete audio amplifier. Resistors are rated at ¼ or ½ watt. The dc working voltages are given for all the electrolytic capacitors. The IC can be either a National Semiconductor LM377 or a Radio Shack RS377 integrated circuit. The ground pins of the IC should be connected to a copper strip, which serves as a heat sink. If the amplifier is assembled on a printed-circuit board, sufficient copper foil should be left around the ground pins to serve as a heat sink.

An IC Audio Amplifier for the Ham Shack

A complete audio amplifier for the ham shack, using the 386 integrated-circuit device, can be assembled in a minimum of time. This amplifier can be powered by a single 9-volt battery or, if desired, an ac power supply can be incorporated in the circuit for 120-Vac operation. Figure 7.18 shows the amplifier circuit and the optional power supply is shown in Fig. 7.18B. A photograph of the amplifier using the 386 IC is shown in Fig. 7.19.

The amplifier and optional ac power supply can be assembled on perforated board or on a "home-brew" printed-circuit board. If a perforated board is used, the audio-amplifier integrated circuit can be mounted in an IC socket. Safety precautions should be observed when assembling and testing the optional ac power supply. All exposed terminals at ac line-voltage potential should be covered with electrical tape. Be sure to use the grounding pin of the 3-wire ac plug. The other end of the ground wire should be connected to the chassis or cabinet of the power supply.

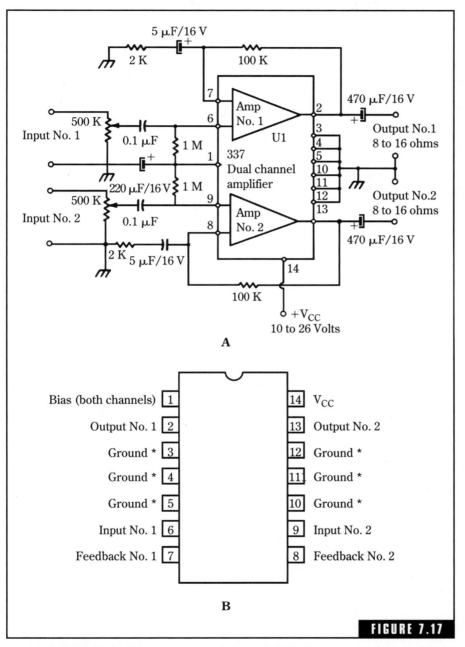

A two-watt dual-channel audio amplifier using a 377 IC. (A) Schematic diagram. (B) Pin connection for IC.

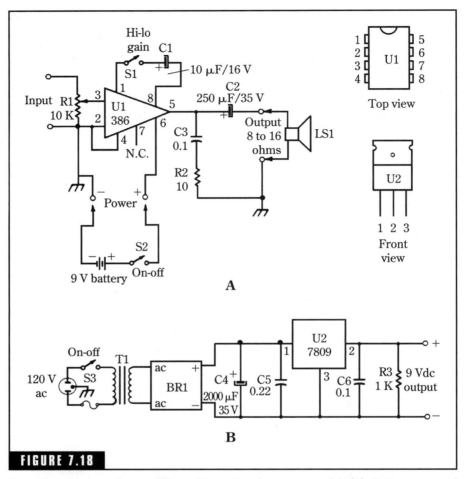

FIGURE 7.18

General-purpose audio amplifier with optional power supply. (A) Battery-powered audio amplifier. (B) Optional ac power supply.

The switch shown in the upper-right of Fig. 7.18 is used to control the gain of the amplifier. It can be set to provide a gain of 20 or 200. Note that a power switch is required for permanent installation. The switch on the volume control (upper-left) can be used for this purpose.

This amplifier can be used to amplify the output from QRP rigs to power a loudspeaker. The amplifier can also be used as a signal tracer for servicing ham receivers, particularly the audio stages. If desired, an RF probe can be constructed that allows signal tracing in the RF or IF stages of a receiver. A recommended circuit for the RF probe is given in

Fig. 7.20. The probe may be mounted in a convenient housing such as a plastic tube. When signal tracing in RF circuits, connect the alligator clip to the nearest ground in the circuit. You can then use the probe to follow an RF signal through the receiver and obtain a rough indication of the gain of each stage. Remember, the RF signal must be audio modulated. Otherwise, there will be no audio signal to detect.

The Oscillator: An Amplifier with Feedback

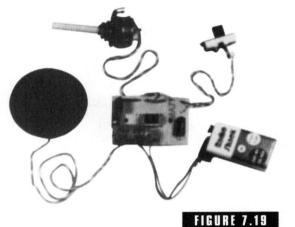

Photograph of an audio amplifier using the circuit shown in Fig. 7.18A.

The oscillator, a vital circuit in all amateur receivers and transmitters, is simply an amplifier with a feedback network connected between the output and input terminals. The amplifier can be operated in a manner similar to that of Class-A, Class-B, or Class-C amplifier characteristics.

The function of an oscillator is to generate an ac signal having a specified frequency and waveform. Most oscillators in radio communications equipment generate a sinusoidal or sine waveform. Oscillators can be designed to operate at frequencies ranging from subaudio (less than 20 Hz) to the top end of the radio frequency spectrum. Oscillators employ transistors or tubes as the active amplifying devices.

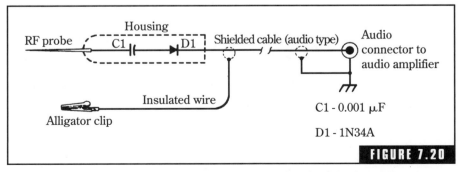

FIGURE 7.20

An RF signal tracer that can be used with the audio amplifier circuit in Fig. 7.18A.

The Basic Oscillator Circuit

Any oscillator can be reduced to three basic elements—an amplifier, a frequency-determining feedback network, and a source of dc power. A block diagram for the basic oscillator circuit is given in Fig. 7.21.

The amplifier section, either a vacuum tube or a transistor, provides sufficient gain to permit sustained oscillation. This means that the ac output signal from the amplifier compensates for the losses in the feedback network and provides a suitable input signal to the input terminals of the amplifier. The amplifier can be operated in the Class-A, -B, or -C modes of operation.

The feedback network contains inductive, capacitive, or resistive elements—or a combination of these elements—for determining the frequency. In crystal oscillators, the crystal element acts as a high-Q LC tuned circuit. The feedback network, in conjunction with the amplifier, must satisfy two basic requirements for sustained oscillations:

- An ac energy from the output terminals must be fed back to the input terminals in the correct phase. Otherwise, the feedback signal would cancel any ac input voltage and oscillation would not occur. This in-phase feedback is called *positive* or *regenerative* feedback.

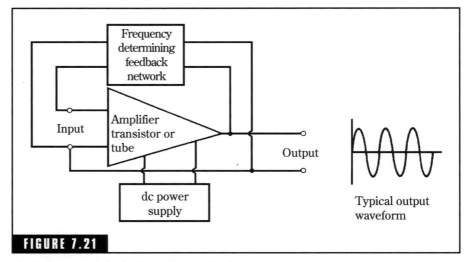

FIGURE 7.21

Block diagram of the basic oscillator circuit.

■ The amplitude of the feedback energy must be sufficient to overcome losses in the feedback network. The gain of the amplifier must be adjusted to provide this additional energy.

For optimum performance, the oscillator requires a well-filtered and stable source of dc power. For example, if the dc power contains an excess amount of 60- or 120-Hz ac power (from half-wave or full-wave rectifiers), this ac ripple signal will modulate the output signal of the oscillator. Such an effect is readily noticeable on CW and radio-telephone transmissions.

Oscillator Characteristics

Oscillators used in radio communications equipment must possess adequate frequency stability and constant amplitude characteristics. The radio-frequency oscillators used in amateur HF transmitters and receivers are designed for frequency stabilities on the order of 100 Hz or less for carrier frequencies up to about 30 MHz. This is equal to about one part in 300,000 or variations of less than 0.00033 percent. In many instances, crystal-controlled oscillators are employed to achieve this extreme stability.

Oscillator amplitude stability, the ability of an oscillator to provide a constant-amplitude output waveform, is important in both transmitters and receivers. The performance of other circuits within this equipment depends on the stability of oscillator amplitude and frequency.

The Tuned LC Oscillator

Many high-frequency oscillators that operate at frequencies above 20 kHz employ tuned circuits consisting of inductors and capacitors. When connected in a parallel configuration, the LC circuit has a natural resonant frequency determined by:

$$f_o = \frac{1}{2\pi \sqrt{LC}} = \frac{0.159}{\sqrt{LC}} \qquad \textbf{(Eq. 7.10)}$$

where
f_o is the resonant frequency in hertz
L is the inductance in henries

C is the capacitance in farads

π is 3.1416

Figure 7.22A shows an ideal LC tuned circuit that has no resistance in the inductor or in the interconnecting wiring. However, all inductors and associated wiring have some inherent resistance that dissipates the energy in the LC circuit.

When electrical energy is supplied to the ideal parallel LC tank circuit, current alternately flows between the inductor and capacitor at a rate corresponding to the resonant frequency given in Equation 7.10. For example, if the capacitor is initially charged, it discharges into the inductor. The resulting current flow through the inductor generates a magnetic field. As the current drops to zero, the magnetic field starts to collapse on itself. The electrical energy stored in the magnetic field generates a voltage that is equal but opposite in polarity to the original voltage impressed across the capacitor. This voltage charges the capacitor. When fully charged, the capacitor again starts to discharge back into the inductor, starting a second cycle of operation.

In the ideal LC tank circuit, these oscillations would occur indefinitely and would produce a constant-amplitude waveform. Figure 7.22B illustrates the voltage waveform from an ideal LC tank circuit. However, a finite amount of resistance exists in any LC tank circuit. This resistance absorbs power and causes the oscillations to eventually drop to a zero amplitude. This effect is known as *damped-wave oscillations* and is shown in Fig. 7.22C.

In addition to ohmic losses in the LC tank circuit, the oscillator must furnish RF power to other circuits within radio communications equipment. The amplifier portion of an oscillator supplies this additional RF power for overcoming these losses and maintaining a constant output.

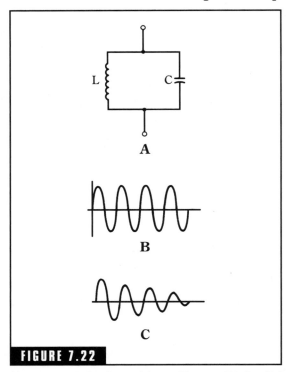

FIGURE 7.22

The LC tuned circuit used in oscillators. (A) Parallel LC tuned circuit. (B) Ideal oscillations. (C) Damped oscillations.

Oscillator Circuits

There are many types of oscillators that have been developed for specific applications. Let's look at several basic types used in amateur radio equipment.

THE HARTLEY OSCILLATOR

One of the earliest oscillators used in radio com-munications equipment, the Hartley oscillator employs a resonant LC tank circuit in the feedback network. Figure 7.23 shows a basic Hartley oscillator circuit using an npn transistor in the amplifier section. Feedback is accomplished by the use of a tapped inductor which provides inductive coupling from the output end of the inductor to the input end. The output end of the inductor is connected to the collector terminal of Q1 via the dc-blocking capacitor, C3. The other end of the inductor provides an input signal to Q1 via dc-blocking capacitor C2. The oscillator frequency is determined primarily by the L1C1 tank circuit. Capacitor C4 couples the output signal to the intended load.

Variable capacitor C1 allows the oscillator frequency to be varied within specific limits. Oscillator circuits can be designed for a two-to-one or a three-to-one change in oscillator frequency. For example, one oscillator might be designed for a 3- to 6-MHz frequency range while another design might call for a 3- to 9-MHz range.

Resistors R1 and R2 form a voltage-divider bias network for developing base bias for Q1. This bias can be adjusted for operation similar to that of Class-A, -B, or -C amplifiers. Class-A type of operation results in good waveform purity and the output signal is relatively free of harmonic content. This type of oscillator is used for the local oscillator in receivers or for the master oscillator in transmitters. Class-C type of operation provides the best efficiency and is used when purity of waveform is not important.

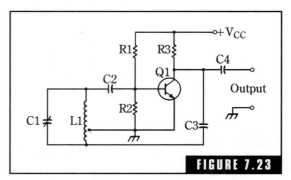

FIGURE 7.23

Hartley oscillator using an npn transistor.

THE COLPITTS OSCILLATOR

Like the Hartley oscillator, the Colpitts oscillator was developed in the early days of radio and uses a resonant LC tank circuit

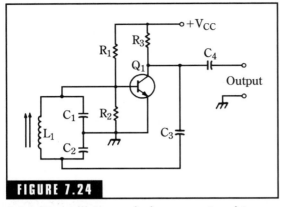

FIGURE 7.24

A Colpitts oscillator employing an npn transistor.

in the feedback network to establish the frequency of oscillation. A capacitive voltage-divider network is used to feed a portion of the amplified output signal to the input terminal of the amplifier. In the Colpitts oscillator, inductive tuning is normally used to vary the frequency of oscillation instead of variable capacitors. Although two variable capacitors could be used in this circuit, such an arrangement is normally not used due to the need for ganged capacitors of different values. Figure 7.24 shows a circuit for the Colpitts oscillator employing an npn transistor.

The frequency of oscillation for the Colpitts oscillator, determined primarily by L1, C1, and C2, is given by:

$$f_o = \frac{1}{2\pi \sqrt{L_1 \times \dfrac{C_1 \times C_2}{C_1 + C_2}}} = \frac{0.159}{\sqrt{L_1 \times \dfrac{C_1 \times C_2}{C_1 + C_2}}} \tag{Eq. 7.11}$$

where

f_o is the frequency of oscillation in Hz

L_1 is the inductance of the tank circuit in henries

C_1 and C_2 are the capacitance of the tank circuit in farads

A more accurate determination of the frequency of oscillation would include the effects of the internal capacitances of the transistor and stray capacitances and inductances in the circuit.

Capacitor C3 couples a portion of the ac output signal back to the L1C1C2 tank circuit in the proper phase and amplitude. The voltage appearing across C1 as a result of this feedback action is fed to the base of Q1. This causes a second cycle of the ac signal to be amplified by Q1. The base bias for Q1 is established by resistors R1 and R2 for operation similar to Class-A, -B, or -C amplifiers. Capacitor C4 couples the ac output signal to the intended load.

The modified Colpitts oscillator shown in Fig. 7.25, sometimes referred to as the Clapp oscillator, uses a series tuned LC circuit. This arrangement allows the use of a single variable capacitor for changing the oscillator frequency. Capacitors C2 and C3 are large in comparison to variable capacitor C1 and help to shunt any effects of external capacitance such as the base-to-emitter capacitance of the transistor. The frequency of oscillation is determined primarily by L1 and C1 and can be calculated using Equation 7.10. This modified Colpitts oscillator is extremely stable and is frequently used in amateur radio communications equipment.

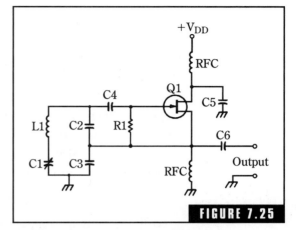

FIGURE 7.25

Modified Colpitts or Clapp oscillator employing an n-channel JFET.

The Colpitts oscillator produces an excellent sine-wave output signal both in terms of constant amplitude and frequency stability. This oscillator is used in receivers, signal generators, and other applications where a stable, variable source of high-frequency energy is required. Colpitts oscillators can be designed to operate as frequencies ranging from subaudio to the RF range of the frequency spectrum.

QUARTZ-CRYSTAL OSCILLATORS

Oscillators that employ quartz crystals as the frequency-determining element exhibit a high degree of frequency stability. Many types of crystal oscillator circuits, using transistors or vacuum tubes, have been developed for specific applications. Some of the most popular crystal oscillator circuits are the Colpitts, Pierce, and Miller. These circuits are generally used for fundamental frequencies up to about 20 MHz and overtone or harmonic frequencies up to about 200 MHz. To achieve higher frequencies in the VHF and UHF spectrum, frequency-multiplying amplifiers are used ahead of the crystal oscillator.

The ideal frequency stability of conventional crystal oscillators is about one part in one million, i.e., the oscillator frequency will not vary more than 1 Hz in 1 MHz. However, crystal aging, changing temperature, and varying dc supply voltages will reduce this stability to about one part in one hundred thousand (or about 10 Hz in 1 MHz). This degree of stability is adequate for most amateur radio applications.

Because varying temperature accounts for most of the frequency variations in a crystal-controlled oscillator, the crystal element can be enclosed in a temperature-controlled oven. Most of the frequency variations can be eliminated by maintaining a constant temperature within one degree centigrade. Some precision crystal oscillators used in a laboratory environment feature temperature control within 0.1 degree centigrade or less. Frequency stabilities of these precision oscillators will approach one part in one billion or better.

Quartz crystals. Small wafers or slabs cut from a block of quartz crystal act as resonant tuned LC circuits. When used in the feedback networks of oscillators, these crystals generate a stable single frequency of oscillation. The physical layout of a typical oscillator crystal and its equivalent electrical circuit are given in Fig. 7.26.

Crystals exhibit extremely high values of Q (up to 10,000) and also exhibit stable resonant-frequency characteristics. Each crystal oscillates at a fundamental frequency or at specified overtone frequencies. If different frequencies are required from the same oscillator, a separate crystal for each frequency and a switching circuit is required.

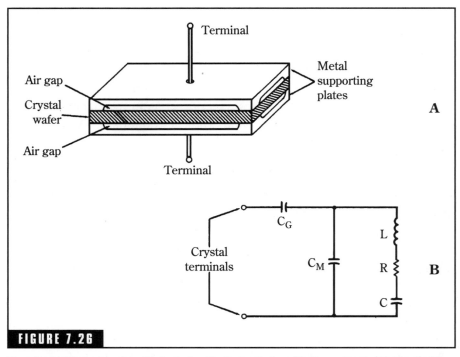

FIGURE 7.26

The physical layout and equivalent circuit of a typical oscillating crystal. **(A)** Physical layout. **(B)** Equivalent electrical circuit.

The frequency of oscillation depends primarily on the thickness of the crystal wafer. The capacitance of the crystal holder and the associated oscillator circuit will cause the frequency of oscillation to vary slightly. Standard crystal oscillator circuits have been established to maintain an identical frequency when installing a replacement crystal in the circuit. Also, trimmer capacitors are sometimes included in the oscillator circuit to adjust the frequency precisely to the required value.

In the equivalent electrical circuit shown in Fig. 7.26B, capacitance C_G is the effective series capacitance caused by the air gap when the metal plates are not touching the crystal wafer. This capacitance is not considered in some analyses of crystals. Capacitance C_M is the capacitance of the crystal holder and must be considered in the circuit design of crystal oscillators. Elements R, L, and C represent the inherent electrical equivalent of the mechanical oscillating characteristics of the crystal wafer. The value of C_M is much greater than that of crystal capacitance C by a factor of up to 100.

How crystals work. Crystal operation is based on the piezoelectric effect. When physically stressed, the crystal generates a voltage that is proportional to the amplitude and frequency of the physical force. An example of a device using this principle is the crystal microphone found in some amateur transmitters. Sound waves striking a diaphragm in the microphone cause it to vibrate. These vibrations are applied to a crystal element that, in turn, generates an ac voltage proportional to the amplitude and frequency of the sound waves. The reverse condition, placing an ac voltage across the crystal wafer, results in electromechanical vibrations or oscillations. The ac voltage generated by the crystal can be used to drive the oscillator at the required frequency.

Many crystal materials found in nature or produced by commercial processes, exhibit the piezoelectric effect. Three such materials, Rochelle salts, tourmaline, and quartz, are used in electronics applications. Rochelle salts produce the highest piezoelectric activity, i.e., the highest ac voltage for a given mechanical stress force. However, this material is not suitable for crystal oscillator circuits due to poor temperature, humidity, and mechanical characteristics. Rochelle salts crystal elements are used primarily in phonograph cartridges, microphones, and headsets.

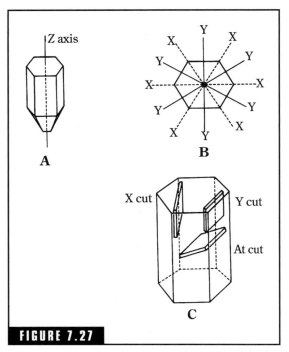

Quartz crystal structure and cut orientation. (A) Side view. (B) Top view. (C) Three of the more common cut orientations for crystal elements.

Both tourmaline and quartz crystals possess excellent operational characteristics for crystal oscillator circuits. Although tourmaline shows a lower piezoelectric activity than quartz, crystals made from this material will oscillate at frequencies up to about 100 MHz. Tourmaline, a semiprecious material, is rarely used in crystal oscillators due to the high cost involved. Quartz crystals are used almost exclusively in crystal oscillator circuits. Quartz crystals are also used in filter circuits at RF or IF frequencies. This application is discussed in Chap. 9.

How crystals are produced. Crystal wafers are cut from a block of quartz crystal material in one of several different planes or cuts. Quartz crystal material in its natural form has a hexagonal or six-sided shape as illustrated in Fig. 7.27A. For reference purposes, the crystal is assigned X, Y, and Z axes. The Z or optical axis extends from top to bottom. The X and Y axes, both at right angles to the Z axis, describe the orientation of the hexagonal plane. The X axes pass through the hexagonal points and the Y axes are perpendicular to the faces of the crystal. Figure 7.27B shows a top view of the crystal and the orientation of the X and Y axes.

Crystal wafers or elements are cut in many planes along the Z axis or at various angles with respect to the Z axis. Each family of cuts exhibits a particular set of operational characteristics in terms of piezoelectric activity (or ease of oscillation), temperature effects on frequency stability, and other factors. Some of the more common cuts, the X cut, the Y cut, and the AT cut, are illustrated in Fig. 7.27C. For example, the X cut, which was developed in the late nineteenth century, possesses high piezoelectric activity but is sensitive to changes in temperature. The frequency range of the X cut is approximately 40 to 350 kHz. The AT cut, on the other hand, possesses a low level of piezoelectric activity but exhibits excellent operational characteristics in terms of frequency stability with changing temperature and frequency of oscillation up to about 200 MHz. The AT-cut crystal is widely used in amateur and other radio communication equipment.

Crystals are available with fundamental or overtone frequencies ranging from less than 4 kHz to over 200 MHz. Many military and commercial specifications have been developed to assure precise manufacture and assembly of these devices. Crystals produced for the military and other government agencies normally are covered by the CR-XX/U designators. CR stands for *crystal resonator,* and XX indicates the number or letters assigned to a specific series of crystals. For example, the CR-5A/U designator refers to a family of crystals with a frequency range of 2 to 10 MHz. Many commercial crystals can be referenced to an equivalent of the government CR-XX/U designators.

Crystal elements are mounted or installed in a variety of holders. The holders are designed for installing the crystal in special crystal sockets or for direct soldering in printed-circuit boards. Again, almost all types of crystal holders are covered by government specifications bearing an FT-XXX or HC-XX/U designator. Outlines of some of the more popular crystal holders are shown in Fig. 7.28. In some instances, the type of holder indicates the range of frequencies available. For example, the HC-6/U holder might contain crystals with frequencies from 200 kHz to over 200 MHz.

The frequency of oscillation is normally stamped or marked on commercial crystals and many crystals produced for the government. However, some government crystals can be marked in terms of channel numbers. Because these channel numbers generally apply to a particular radio set, it is difficult to equate channel numbers with specific frequencies. Watch for this limitation if you plan to purchase surplus crystals or "rocks." Some typical crystals used in amateur radio equipment are shown in Fig. 7.29.

The frequency tolerance for most crystals will usually be specified in the range of 0.0005 to 0.02 percent. Tolerances for amateur crystals are normally specified from 0.002 to 0.005 percent in order to reduce the costs. Many crystal oscillator circuits include a trimmer capacitor for "zeroing-in" the crystal frequency to the required frequency.

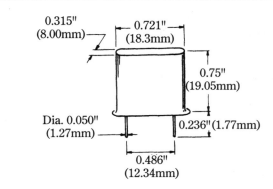

Typical frequency range: 200 kHz to 200 MHz

Installation: For installation in two-pin crystal socket

A

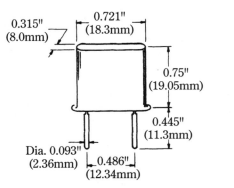

Typical frequency range: 200 kHz to 20 MHz

Installation: For installation in two-pin crystal socket or in eight-pin octal socket.

Note: FT-241 holder is similar in size and has same type of pins for installation in two-pin crystal socket or eight-pin octal socket.

B

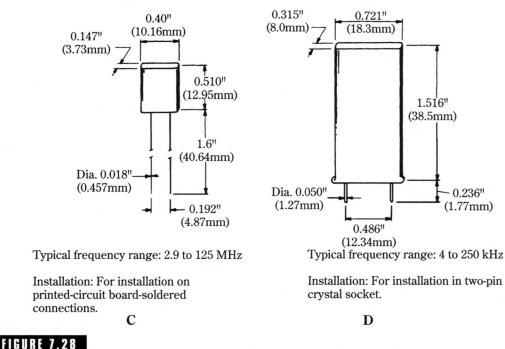

Typical frequency range: 2.9 to 125 MHz

Installation: For installation on printed-circuit board-soldered connections.

C

Typical frequency range: 4 to 250 kHz

Installation: For installation in two-pin crystal socket.

D

FIGURE 7.28

Some typical types of crystal holders used in amateur radio equipment. (A) HC-6/U type holder. (B) HC-17/U type holder. (C) HC-18/U type holder. (D) HC-13/U type holder.

FIGURE 7.29

Typical types of crystals used in amateur radio applications.

The Pierce crystal oscillator. One of the simplest oscillator circuits, the Pierce oscillator is usually operated in the low and medium radio-frequency ranges. This oscillator is used for applications requiring moderate output power and good frequency stability. The ac output waveform is an approximation of a sine wave.

Requiring no LC tuned circuit, this simple crystal oscillator is easy to assemble and operate. No adjustments other than a trimmer capacitor are required for proper operations. Figure 7.30 shows a Pierce crystal oscillator circuit using an n-channel FET such as the 2N3819. The crystal serves as a tuned feedback network from the drain to the gate. If no minor adjustment of the oscillator frequency is required, capacitor C1 can be any fixed value from 0.001 to 0.01 μF. The value is not critical because the purpose of this capacitor is to provide a low-impedance path for the ac signal while blocking the dc supply voltage from the crystal element. The ac output signal is taken from the drain terminal via dc blocking capacitor C2. The

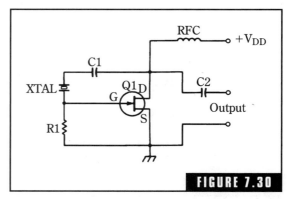

FIGURE 7.30

The Pierce crystal oscillator.

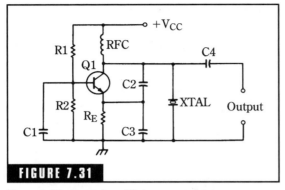

FIGURE 7.31

The colpitts crystal oscillator.

value of this capacitor should be approximately 50 to 100 pF for minimal loading effects on the oscillating circuit. The gate bias resistor, R1, can range from approximately 47 k to 100 k. The radio-frequency choke (RFC) isolates the oscillator from the dc power supply and is normally 2.5 mH.

The Colpitts crystal oscillator. The Colpitts crystal oscillator, illustrated in Fig. 7.31, is an excellent high-frequency oscillator with extremely stable frequency characteristics. This circuit is used at low, medium, and high radio frequencies for receiver local oscillators, transmitter master oscillators, and other applications requiring stable operating frequencies.

The ac feedback circuit consists of the capacitive voltage divider, C2 and C3, and the crystal element. A portion of the ac signal developed at the collector of Q1 provides regenerative feedback to the emitter. The amplitude of this ac feedback signal is determined by the ratio of C2 and C3. The resonant-frequency characteristics of the crystal controls the frequency of the ac signal developed at the collector. The ac output signal is taken from the collector terminal via dc blocking capacitor C4. Resistors R1, R2, and R_E establish the operating bias for the base of Q1. Although a bipolar npn transistor is used in this circuit, FETs or vacuum tubes can be used in the Colpitts crystal oscillator with appropriate circuit changes.

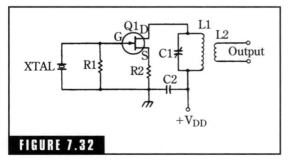

FIGURE 7.32

The Miller crystal oscillator.

The Miller crystal oscillator. The Miller crystal oscillator, popular with many designers and builders of amateur radio equipment, provides high signal output power with low crystal excitation current. Used with either vacuum-tube or transistor crystal oscillator circuits, the excitation or feedback current is extremely small and results in more reliable operation.

This oscillator circuit is often used with overtone crystals for frequencies well above 20 MHz.

The Miller crystal oscillator, illustrated in Fig. 7.32, employs an LC tuned network in the output circuit. For proper operation, the L1 C1 tuned circuit is tuned to a frequency slightly higher than the resonant frequency of the crystal. Link coupling, L2, is used to inductively couple the output of the oscillator to the required load.

Introduction to Radio Transmitters

The radio transmitter, one of the basic pieces of equipment in the amateur radio station, generates signals at frequencies in the amateur bands. When coupled to a transmission line and antenna, this energy produces electromagnetic radiation, or "radio waves." Transmitters may be simple or complex, single band or multiband, and have a variety of modulation schemes [continuous wave (CW), amplitude modulation, single sideband, frequency modulation, or a combination thereof].

Radio transmitters, as well as radio receivers, are generally classified in terms of the frequency bands they cover. For example, a transmitter designed to operate in the HF band, 3 to 30 MHz, requires design and construction techniques that are generally different than those used in VHF and UHF frequency bands. As a rule, construction of HF transmitters and receivers are easy when compared to those operating in the VHF and higher bands. The reason for this is that the size of components and length of interconnecting conductors become smaller as the radio frequency of operation increases. Finally, at microwave frequencies, the size of components and interconnecting wires is so small that special construction techniques are required. In general, the beginning amateur operator can learn to construct and operate simple HF transmitters and receivers. Fortunately, these simple radios can operate in the popular CW HF bands in the QRP, or low-power mode. Many amateurs employ low-power rigs (5 watts or lower) to make long-range contacts over thousands of miles.

Radio transmitters suitable for Technician Class operation in HF bands may range from simple, low-power, crystal-controlled rigs to the more elaborate transceivers. Power outputs of low-power transmitters will vary from less than 1 watt to about 10 watts while the high-power transceivers range from about 50 to 300 watts. Older types of equipment employ vacuum tubes or a combination of tubes and transistors to perform the functions of oscillation, amplification, and modulation. Most new amateur radios are fully solid-state and are generally designed for 12-V dc operation.

You have a number of options for "getting on the air." Many designs for simple "home-brew" CW/phone transmitters and receivers are available in amateur publications, such as handbooks and magazines. Transmitters and transceivers, many in kit form, are available from amateur radio suppliers. The Ramsey Model QRP-40 40-Meter Transmitter, shown in Fig. 8.1, is available in kit form for easy construction.

FIGURE 8.1

The Ramsey Model QRP-40, 40-Meter (7-MHz) CW Transmitter. This 1-watt, 40-meter CW transmitter features two-channels (a choice of two transmitting frequencies), crystal controlled with variable tuning of up to 5-kHz around each crystal; a built-in antenna transmit-receive switch; and operation from an external 12 Vdc power supply (or batteries). This versatile transmitter is available in kit form from Ramsey Electronics, Inc. *Ramsey electronics, Inc., 793 Canning Parkway, Victor, NY 14564, Telephone: (716) 924-4560*

Building this kit can provide good experience. Figure 8.2 shows a simple, easy-to-build HF receiver kit that covers the 80-meter, 40-meter, 30-meter and WWV, 20-meter, and 15-meter bands. After you finish this type of kit, you may want to investigate transceiver kits for other amateur bands on HF and VHF bands. With these simple "home-brew" radios at a low investment, you can get on the air with either HF Morse code or 2-meter FM communications capability.

Used amateur radio equipment is another excellent and inexpensive way to assemble an amateur radio station. You will find many good bargains through the local amateur radio club, swap-fests, ham auctions, and amateur radio conventions. In some instances, however, you may have to repair or rebuild used equipment for proper operation. Experienced hams at

FIGURE 8.2

The MFJ World Band Radio, Model MFJ-8100. This high performance "short wave" radio is available in kit form for a modest price. In addition to covering the following hf amateur bands (80m, 40f, 30 m, 20m, and 15m), the radio provodes coverage on other international frequencies and the WWV and WWVH time standard stations. This radio in kit form helps the beginner to become familiar with electronic construction practices and testing of electronic circuits

most amateur radio clubs will be glad to help you with technical problems or loan you an instruction manual or schematic for the used equipment. Also, most amateur radio journals carry advertisements listing used amateur radio equipment. Many of the companies involved will check out and repair all used equipment before it is offered for sale.

Another option for assembling your amateur radio station is the purchase of new equipment. Many excellent HF transceivers are available from such companies as Kenwood USA, ICOM, MFJ Enterprises, Ten-Tec, Yaesu, and others. Most of the equipment produced by these companies is capable of single-sideband operation as well as CW—this is a vital factor when the Technician upgrades to General or higher class license.

The No-Code Technician Class ham will be interested in amateur equipment for operation above 50 MHz. Many VHF and UHF radios are available on the amateur market, both new and used. A modern Twin-Band 2-Meter Handheld radio is shown in Fig. 8.3. Comparable mobile radios are available for portable operation or for installation in

FIGURE 8.3

The Standard C528A Twin Band FM Hand-held 2-Meter and 440-MHz Transceiver. This versatile radio covers the amateur 144–148 MHz and 438–450 MHz bands, providing FM communications for both simplex and repeater operation. The features packed into this tiny radio include 40 memory channels, cross-band operation between 2-meter and 70-centimeter frequencies, subaudible tone (CTCSS) for both encode and decode functions, DTMF "touch dial tones" for phone patch operation, and receiver tuning ranges of 130–174 MHz and 248–479 MHz. *Standard Radio Products, Inc., P.O. Box 48480, Niles, Il 60648, Telephone: (312) 763-0081*

automobiles or other vehicles. Also, some companies offer "full-mode" radios which cover all forms of modulation (CW, SSB phone, FM repeater and packet operation, and split-frequency operation for amateur satellites).

This chapter will cover the basic types of radio transmitters with particular emphasis on the CW transmitter. More advanced amateur radio publications are available for specific information of radiotelephone or voice transmitters.

Definitions

Amplitude modulation Amplitude modulation (AM) is the process of varying the amplitude of the carrier radio wave in proportion to the modulating signal. The information being conveyed by the transmitted radio signal, normally CW-coded signals or voice, is directly related to changes in amplitude of the RF signal. Amplitude modulation includes CW, tone modulation, double-sideband modulation, and single-sideband suppressed-carrier modulation.

Backwave Backwave is defined as residual or feed-through radiation from the master oscillator when the final RF amplifier is not keyed to the transmit mode. In some transmitters, the master oscillator is left on during receive as well as transmit modes to improve stability. If the final amplifier stage is not properly neutralized, some of the energy from the master oscillator may be fed through the final amplifier stage and cause a weak signal to be radiated from the antenna.

Band-pass filter   A circuit that passes a signal band of frequencies and blocks signals with frequencies below and above the band-pass limits.

Continuous-wave modulation   Referred to as CW modulation, this type of radio transmission conveys information by keying the carrier wave on and off. Thus, the keyed dits and dahs are used to form characters of the International Morse code set. Sometimes CW modulation is called radiotelegraphy, a term left over from the early days of radio.

Crystal oscillator   An RF oscillator that uses a piezoelectric crystal to control the frequency of the oscillator within very accurate limits. See Chap. 7 for more details.

CW   CW, or continuous wave, is the name for emissions produced by switching a transmitter's output on and off.

Detector   A circuit (or stage) in a receiver that detects the audio modulating component of an RF modulated signal. A detector is found in every receiver.

Deviation   Usually refers to a frequency deviation of the carrier signal as produced by an FM transmitter. Over-deviation of an FM transmitter's carrier can produce "out-of-channel" emissions.

Frequency discriminator   A circuit used in an FM receiver to produce an audio signal contained in the FM-modulated RF carrier signal.

Frequency modulation   Commonly referred to as FM, this is a process whereby the frequency of the carrier wave is varied in accordance with the instantaneous amplitude of the modulating signal. The frequency of the modulating signal determines the rate or number of times that the carrier frequency varies each second. Frequency-modulation modes used in amateur radio include frequency-shift telegraphy and radiotelephone.

Frequency stability   As related to transmitters, frequency stability describes the ability to maintain a constant carrier frequency under such conditions as on-off keying, varying power-line voltages, and changes in temperature. For example, poor frequency stability in CW transmitters reduces intelligibility of copy and may cause out-of-band transmissions.

Harmonic radiation   Harmonic radiation is an undesired signal produced in transmitters. The frequency of the harmonic radiation may be a multiple of the fundamental or desired signal. Modern HF transmitters usually have a built-in low-pass filter in their RF output circuits.

Key chirps   Usually caused by poor frequency stability, key chirps occur as a result of changes in the carrier frequency when the transmitter is keyed on. The sound of the CW signal has a distinct "chirp" at the instant the transmitter is keyed. In order to minimize or eliminate this problem, some CW transmitters are designed with a constant-on "free-running" RF oscillator and the carrier wave is keyed on and off in the RF amplifier stages.

Key clicks   Unlike key chirps, key-click problems are encountered primarily with switching on and off the RF amplifier stages. This sudden transition from off to on and from on to off can cause annoying transient signals to appear on the carrier wave. Usually, key clicks can be eliminated with a key-click filter installed across the key terminals. A related problem in keying some CW transmitters is that of unintentional radiation of a small portion of the carrier wave during key-up conditions. This is caused by a small amount of the RF energy from the RF oscillator leaking through the unkeyed RF amplifier stages to the antenna terminals. Adequate RF shielding, decoupling networks, and proper neutralization of the first RF amplifier will generally clear up this type of problem.

Modulation   A term used to describe the process of combining an information signal (such as a voice or phone) with a radio signal.

Phone emissions   This refers to telephony or voice signals usually transmitted as AM, FM, or SSB emissions.

Product detector   A circuit used in a single-sideband (SSB) receiver to detect or demodulate the SSB signal to produce an audio signal.

RF amplifier   An integral part of most transmitters, one or more RF amplifiers are used to boost the power of the carrier wave to usable levels. The RF amplifier may use tubes, transistors, or a combination of both as the active amplifying devices.

RF carrier   The unmodulated RF signal or wave generated by a transmitter is referred to as the carrier or carrier signal. Normally the carrier is an RF signal of constant amplitude and frequency. The carrier can be modulated either in terms of amplitude or frequency changes in order to convey information to the intended receiving station.

RF oscillator Sometimes referred to as the master oscillator, this section of the transmitter generates the initial carrier signal prior to amplification of conversion. Most transmitters employ a variable-frequency oscillator for controlling the frequency of the output carrier wave. Other transmitters use a crystal oscillator for the master oscillator function.

Single-sideband (SSB) signals Single-sideband signals may be either upper (USB) or lower sideband (LSB) and convey the original audio signal used to modulate the SSB transmitter. Normally, only one set of signal frequencies, USB or LSB, is transmitted by the transmitter.

Splatter interference sideband When a transmitter's output carrier is overmodulated by the modulating signal, RF signals of other frequencies, or "splatter signals," are produced.

Spurious emissions Spurious emissions are those unintentional radiations from electronic equipment such as transmitters, which can cause interference to other electronic equipment. For example, an HF amateur transmitter may inadvertently radiate interfering signals in the television broadcast frequencies and cause television interference (TVI). Usually the frequency of a spurious emission is different from that of the carrier and may be completely out of the amateur bands. Some spurious emissions can be eliminated by installing filter networks in the transmission line.

Superimposed hum Poor filtering in an ac power supply for a transmitter can cause 60 or 120 Hz hum to be mixed with the output carrier signal. The carrier signal is modulated with this low-frequency audio signal and produces an objectionable sound to the listener. The problem can usually be cleared up by installing adequate filter capacitors in the power supply.

VFO The variable frequency oscillator (VFO) circuit is used to help produce the transmitter carrier signal output.

The CW Transmitter

Continuous-wave (CW) transmitters are used to transmit information with a series of short and long pulses of radio-frequency (RF) energy. To accomplish this, the transmitter's carrier RF signal is turned on and off by a hand or automatic keyer. These on-off transmissions convey information in the form of International Morse code characters, a universal language among amateur radio operators worldwide.

The CW transmitter is relatively simple to build and operate. With one or two transistors (or vacuum tubes), the Technician operator can assemble a CW transmitter capable of contacting other hams at distances of hundreds to thousands of miles.

The earliest amateur transmitters, described in Chap. 1, utilized spark gaps for generating damped radio waves. These old transmitters sometimes produced only a few watts of power spread over a wide band of frequencies. This caused considerable interference with other amateur stations and commercial "wireless" facilities. The introduction of the vacuum tube resulted in constant-amplitude carrier signals with good frequency stability. The development of crystal oscillators and highly stable variable-frequency oscillators improved the frequency stability of transmitters to a fine degree. Today, the transistor has all but replaced the vacuum tube for generating RF energy.

A simple CW transmitter, illustrated by the block diagram in Fig. 8.4, consists of an RF oscillator, a hand key, and a dc power source. Many amateur operators launched their careers in amateur radio using a one-tube or one-transistor transmitter. Today, some amateur operators employ a single-transistor CW transmitter for challenging contacts with other hams. For optimum frequency stability, these simple transmitters are crystal controlled.

The one-transistor transmitter has the disadvantage of low power output. Also, changes in loading effects from the transmission line and antenna are reflected back into the oscillator circuit, which can cause frequency stability problems. Even the swinging of the antenna in the wind can cause changes in frequency.

The Master Oscillator Power Amplifier Transmitter

A significant improvement over the single-stage RF oscillator type of transmitter is the master oscillator power amplifier (MOPA). This type of transmitter, shown in Fig. 8.5, has separate stages for oscillator and amplifier functions. Greater frequency stability as well as higher power output can be obtained from this circuit. The RF power amplifier isolates the RF oscillator from external loading effects in addition to providing power amplification of the RF signal. An LC-tuned oscil-

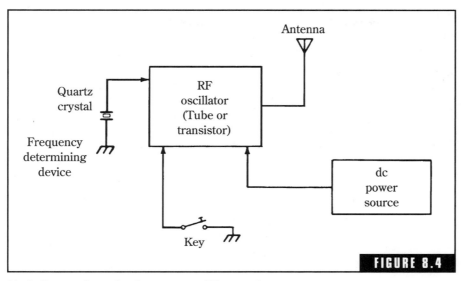

FIGURE 8.4

Block diagram for a simple one-stage CW transmitter.

lator can be used with this type of circuit for varying the frequency of the carrier signal.

A schematic for a modified crystal-controlled MOPA CW transmitter, developed by Ramsey Electronics Inc., is given in Fig. 8.6. This CW transmitter, capable of providing about 1 watt of output power and switchable in frequency between two crystals, incorporates a driver stage between the crystal oscillator and final amplifier stages to improve RF frequency and keying stability. The QRP-40 VXO Transmitter is available in kit form from Ramsey Electronics Inc., 793 Canning Parkway, Victor, NY 14564 [telephone: (716) 924-4560; internet address: www.ramseyelectronics.com]. The kit includes all required parts, including a predrilled PC board and a complete manual for assembly, test, and operation. This PC board permits easy modification of the existing circuit and/or installation of additional circuits for later upgrades. A plastic case and knob kit is also available to house the transmitter for easy installa-

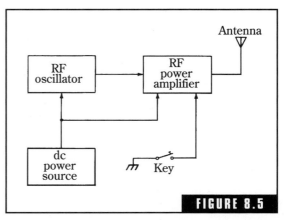

FIGURE 8.5

The master oscillator/power-amplifier (MOPA) CW transmitter.

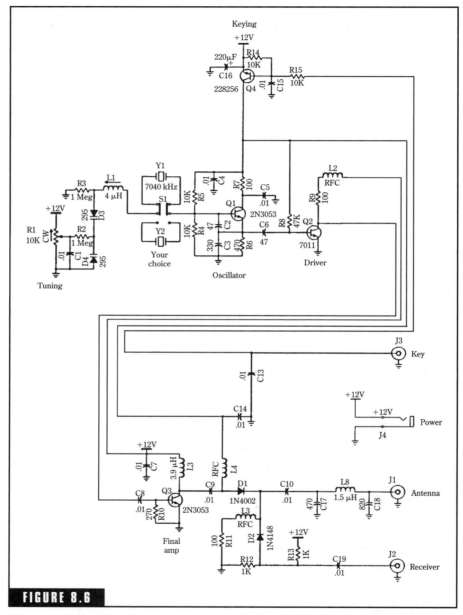

Schematic diagram for the Ramsey Electronics Model QRP-40 CW Transmitter. This easy-to-build CW transmitter provides about one watt of output RF power in the 40-meter band. *Ramsey Electronics Inc.*

tion in your "ham shack." (Figure 8.1 shows a photograph of this CW transmitter with the top cover removed.)

Functional data taken from the Ramsey Model QRP-40 Transmitter's instruction manual shows the following circuit description:

Transistor Q1 is configured in a crystal oscillator circuit that can be switched to two separate frequencies of operation. Switch S1 selects one of two internally installed crystals. The R1/D3/D5/L1 varactor-controlled series resonant circuit establishes the crystal oscillator frequency. Adjusting R1 allows the crystal oscillator frequency to be varied up to 5 kHz about the crystal frequency. This feature permits you to adjust your operating frequency away from other existing signals on the air.

The output RF signal from Q1 is fed to the Driver Q2, which acts as a buffer between the crystal oscillator stage and the final amplifier stage. Q3 is a Class C RF amplifier that amplifies the RF signal to about 1 watt of output power. The C7/L6/C18 "pi" output network forms a low-pass filter that matches the low impedance output of Q3 to the antenna impedance of about 50 ohms. Diodes D1/D2, along with resistors R11/R12/R13, and the L5 RFC "RF choke" form an electronic transmit-receive switching network which switches the antenna to either the transmitter or the receiver.

Dc power to the transmitter is applied through an electronic keying circuit employing transistor Q4 and a biasing network. When the telegraph or hand key is pressed down to transmit the dots and/or dashes of Morse code:

1. Q4 conducts +12 VDC to Oscillator Q1 and applies a positive bias to the base of Driver Q2 through R8. This turns on the crystal oscillator stage and applies a low-level RF signal to the driver stage. Q2 amplifies the RF signal and applies it to the base of the Q3 Final Amplifier for amplification to about 1 watt of RF power. Note that +12 VDC power to the buffer and final amplifier stages is always present, eliminating keying action which could produce clicks and chirps.

2. Q4 also conducts +12 VDC through choke L4 to the anode of Diode D1, thus conducting the amplified RF signal power to the antenna through the low-pass filter section. The +12 VDC from Q4 also

applies a negative bias to the cathode of D2, which blocks the high-level RF signal from the transmitter from entering the receiver. (Actually, a tiny amount of power does reach the receiver, permitting the operator to hear the transmitted signal and ensure that the receiver is tuned to the frequency of the transmitted signal.)

3. When the telegraph key is released—completing the intended transmission of each dot or dash of the Morse code—the +12 VDC power is removed from the oscillator stage, the driver stage, and Diode D1. This turns off the RF signals and opens the D1 diode. The +12 VDC applied to R13 can now provide a positive bias to Diode D2, permitting the signals intercepted by the antenna to be applied to the receiver.

Amplitude-Modulated Transmitters

The Technician Class operator is authorized single-sideband (SSB) voice modulation on a portion of the 10-meter band as well as all amateur modes of operation on 6 meters and above. The General and higher class operators are authorized voice modulation modes on specified portions of all amateur bands from 160 meters up. This increased scope of operation is a powerful incentive for continuing to improve your knowledge of amateur radio and upgrade to the General Class license. In this chapter, we'll look at basic AM and SSB radio telephone transmitters for operation in the HF band and frequency modulation (FM) transmitters used primarily in the VHF and above.

Conventional AM transmitters

Amplitude modulation is produced when the amplitude of a radio carrier wave is varied in proportion to a modulating signal such as an audio or voice signal originating from a microphone. Figure 8.7 shows a block diagram of an AM transmitter. The only observable difference between the CW transmitter and the AM transmitter is the addition of a speech or audio modulator, which is capable of varying the amplitude of the output FR carrier wave. In fact, an "outboard" modulator could be added to the QRP CW 40-meter transmitter in Fig. 8.6 for phone operation. Although most hams today use SSB modulation, some hams prefer AM phone on 75- and 40-meter bands.

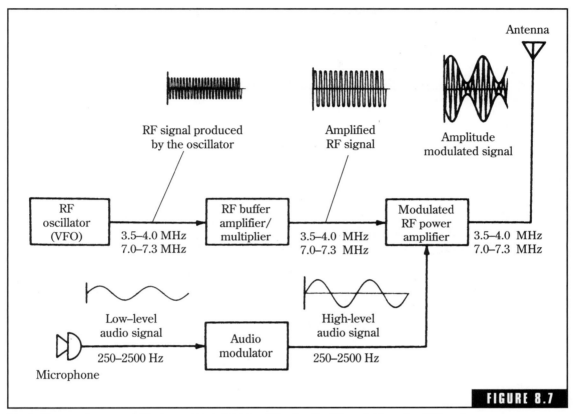

Block diagram for an amplitude modulated transmitter for the 75-and 40-meter HF bands.

Here's how the AM transmitter in Fig. 8.7 works. Note that the RF sections of this transmitter (VFO RF oscillator, RF amplifier/multiplier, RF power amplifier) are similar to the MOPA CW transmitter in Fig. 8.6. With the addition of a telegraph key circuit, the transmitter in Fig. 8.7 is capable of either CW or AM phone operation on all HF bands between 3 and 30 MHz. The speech modulator amplifies the low-level audio signals from the microphone, producing a high-level audio signal. This high-level signal is applied to the RF power amplifier stage, causing the amplitude of the carrier wave to be varied from nearly zero to about twice the amplitude of the unmodulated wave. If the modulating signal is applied to the modulation stage output circuit of the RF power amplifier stage, this is referred to as *high-level modulation.* (High-level modulation for solid-state transmitters is accomplished when the audio modulating signal is applied to the collector circuit of the final RF power transistor stage.) *Low-level modulation* in solid-state transmitters is produced when the audio modulation signal is applied to the base (or gate) of the RF power amplifier transistor stage.

The power level of the audio modulating signal for high-level modulation must be approximately one-half that of the carrier wave power. For example, a 100-watt AM transmitter requires 50 watts of modulating power in order to produce 100 percent modulation.

The RF carrier wave power represents about two-thirds of the total RF power being radiated from the antenna. The remaining one-third of the RF power is contained within the two sidebands. Because either sideband contains identical information relating to the original modulating signal (i.e., the modulating signal such as voice), the effective transmission of the voice information by single sideband requires one-sixth of the original power. For all practical purposes, a 20-watt single-sideband transmitter is as effective as a 120-watt AM transmitter!

AM carrier and sideband signals

The generation of AM signals results in a carrier wave and two identical sets of sideband signals spaced on opposite sides of the carrier-wave frequency. Figure 8.8 illustrates the frequency spectrum of a typical AM signal in the 80-meter band. The carrier wave frequency, generated by the VFO, is positioned at 3923.0 kHz. Voice signal frequencies from 250 to 2500 Hz are fed to the modulator for amplification to the required power level. When applied to the RF power amplifier stage, this produces sideband signals below and above the carrier frequency. The set of lower sideband signals extends down to 3920.5 kHz. In terms of decreasing carrier frequency, the lower sideband appears as an inverted set of frequencies, with the lowest voice frequencies (around 250 Hz) adjacent to the carrier frequency. The upper sideband extends up to 3925.5 and appears as a noninverted set of voice frequencies ranging from the lowest frequencies (around 250 Hz) to the highest.

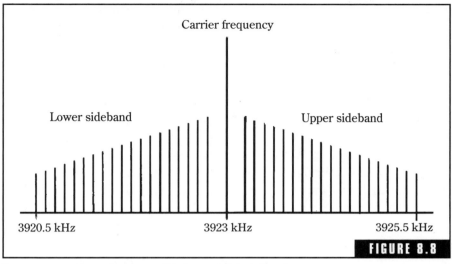

Frequency spectrum for an amplitude-modulated signal.

Audio modulation requires a bandwidth of twice the highest frequency of the modulating signal. For speech signals ranging up to about 2500 Hz, the AM modulating bandwidth is about 5 kHz.

Single-Sideband Modulation

The disadvantages of AM modulation in terms of wide bandwidths and power lost in transmitting the carrier wave and both sidebands can be eliminated by single-sideband (SSB) transmissions. As noted earlier, the transmitter carrier wave does not contain the information being conveyed by the person talking into the microphone. All of this information goes into the two sidebands. Thus, only one sideband needs to be transmitted. This is the theory behind SSB transmission.

Up to a point, the SSB transmitter is almost identical to the AM transmitter. A carrier wave and two sidebands are generated in the initial stages of an SSB transmitter. Figure 8.9 shows the block diagram for a filter type of SSB transmitter. The balanced modulator stage removes or suppresses the carrier wave, leaving only the two sidebands. One of these sidebands is removed by a very sharp filter. The remaining sideband, which appears as a set of audio frequencies converted to intermediate RF frequencies, can be heterodyned with an

RF oscillator to produce the desired RF transmission frequency. After sufficient amplification, the modulated SSB wave is fed to the antenna for radiation into space.

The SSB transmitter in Fig. 8.9 covers the 80-meter amateur band and produces a lower-sideband output signal. A fixed-frequency oscillator, V1, generates an RF carrier wave of 456 kHz. Speech signals from the microphone are amplified by the speech amplifier stage, V2. When both the 456-kHz carrier wave and speech signals (nominally 250 to 2500 Hz) are applied to the balanced modulator, the mixing action removes the 456-kHz signal and produces modulated lower and upper sidebands. The lower sideband frequencies extend from 453.5 kHz to 455.75 kHz, representing the 250- to 2500-Hz speech modulating signals. Conversely, the upper sideband contains frequencies from 456.25 kHz to 458.5 kHz. Figure 8.10 shows the frequency spectrum of the output of the balanced modulator containing the two sidebands. This is a form of modulation known as double-sideband suppressed-carrier modulation. If applied directly to the mixer stage (V4 in Fig. 8.9) this

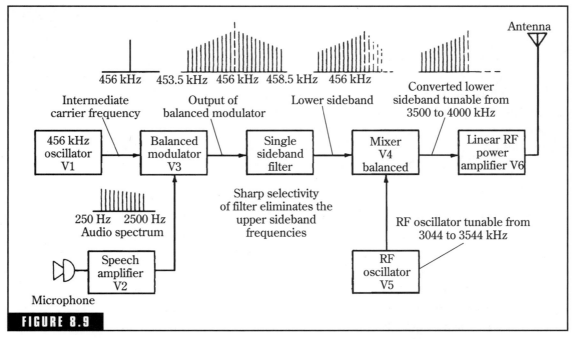

FIGURE 8.9

Block diagram for a lower-sideband SSB 80-meter transmitter.

would result in transmission of a double-sideband suppressed-carrier signal. Some of the earlier sideband transmitters employed this mode of transmission.

When applied to the sharp bandpass filter, the upper sideband is suppressed, leaving only the lower sideband. Either electromechanical or quartz-crystal filters are used to obtain the required sharp bandpass characteristics.

The lower sideband is heterodyned to the desired frequency of operation within the 80-meter band by the mixer stage, V4, and a variable frequency oscillator, V5. In order to cover the 80-meter band, the VFO must be able to produce a signal between 3044 kHz and 3544.0 kHz.

For upper-sideband operation, the frequency of the fixed frequency oscillator, V1, is changed to 453.5 kHz. Many SSB transmitters employ a crystal oscillator for this stage with a capability for switching between two crystals to provide the lower- or upper-sideband mode of transmission. Changing this oscillator frequency to 453.5 kHz shifts the lower sideband below the passband of the sharp filter and centers the upper sideband within the passband of the filter.

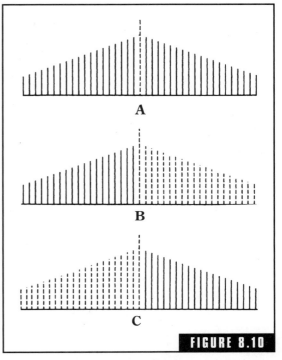

FIGURE 8.10

Forms of amplitude-modulated signals. (A) Double-sideband suppressed-carrier modulation. (B) Lower-sideband modulation (upper-sideband and carrier are suppressed). (C) Upper-sideband modulation (lower-sideband and carrier are suppressed).

Operation for the other amateur HF bands (40 to 10 meters) can be obtained by changing the frequency coverage of the VFO and the tuned LC networks in the linear RF power amplifier. Most amateur HF SSB transmitters contain switching networks for quickly changing to any desired band of operation. Figure 8.10B and C illustrates the frequency spectrum of both lower- and upper-sideband transmissions.

The last stage of the SSB transmitter, the linear RF power amplifier, must possess two basic characteristics. First, the input tuned network must pass only the desired RF frequencies from the mixer stage. The

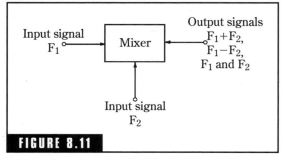

FIGURE 8.11

Basic action of a mixer circuit.

output of a mixer contains both the sum and difference frequencies of the two input frequencies. In Fig. 8.9, with the VFO set at 3432.75 kHz, the mixer provides two sets of frequencies, one set centered at 3888.75 kHz and a second set centered at 2976.75 kHz. The latter set of frequencies clearly does not fall within the authorized amateur frequencies. Figure 8.11 illustrates the action of a frequency mixer.

The second characteristic of the RF power output amplifier is that it must provide for uniform or linear amplification. Linear amplifiers in subtransmitters are usually operated in Class-B or -AB modes in order to obtain linear amplification.

The power output of SSB transmitters is normally specified in terms of peak envelope power, or pep. This is a practical way to rate SSB transmitter power because no carrier wave is radiated and during conditions of no modulation, the transmitter does not produce an RF output. Average power is difficult to measure due to the differing characteristics of voice signals. Also, dc ammeters or voltmeters cannot provide an accurate indication of the power output. An approximation can be made by multiplying the indicated average power by a factor of two. For example, if the plate-current meter of an SSB transmitter indicates about 100 mA during modulation with a plate supply voltage of 750 volts dc, the average power input to the final stage is approximately:

$$P_{avg} = V_P \times I = 750 \times 0.1 = 75 \text{ watts}$$

The pep produced by the SSB transmitter is approximately twice the indicated average power, or about 150 watts in this example.

The effective "talk power" of SSB transmitters is superior to that of conventional AM transmitters with the same-rated power output levels. In the example just given, the final RF amplifier stage has a power dissipation of 75 watts with a peak effective power of about 150 watts. An AM transmitter of similar power ratings will produce about 75 watts of power during periods of no modulation. When modulated, the total power of the carrier and the two sidebands is approximately

110 watts. Only one-sixth of this power, or about 18 watts, is contained in each of the sidebands. This illustrates the tremendous advantage of SSB transmitters over AM transmitters. Along with the advantage of requiring less bandwidth, it is no wonder that SSB is the most popular mode of voice communication in the crowded HF amateur bands.

Most amateur equipment available for the HF bands consists of multipurpose transceivers, or combination transmitter and receiver units. The transceiver combines the use of common circuits such as the master oscillator and common RF, IF, and audio amplifiers. This makes possible a more compact unit as well as reducing the overall cost of an amateur radio station. Figure 8.12 shows a modern, all solid-state HF transceiver capable of CW and SSB operation over all authorized amateur bands from 80 to 10 meters. The 100-watt peak input transmitter power and sensitive receiver characteristics make this transceiver and similar models popular with both beginning and experienced hams. Specifications and pricing information on this transceiver and similar equipment are available in most amateur magazines such as *CQ, Ham Radio, QST*, and *73*.

FIGURE 8.12

The ICOM 728 All-Mode HF Transceiver. This high-performance amateur HF transceiver covers all the amateur bands between 1.8 to 29.7 MHz; the receiver section provides general coverage from 30 kHz to 30 MHz. All mode operation includes CW, SSB (USB and LSB), and FM modulation. Dual VFOs and 26 memory channels provide for split or cross-band operation. Operating accessories include an automatic antenna tuner, dc power supply, desk microphone, and a 500-watt HF linear amplifier. *ICOM America Inc., 2380 116th Avenue N.E.; Bellevue, WA 98004; Telephone: (206) 454-7619*

Frequency Modulation

Another form of RF modulation used by amateurs is that of frequency modulation (FM). Here the frequency of the RF carrier wave is varied within limits by an audio (or other low-frequency signal such as RTTY or data) signal waveform. FM provides certain advantages over AM modulation systems. However, those advantages come at a price.

Radio communications via FM systems offer clear, noise-free reception of voice and other amateur FM signals. Even in the presence of high-ambient electrical noise and interference, FM communications sytems provide virtually crystal-clear signal reception. These characteristics are ideal for amateur repeater operation in the VHF and UHF where many ham radios are installed or used in automobiles and other vehicles.

Another teature of FM communications is the "capture effect." An FM receiver will always "lock on" to the strongest signal and ignore weaker ones. This is excellent for a two-way contact with a ham friend. However, it prevents another ham with a weaker signal from calling you. This is the reason why airport control-tower operators use conventional amplitude modulation radios for contacting and directing aircraft activity. In this instance, the control-tower operators can hear weaker signals from more distant aircraft.

On the negative side, FM systems normally require more bandwidth than similar AM systems. One reason for this is that the FM modulation process generates a series of sideband signals extending well past the frequency of the modulating audio signal. For example, FM broadcast stations require a bandwidth of 150 kHz to transmit a high-fidelity sudio signal of 50 to 15,000 Hz. This results in a +/− 75 kHz deviation. Amateur radio FM transmitters use a "narrow-band" FM modulation system, which generates sidebands up to +/− 5000 Hz when modulated by a voice signal with a frequency range of 300 to 3000 Hz. The total RF bandwidth required to transmit this FM signal is given by

$$BW = 2 \, (M + D) \, \text{Hz} \qquad \textbf{(Eq. 8.1)}$$

where

BW is the required RF bandwidth

M is the maximum modulation frequency in Hz

D is the peak deviation in Hz

The wide-bandwidth FM RF signals become distorted when transmitted via skywave propagation modes. This is due to the changing phase of the different frequency components of the transmitted signal. This limitation prevents use of FM modulation methods at frequencies below the amateur 10-meter band.

FM transmitters

Modern FM transmitters employ several methods of generating FM signals. A popular circuit is the "direct FM" modulator, which employs a voltage-controlled oscillator (VCO). Conventional LC RF oscillators use an LC resonance circuit (refer to the "The Tuned LC Oscillator" in Chap. 7). If the capacitance of the capacitor in the LC resonant circuit can be varied by an audio signal voltage waveform, this change in capacitance will vary the frequency of the RF oscillator. The amount of change of frequency from this circuit is relatively small. By designing the VCO oscillator for a low RF frequency and using frequency multiplier stages, the center frequency of the AM signal can be raised to the desired VHF or UHF channel. Some FM transmitters employ frequency synthesizer circuits to generate and maintain accurate RF channel frequencies. A block diagram of a typical VHF modern FM transmitter is given in Fig. 8.13 for a simple, single-channel VHF FM transmitter. Audio (voice) signals generated by the microphone are amplified by the AF amplifier and applied to the VCO. These audio signals cause the frequency of the VCO to vary as a function of the original voice signals. The amount of frequency change (or deviation) is a function of the voice signal's amplitude, and the number of frequency swings is a function of the original voice frequency. The frequency changes at the output of the VCO stage are small. Frequency multipliers (harmonic generators) increase the VCO's center frequency (eightfold in this example) to the VHF channel's authorized operating frequency (147.2 MHz) while increasing the final maximum frequency deviation to 5000 Hz. In this example, the maximum output of the FM transmitter reuqires a bandwidth of 16 kHz. Channel spacing and frequency allocations for FM operation on the amateur bands are established by a regional frequency coordinator group made up of amateur operators.

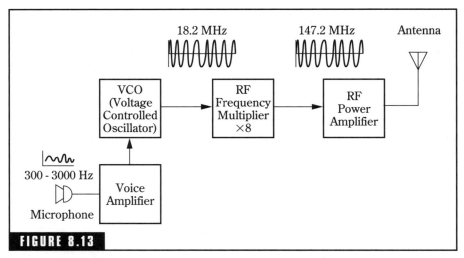

FIGURE 8.13

Block diagram for a simple, single-channel VHF FM radiotelephone transmitter.

Transmitter Performance Tests

The FCC requires that all licensed amateur operators monitor their transmitters to ensure that emissions are always within the authorized frequency and power limits. This means that you must have the capability to determine the performance of your transmitter at any time. This section covers some of the basic methods and test equeeipment used in checking the transmitter for proper operation.

Transmitter frequency measurements

Several types of equipment are available for determining the output frequency of the transmitter. In addition, this equipment will help to detect the presence of spurious radiation outside of the authorized amateur bands.

An accurately calibrated HF receiver is one of the simplest approaches to monitoring transmitter output signals. One way to ensure the calibration accuracy of the receiver involves the use of a frequency marker or frequency standard. If the receiver does not have this capability, an external frequency standard can be assembled using the crystal oscillator circuits described in Chap. 7.

A general-coverage receiver is useful in determining the presence of spurious signals outside of the amateur bands. For example, a 0.5-

to 30-MHz receiver is capable of tuning to spurious signals within this frequency range. However, if the receiver is limited to amateur band coverage, it cannot be used to detect spurious signals outside of the amateur bands. Other means, such as an absorption wave meter, may be employed for this purpose.

Frequency counters

The development of digital integrated-circuit chips, such as the 7400 TTL series, has made possible low-cost, highly accurate frequency counters. When calibrated with the precise frequency transmissions from the National Bureau of Standard's WWV and WWVH transmitters, the digital frequency counter can exhibit errors of less than 100 Hz in the amateur HF bands. Figure 8.14 shows a typical frequency counter.

The sensitivity of most frequency counters is on the order of 50 to 100 mV (millivolts). This allows you to check the frequency of the low-power oscillator as well as other circuits. However, the output of a high power amplifier should never be connected directly to a frequency counter. The high RF voltages can burn out the input circuit of the counter. When making frequency measurements on a high-power amplifier, a small 2- to 3-turn coil can be used to pick up sufficient RF signal voltage for the counter's input circuit. Many modern trnsmitters and transceivers have an external jack for connecting a frequency counter. Also, the newer "hand-held"frequency counters have a small whip antenna to "sniff" out the frequency signals.

Frequency digital displays are now incorporated in most amateur radios to permit frequency monitoring. This permits a continual display of the operating frequency.

FIGURE 8.14

The MFJ Model-346 Frequency Counter is a precision counter for measuring RF frequencies up to 600 MHz. This portable, battery operated unit features a 10-digit LCD frequency display within 0.1 Hz resolution and a high accuracy of 1 ppm (parts per million). *MFJ Enterprises, Inc..*

Transmitter power measurements

Paragraph 97.313 (Transmitter Power Standards) of the FCC Part 97 Rules and Regulations (see Chap. 2) states the RF power limitations of an amateur radio station. Thus the amateur operator needs some capability to measure the RF power output to the antenna.

The inherent design of modern transmitters and transceivers provides an indication of the expected RF power output. Thus a 2-meter FM, 3-watt Handie-talkie will provid a power out of approximately 3 watts. The same is true for a 100-watt HF transceiver. The power output will usually never exceed 100 watts for any amateur band. However, the power output of some transceivers in the higher RF bands may drop below the stated 100 watts.

Many of the modern RF power amplifiers in the 600- to 1000-watt category use transmitting-type vacuum tubes. These tubes may be triodes or screen-grid tetrodes. (Yes, these types of tubes are still being manufactured.) The power output of these amplifiers is defined as the sum of the plate power and the screen-grid power (for tetrodes) during key-down conditions. The output power of solid state (transistor) power amplifiers can be calculated in a similar manner if access to the collector voltage and current can be determined.

Example. During the key-down operation, a tube-type RF power amplifier has the following characteristics:

dc plate voltage = 600 Vdc

dc plate current = 300 mA dc

dc screen voltage = 300 Vdc

dc screen current = 10 mA dc

filament current = 500 mA ac

Compute the power output of the RF power amplifier and determine if this power is within the legal limits for the HF bands authorized for the Technician operator with 5 WPM certification. This power limit is 200 watts pep. Note: All other amateur operators, including Technician operators, must observe this power limitation when operating on these band segments.

If you have wondered how the output power of RF transistor power amplifiers can be determined, the theoretical answer is relatively simple. Simply multiply the transistor's collector current by the Vcc supply voltage—the answer will be in watts. However, you may experience difficulties in measuring the collector current. Also, the compact construction may invite accidental shorting of components and cause costly transistors to be destroyed. Don't attempt this procedure unless you are an experienced technician.

Solution.

A review of the data given in this example shows that the filament power is not considered when calculating RF power output. Accordingly, this data will be ignored.

Step 1. Calculate the plate power:

$$P_p = V_p \times I_p \qquad \text{(Eq. 8.2)}$$

where

P_p is the plate input power in watts
V_p is the dc plate voltage in volts
I_p is the dc plate current in amperes

Thus,

$$P_p = 600 \text{ V} \times 0.3 \text{ A} = 180 \text{ watts}$$

Step 2. Calculate the screen grid power:

$$P_s = V_s \times I_s$$

where
P_s is the screen power in watts
V_s is the screen grid voltage in volts
I_s is the screen grid current in amperes

Thus,

$$P_s = 300 \text{ V} \times 0.010 \text{ A} = 3 \text{ watts}$$

Step 3. Add the values obtained from Steps 1 and 2 to obtain total output power for the RF power amplifier:

$$P_{total} = 180 + 3 = 183 \text{ watts}$$

FIGURE 8.15

The Bird Model 43 Directional "THRULINE" Wattmeter. This versatile, highly accurate RF wattmeter is extremely popular with amateurs for measuring RF power levels. The Model 43 features a 50-ohm impedance; direct meter reading in watts; low insertion VSWR of less than 1.05 to 1; and plug-in elements for power levels from 100 milliwatts to 10,000 watts full scale and a frequency range of 0.45 to 2300 MHz. *Bird Electronic Corporation, 303 Aurora Road, Salon, OH 44139-2794, Telephone: (216) 248-1200*

Both dc voltmeters and ammeters (or a multimeter set up for dc measurements) can be used to monitor the voltages and currents associated with the operation of the final RF power amplifier stage of the transmitter or transceiver. Many transmitters and transceivers incorporate a dc or RF ammeter for continual monitoring of the final amplifier current during "tune-up" and normal operation. Commercial RF watt meters for "in-line" connection with the antenna transmission line or with dummy loads are available for hams who regularly use high-power RF power amplifiers. Also, when amateur operators want to experiment with RF circuits and antennas, they will want to continually monitor power output levels. Figure 8.15 shows an in-line RF wattmeter which provides true peak power measurements of CW, AM, FM, and SSB modes of operation. Many antenna tuners (covered in Chap. 10) incorporate power meters for monitoring output RF power levels. Finally, designs for "home-brew" construction of RF power meters are available in amateur publications.

Introduction to Radio Receivers

The antenna connected to the amateur radio receiver in your ham shack may collect hundreds to thousands of radio signals at any one time. The frequencies of these signals may range from the AM broadcast band and below, to the VHF and UHF FM and TV broadcast bands and above. In between these extreme frequency limits are amateur, government, commercial, maritime, and other communications services. The amplitudes of these signals being fed into the antenna terminals of the receiver vary from less than 1 microvolt to many millivolts. Many amateur contacts are made with received signals on the order of a few microvolts.

The purpose of the amateur receiver is twofold: select, amplify, and demodulate the desired radio signal; and reject all other signals. The old amateur expression "If you can't hear 'em, you can't work 'em" is most appropriate for this discussion.

Amateur receivers are available as separate pieces of equipment or as part of a transceiver. Many excellent commercial receivers, new or used, are available to help get you on the air as soon as possible. Figure 9.1 shows a deluxe amateur HF and VHF, all-mode transceiver—a complete amateur station.

FIGURE 9.1

The Icom IC-746 all-mode hf/vhf transceiver. This modern amateur radio transceiver provides a remarkable coverage of the following amateur and related radio frequencies. Receive coverage: 30 kHz–60 MHz and 108 MHz–174 MHx; transmit coverage: all amateur bands from 160 meters to 2 meters. The transmit function provides 100 watts pep on all hf bands and the vhf bands of 6 meters and 2 meters. All-mode operation covers CW, AM, USB, LSB, and FM. Other valuable operator features include digital signal processing (DSP) for selecting desired signals while rejecting unwanted signals; a "band scope" that provides a large multifunction LCD display displaying band conditions around a displayed frequency, received signal amplitude display, and transmission modes; and a built-in hf automatic antenna tuner. The transceiver operates on 13.8 vdc, 20 amperes, maximum power source (battery or external ac power supply) and can be operated as a mobile or fixed station radio.

Sometimes you will find used receivers, in good operating condition, at local ham swap-fests or conventions. Replacement transistors, vacuum tubes, and other parts can usually be obtained from local electronics stores, swap-fasts, and advertisements in amateur radio journals. These old "boat anchors" can give you years of good service, either as the primary receiver or as a back-up receiver in your ham station.

Amateur receivers in kit form are available from many commercial suppliers. Figure 9.2 shows the HR-40, a 40-meter, single-band CW/SSB receiver kit offered by Ramsey Electronics Inc., 793 Canning Parkway, Victor, NY 14564 [telephone: (716) 924-4560]. Other receiver kits are available for HF and VHF amateur bands. A receiver kit , along with a transmitter kit, is an exciting as well as inexpensive way to get on the air as soon as possible.

Definitions

The following definitions of terms related to receivers are important to both the beginner and the licensed amateur radio operator. The Technician exam will include questions on the theory and operation of amateur radio receivers.

Beat frequency oscillator (BFO) The BFO stage in a receiver generates a high-frequency signal that is mixed or heterodyned with the incoming modulated carrier, such as a CW signal. This mixing action produces an audio beat frequency signal that can be used to drive a loudspeaker.

Demodulator (or detector) The demodulator or detector stage in a receiver separates the audio signals from an incoming modulated carrier signal. The key component in a demodulator or detector circuit is usually a silicon diode, a nonlinear device. Older tube radios use a vacuum-tube diode.

Frequency-selective networks Frequency-selective networks are used in receivers to pass or reject RF, IF, or audio signals. These networks may consist of parallel or series LC circuits, quartz crystal filters, mechanical filters, or a combination of these devices.

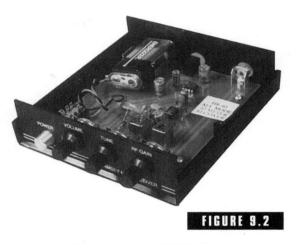

FIGURE 9.2

Ramsey Model HR-40 40-Meter (7 MHz) All Mode Receiver. The HR-40 is a single band, direct conversion receiver for receiving SSB, CW, RTTY, and AM signals on the popular 40-meter band. Available in kit form, this receiver features the Signetics NE-602 Oscillator-Mixer IC for outstanding sensitivity and efficient operation; smooth varactor diode electronic tuning over any 250 kHz segment of the 40-meter band; front panel RF gain, tuning, volume, and on-off controls; an LM386 audio amplifier IC for providing crisp clean audio to external headphones or speaker; and an internal transistor 9-volt battery. *Ramsey Electronics, Inc.*

Intermediate-frequency (IF) amplifiers The IF amplifier in superheterodyne receivers is a fixed-frequency, high-gain amplifier that exhibits sharp selectivity. The IF amplifier is usually connected between the mixer and demodulator stages of the receiver. Some receivers employ two or more stages of IF amplification.

Local oscillator The local oscillator is used in superheterodyne receivers to generate a high-frequency signal. This signal is mixed with the incoming modulated carrier signal, producing the IF signal. Local oscillators may be tunable or fixed-frequency, depending on how they are used in the receiver.

Product detector   The product detector is incorporated in some CW and SSB receivers to demodulate the incoming modulated carrier signal.

Receiver   In communications terms the receiver is a device for receiving selected radio signals from an antenna and processing these signals for the required output form.

Receiver overload   This condition occurs when the antenna collects a strong signal and desensitizes the receiver and the desired signal is distorted or no longer heard. Receiver overload can occur regardless of the frequency the receiver is tuned to. Front-end overload is another name given to receiver overload.

Radio-frequency (RF) amplifier   The RF amplifier used in most receivers amplifies the incoming RF signal from the antenna. The RF amplifier provides for some selectivity and limited gain characteristics.

Superheterodyne receiver   The superheterodyne receiver combines the incoming RF signal with the output from a local oscillator to produce an intermediate-frequency signal. The IF signal is amplified , demodulated, and converted to an appropriate output signal, such as the audio signal required by headphones or speakers.

Tuned radio-frequency (TRF)   receiver An early form of radio receiver, the TRF receiver employs one or more stages of RF amplification followed by a demodulator and audio amplifier. Usually, the TRF receiver does not provide the gain and selectivity offered by the superheterodyne receiver.

Receiver Basics

Amateur radio receivers must be able to pick up extremely weak signals, separate these signals from noise and the many strong signals always present on the air, have adequate frequency stability to hold to the desired signals, and demodulate the radio signals for reproduction.

In general, receiver performance can be rated in terms of sensitivity, selectivity, and frequency stability. Additional factors related to the operational capability of the receiver include frequency coverage, demodulation or detector modes (CW, AM, SSB, or FM), frequency readout, and power requirements.

Although receivers vary in complexity and performance, they all have four basic functions: reception, selection, demodulation, and reproduction. These functions are illustrated in the simple receiver

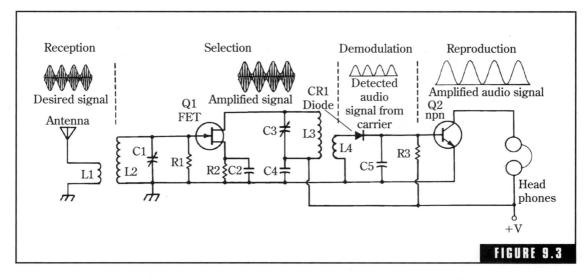

FIGURE 9.3

A simple "tuned radio frequency" (TRF) receiver that incorporates the four basic functions of reception, selection, demodulation, and reproduction.

shown in Fig. 9.4. Some of these functions employ amplifiers to boost the signals to usable levels.

Reception

Reception is the ability of the receiver to process weak signals being collected by the antenna. This is related primarily to the sensitivity of the receiver. Most receivers contain several amplifier stages to produce the large amounts of gain required to boost the signals from levels of microvolts to volts. Thus, the total gain of a receiver may be on the order of a million to one.

Factors influencing receiver sensitivity include electrical noise generated in the antenna circuits and within the receiver, particularly in the front-end or RF stages. Electrical noise is produced by thermal agitation of molecules in a conductor and within amplifying devices such as transistors and tubes. In general, electrical noise is independent of frequency and increases with temperature and bandwidth.

Any noise present at the antenna terminals of a receiver is amplified along with the signal being received. Thus, there is a limit to the sensitivity that a receiver can possess. Most modern receivers have a sensitivity of 1 microvolt or less. Many receiver specifications define

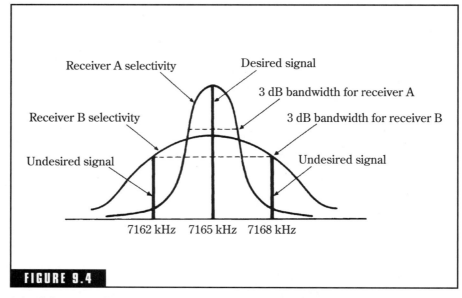

FIGURE 9.4

Selectivity curves for two HF receivers.

the sensitivity in terms of microvolts for a 10-dB (S + N)/N (signal plus noise/noise). For example, a sensitivity of 0.5 microvolt for a 10-dB (S + N)/N simply means that the amplitude of the noise generated in the receiver is about 10 dB below a signal level of 0.5 microvolt. Stated in an alternate manner, the "noise floor" of the receiver is about 0.16 microvolts. Any signal below this level will be masked by the noise and will not be heard in the speaker.

For HF frequencies below about 20 to 30 MHz, external noise (atmospheric and man-made) is usually well above internal receiver noise. Most HF receivers have an adequate sensitivity to handle signals in this frequency range.

Selection or selectivity

Selection or selectivity is the ability of the receiver to tune to a desired signal and reject unwanted signals with frequencies other than that of the desired signal. A receiver employs high-Q LC tank circuits or special crystal filters to pass only a narrow band of frequencies. Thus, the passband or bandwidth of the receiver is all-important in rejecting unwanted signals. Figure 9.4 illustrates the bandwidths of two receivers. Both receivers are tuned to the same signal at 7165 kHz.

Unwanted CW signals are also present at 7162 and 7168 kHz. The sharp selectivity of receiver A effectively rejects the two unwanted signals. Receiver B has a broad selectivity that passes all three signals, causing interference to the operator.

The selectivity of many modem CW receivers is on the order of several hundred hertz. This is adequate for rejecting most signals close to the frequency of operation. All SSB and AM receivers should have a selectivity of about 2500 Hz for amateur voice operation. Many CW/SSB receivers and transceivers employ switchable filters to permit optimum reception of either type of signal.

Demodulation or detection

Demodulation or detection is the process of separating the low-frequency intelligence (usually an audio signal) from the incoming modulated carrier signal. Solid-state devices or vacuum tubes are used in receiver detector circuits.

The two more common types of detectors used in HF receivers are the diode detector for CW and AM and the product detector for CW or SSB signals. Figure 9.3 illustrates the diode detector. A product detector is shown in Fig. 9. 5. The n-channel FET serves as a mixer where

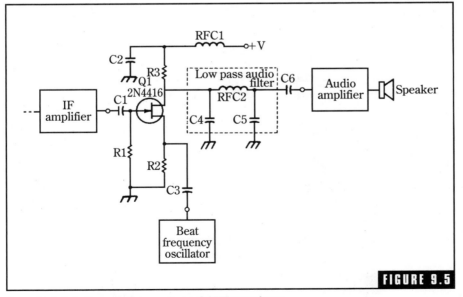

FIGURE 9.5

A product detector using an n-channel JFET transistor.

the IF signals are mixed with the audio-frequency signal from the BFO. The low-pass audio filter consisting of C4RFC2C5 prevents the IF or BFO signals from entering the audio amplifier. The product detector is useful for demodulating either CW or SSB signals.

Reproduction

The final process in receiving a radio signal is to convert the detected signal to an audio signal to drive a speaker or headphones. Usually a high-gain audio amplifier is used to amplify the weak signals from the detector stage. The output of the audio amplifier is then fed to either a speaker or headphones for reproduction.

Most HF receivers have an internal speaker and an output jack for headphone operation. The simple single-stage audio amplifier shown in Fig. 9.3 is suitable for headphone operation. A two- or three-stage audio amplifier is usually required for driving a speaker.

Simple Receivers

The first radio receivers were simple devices consisting of nothing more than a tuned circuit, crystal detector, and headphones. Figure 9.6 shows a typical crystal set that is capable of receiving local AM broadcast stations. Parts for this simple receiver can be obtained from most local electronic parts houses or from discarded AM tube-type or transistor radios. Kits such as the Radio Shack Crystal AM Radio Kit, Catalog No. 28-177, are also available for constructing a crystal set.

The crystal set cannot properly demodulate CW or SSB signals. Also, the sensitivity and selectivity of this type of circuit is not adequate for amateur operation. Additional sensitivity can be obtained by adding an audio amplifier to the output of the crystal set.

Tuned radio-frequency (TRF) receivers

Receiver sensitivity and selectivity can be improved by the addition of one or more stages of radio-frequency amplification. This type of receiver is called the TRF receiver. Figure 9.3 is basically a TRF receiver design. Many commercial AM receivers built in the 1920s and 1930s used the TRF approach. Some of these receivers employed two to four stages of RF amplification to obtain the required sensitivity and selectivity.

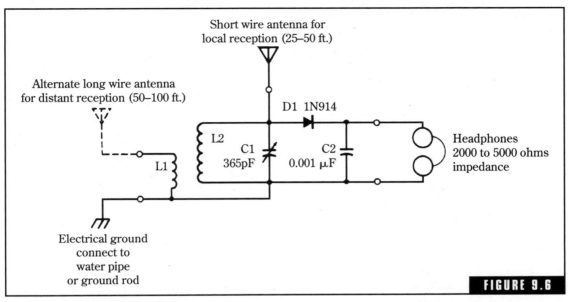

A simple crystal receiver for the beginner.

Direct-conversion receivers

The direct-conversion receiver, shown in Fig. 9.7, is a simple and popular approach for receiving CW and SSB signals. The received signal is applied to the detector stage along with an RF signal generated by the HF oscillator. The frequency of the HF oscillator signal is slightly higher (or lower) than the frequency of the incoming signal to produce an audio beat frequency. For example, if the incoming signal frequency is 7155.0 kHz and the HF oscillator is tuned to 7155.4 kHz, a 400-Hz audio signal is produced by the mixing action in the detector. This 400-Hz signal is applied to the high-gain audio amplifier through a very narrow bandwidth audio filter.

Selectivity in this type of receiver is accomplished by LC tuned networks ahead of the detector stage and the audio filter between the detector and audio amplifier.

Many amateur designs for the direct conversion receiver have been developed over the past few years. Many of these designs employ the Signetics NE602 Oscillator-Mixer IC and the LM386 Audio Amplifier IC. The popular Ramsey HR Series Receiver Kits for 20, 30, 40, and 80 meters employ these two ICs in a direct conversion circuit.

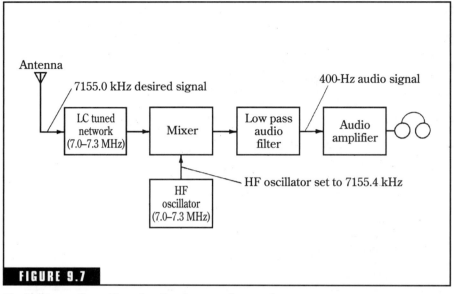

FIGURE 9.7

Block diagram for a 40-meter direct conversion receiver.

Figure 9.8 shows the circuit for the HR-40 Direct Conversion Receiver for 40 meters. Here's how this circuit works. The incoming signal from the antenna is coupled through C5 and the RF gain control R1, to antenna input transformer L1. (A 10.2-MHz IF transformer, detuned to cover the 40-meter band, can be substituted.) This "tuned" transformer peaks the desired signal and applies it to the NE602's mixer section. The shielded oscillator coil, L2, along with varactor diode D1, R2, R5, C1, C2, C3, and C4, form an oscillator network with the NE602's internal oscillator. Rotating R2 varies the oscillator's frequency over a tuning range of about 250 kHz. The NE602 mixes the incoming RF signal with the signal from the internal oscillator to produce an audio signal at Pin 4.

The output audio signal is coupled to the LM 386 audio amplifier via coupling capacitor C8 and volume control R3. Capacitor C9 boosts the voltage gain of the LM386 to about 50. The high-level audio output is coupled from Pin 5, U2, to external headphones or a low impedance (4 to 8 ohms) speaker through coupling capacitor C12 and output connector J2.NE602.

An internal 9-volt transistor battery (or an external power source of about 9–12 Vdc, 60 mA) is used to power the HR-40. Note that zener diode D2, along with C10 and R4, form a 6.2-Vdc voltage regulator network to improve stability of the NE602's internal oscillator circuit.

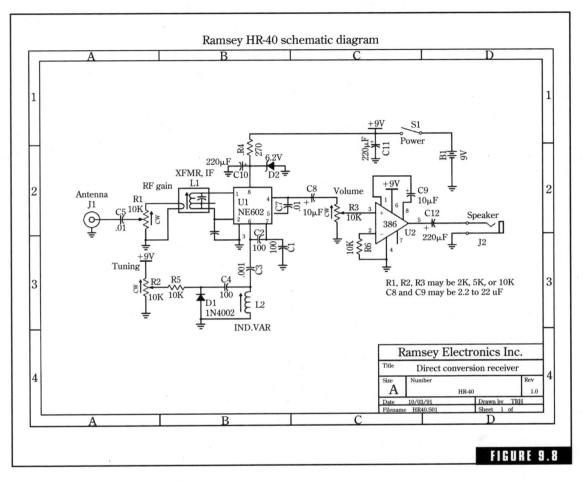

Schematic diagram for the Ramsey HR-40 CW/SSB Direct Conversion Receiver. *Ramsey Electronics Inc.*

The Superheterodyne Receiver

The superheterodyne receiver, developed in the early 1930s, eliminated most of the problems experienced with earlier types of receivers. Today, the superheterodyne receiver is used in virtually all types of radio communications services, including amateur, commercial two-way radio, and AM, FM, and TV broadcast receivers.

The primary difference between the TRF and superheterodyne receivers is the conversion of the incoming RF signal to an intermediate-frequency (IF) signal by the superheterodyne receiver. Figure 9.9

shows a block diagram for a CW superheterodyne receiver. The addition of the IF amplifier stage provides for controlled high gain and narrow bandwidth selectivity at the IF signal frequency.

To understand the operation of the superheterodyne receiver, each stage is discussed in detail. The superheterodyne receiver shown in Fig. 9.9 indicates operation in the 80-meter band. However, any amateur band could have been selected for this example. Band switching circuits in the RF and local-oscillator stages provide for operation on 80 through 10 meters in most amateur HF receivers and transceivers. Sometimes the tuning capacitors in the RF and local-oscillator stages are ganged for single-knob tuning to the desired frequency of operation. For example, the RF stage in Fig. 9.9 can be tuned from 3500 to 4000 kHz while the local-oscillator frequency ranges from 3955 to 4455 kHz. At any frequency within the 80-meter band, the local-oscillator frequency is 455 kHz above that of the incoming RF signal.

When switch S1 in Fig. 9.9 is set to the CW/SSB position, either CW or SSB signals can be received. In the AM position, the beat-frequency oscillator is disabled and the receiver will receive conventional AM signals. An automatic gain control (AGC) amplifier is

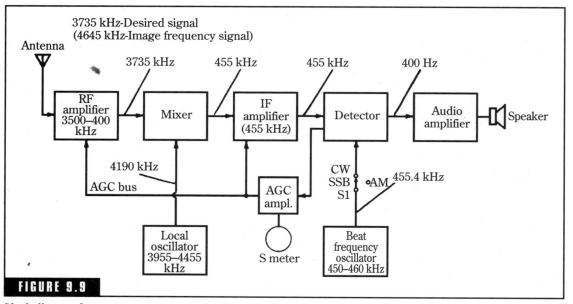

FIGURE 9.9

Block diagram for an 80-meter superheterodyne receiver.

included to compensate for variations in the incoming signal level. The output of the AGC amplifier is a dc voltage that controls the gain of the IF and RF amplifier stages. The S meter is also connected to the AGC amplifier stage to show relative levels of incoming signals.

RF amplifiers

The RF amplifier contains LC tuned circuits that provide for some selectivity and limited gain at the desired signal frequency. The RF amplifier also provides two additional advantages in the superheterodyne receiver. First, it isolates the mixer and local-oscillator stages from the antenna circuits, preventing radiation of the local-oscillator signal from the antenna.

The second advantage of the RF amplifier is that it attenuates unwanted image signals appearing at the antenna terminals. The image signal frequency differs from the desired signal frequency by twice the IF frequency. In Fig. 9.9, the desired signal frequency is 3735 kHz and the image frequency is 4645 kHz (or 3735 kHz plus 2×455 kHz). If allowed to reach the mixer stage, the 4645-kHz image signal would mix with the 4190-kHz local-oscillator signal and produce another 455-kHz signal into the IF amplifier. The ratio of the desired signal level to that of the image frequency signal level is called the image rejection ratio.

Local oscillators

The local oscillator generates a constant-amplitude sine-wave signal whose frequency differs from the frequency of the incoming carrier signal by an amount equal to the IF frequency. The local-oscillator frequency can be positioned either above or below the incoming carrier frequency. This selection is determined by the frequency coverage of the receiver and any requirements for ganged tuning of the RF and local-oscillator stages. Most local oscillators in AM broadcast receivers and the lower bands in amateur HF receivers generate a frequency above that of the incoming carrier frequency.

The tunable local oscillators in superheterodyne receivers must possess adequate frequency stability for the intended operation. For example, if the local-oscillator frequency drifts more than several hundred hertz, this would shift the incoming signal out of the passband of the receiver. For single-sideband operation, a shift of 50 to

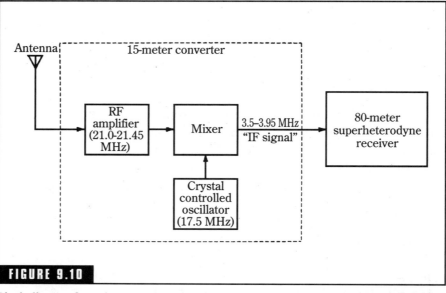

Block diagram for a down-converter that converts the 15-meter band to the 40-meter band.

100 Hz in the local-oscillator frequency can result in unintelligible speech at the output.

Frequency stability in the lower amateur HF bands (80 and 40 meters) can usually be achieved by the use of rigid mechanical construction techniques, temperature-compensating components, and a regulated power supply. However, it is difficult to design LC oscillators for stable operation at frequencies above these bands. Receivers that employ tunable local oscillators in the higher HF bands represent a compromise between operational performance and cost.

One approach to achieving frequency stability in the higher HF bands is the use of crystal-controlled down converters or frequency translators. Figure 9-10 illustrates a typical 15-meter converter that translates the 15-meter band down to 3.5 to 3.95 MHz. In this manner, the 80-meter superheterodyne receiver acts as a variable-IF receiver. Receivers that use this approach are called double-conversion superheterodyne receivers. Most of these types of receivers contain crystal-controlled converters for the amateur HF bands above 40 meters. On 80 and 40 meters, the receiver employs a single-conversion approach. An additional advantage of the double-conversion receiver is improved image rejection.

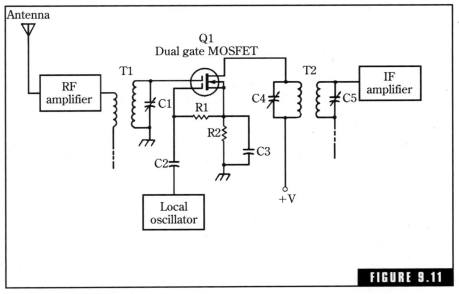

FIGURE 9.11

A dual-gate MOSFET transistor mixer circuit used in many HF receivers.

Mixers

The purpose of the mixer stage is to convert the frequency of the incoming carrier signal to that of the IF amplifier. The action of the mixer is illustrated in Fig. 8.11 in Chapter 8. The mixer produces four major output signals when driven by two input signals of different frequencies, f_1 and f_2:

f_1 = an original input signal frequency,

f_2 = an original input signal frequency,

$f_1 + f_2$ = sum of the two input frequencies,

$f_1 - f_2$ = the difference of the two input frequencies.

Only one of the four mixer output frequencies is used in the superheterodyne receiver, either $f_1 + f_2$ or $f_1 - f_2$. The remaining three mixer output signals can cause interference within the receiver if proper design steps are not taken.

Tubes, diodes, transistors, or special integrated-circuit chips are used as the active elements in mixer circuits. A typical MOSFET transistor mixer circuit is shown in Fig. 9.11. Note that transformer T1 is tuned to the frequency of the incoming signal while transformer T2 is tuned to the IF frequency.

Regenerative-detector receivers

The regenerative-detector receiver, shown in Fig. 9.12 in a vacuum tube circuit, is a simple but highly sensitive receiver. Easy to construct, this circuit has been popular with amateurs since the 1920s and 1930s. Some versions of this receiver employed an RF amplifier stage to improve selectivity and an audio amplifier to boost the audio output for speaker operation. Plug-in coils were used for band switching.

The regenerative-detector circuit employs regenerative feedback from the plate to the grid elements by inductive coupling from L3 to L2. The amount of feedback is controlled by C3. For CW and SSB operation, the detector circuit is made to oscillate at a frequency slightly different from that of the incoming signal.

The regenerative-detector receiver has been largely replaced by the more selective and stable superheterodyne receiver for most amateur operation. However, the simplicity and low cost of the "gennie" regenerative-detector circuit is very attractive to the experimenter and newcomer to amateur radio.

IF amplifiers

The characteristics of the IF amplifier in a superheterodyne receiver are best described in terms of high gain and narrow selectivity. Generally speaking, the gain and selectivity of a receiver are determined by the IF amplifier.

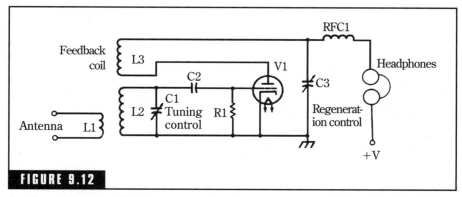

FIGURE 9.12

One form of regenerative-detector receiver.

The selectivity of the IF amplifier must be equal to the bandwidth of the incoming modulated RF signal. If a greater selectivity is present, any adjacent-frequency signal will be passed through the IF amplifier and cause interference to the operator. On the other hand, if the selectivity is too narrow, some of the sidebands contained in the modulated RF signal will be cut. This results in a loss of fidelity when the audio portion of the signal is reproduced by the speaker or headphones.

The optimum selectivity for SSB reception is about 2300 to 2500 Hz. Although some of the higher sidebands associated with speech signals extend beyond 2500 Hz, the loss of these sidebands does not materially affect the sounds or information being conveyed by the transmitting operator.

A selectivity of about 400 to 500 Hz is adequate for CW operation. This narrow selectivity helps to reject any adjacent-frequency CW signal that could interfere with reception of the desired signal.

The more elaborate amateur HF receivers employ two or more stages of IF amplification preceded by a highly selective crystal or mechanical filter. With this arrangement, LC tuned circuits or IF transformers are used for interstage coupling.

The selection of the IF frequency is determined by several factors that include: gain, selectivity, and image signal rejection. For the lower HF amateur bands (80 and 40 meters), 455 kHz is the IF frequency used in many receivers. The IF amplifiers operating at this frequency can provide excellent gain and selectivity. Mechanical filters are available for 455-kHz operation with selectivities of about 400 to 2500 Hz.

Receivers with an IF frequency of 455 kHz generally do not have good image frequency rejection in the higher amateur HF bands (20 meters and above). For operation above 40 meters, IF frequencies between 1.5 and 10 MHz are used to provide for the required image frequency rejection. Crystal filters are employed in the better amateur HF receivers and transceivers to obtain the required selectivity. One popular IF frequency used in many amateur HF receivers and transceivers is 9 MHz. Four- and eight-pole crystal filters are available from commercial sources that have selectivities of 2500 Hz for SSB and 500 Hz for CW modes of operation. The number of poles associated with a crystal filter is related to the number of individual crystal elements used to obtain the desired selectivity characteristics.

Detectors and beat-frequency oscillators

Detection or demodulation is defined as the process of separating the audio-frequency components from the modulated carrier signal. Detectors in superheterodyne receivers are often called second detectors—the mixer stage can be considered as being the first detector.

For conventional AM signals, a diode detector is adequate for removing the audio signals from the modulated IF carrier signal. This process was described earlier in this chapter.

The reception of CW and SSB signals requires the mixing of the modulated IF signal with a signal from the beat-frequency oscillator (BFO). The BFO is used to reinsert a carrier into the IF signal for demodulation purposes. For CW reception, the BFO frequency is adjusted slightly higher or lower than that of the IF frequency to produce an audio beat-frequency signal. For example, if the IF frequency is 455 kHz, the BFO frequency is adjusted to 456 or 454 kHz in order to produce a 1000-Hz beat frequency. This audio signal is then applied to the audio amplifier stage to produce an output in the speaker or headphones.

Automatic gain control

The purpose of an automatic-gain-control circuit is to maintain a constant receiver-output level despite variations in the input signal strength. The HF radio waves being propagated through the ionosphere are alternately attenuated and reinforced by an effect known as fading. This causes the received signal strength at the antenna terminals to vary over a wide range of values. Because the rectified signal voltage at the detector stage is proportional to the amplitude of the received signal, a portion of this voltage can be used to control the gain of the receiver. For receivers employing vacuum tubes or npn transistors in the stages preceding the detector, a negative voltage is fed back to these devices for reducing the gain. Amplifiers and mixers employing pnp transistors require a positive voltage to control the amplification of these stages.

Some receivers, particularly the better solid-state types, include an AGC amplifier to obtain more control over the receiver gain characteristics. These AGC circuits might have different time constants to aid in receiving different types of signals, such as AM and SSB voice and CW signals. The time constant of an AGC circuit describes the length of time that gain control is maintained after the transmitted signal is terminated. For example, without an appreciable time constant in the AGC circuit, the normal pauses in SSB voice transmissions would allow the receiver to immediately resume full gain characteristics. This would produce an annoying burst of noise during the intervals between voice signals.

S meters

The S or signal-strength meters are provided in some receivers and transceivers to indicate relative strength of the incoming signal. Usually a portion of the rectified IF signal from the detector stage is applied to a microammeter or milliammeter. If the receiver has an AGC amplifier, this stage can be used to drive the S meter in addition to controlling the gain of the receiver.

Most S meters are calibrated in terms of S units (S-1 to S-9) up to about the midscale reading. Figure 9.13 shows a typical S-meter scale. One S unit represents approximately a 6-decibel change in received signal strength. A midscale reading, or S-9, is intended to indicate a signal strength of about 50 microvolts. The upper half of the S-meter scale is calibrated in decibels above S-9, usually up to 60 decibels. Thus, a signal strength reading of 60 decibels indicates that the strength of the received signal is 60 dB above 50 microvolts, or about 50 millivolts.

The S meter is seldom accurate because many factors influence its operation. However, it is very useful in determining the relative signal strength of incoming signals. The S meter is also useful in checking or aligning the receiver. In many transceivers, the S meter also serves to indicate the status of transmitter functions such as final RF amplifier current and RF output power.

Interference and Receiver Limitations

Any radio receiver can have difficulty in receiving a desired signal due to three fac-

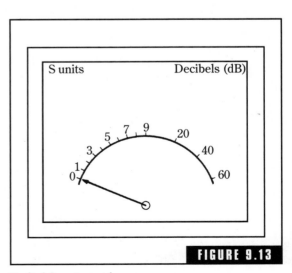

FIGURE 9.13

Typical S-meter scale.

tors: external noise, internal noise, and interfering signals. As stated earlier in this chapter, external noise in the HF frequencies, particularly below about 20 MHz, is much higher than the internal noise generated by HF receivers. Only above approximately 20 MHz does internal receiver noise become a problem in receiving extremely weak signals.

The majority of the internal receiver noise is generated in the first stage, either an RF amplifier or mixer stage. Much research has been expended in developing low-noise circuits and components to reduce internal receiver noise to a minimum level.

External noise can cause problems in receiving weak signals in two ways. First, the noise collected by the antenna can simply mask the incoming desired signal. If the desired signal is near or below the incoming noise level, reception is virtually impossible. Some experienced CW operators can copy CW signals through excessive noise levels. However, voice and other amateur signals are unintelligible under these conditions.

Receiver Overload

Assuming the desired signal is weak but well above external and receiver noise levels, we might still have difficulty in receiving the desired signal. Strong RF signals of widely different frequencies can be collected by the antenna and cause receiver desensitization or front-end overload. For example, strong signals at the image frequency, can cause direct interference with the desired signal as well as reducing the receiver gain by developing a high AGC (automatic gain control) voltage.

Other strong signals, at frequencies completely different from the desired receiving and image frequencies, can produce overloading effects in the first stage. This front-end loading effect reduces receiver gain and the desired signal can be distorted or lost. Some receivers have attenuation networks that can be switched into the input to the first stage. This reduces the overloading effect on the first stage and often permits normal amplification of the desired signal. To check for this type of interference, tune the receiver to different frequencies. If the receiver still exhibits poor gain and distorted audio output signals, this is an indication that a strong signal is causing front-end overloading of the receiver.

Harmonic Signals

Strong undesired signals collected by the antenna might possess a harmonic frequency equal or close to the desired signal frequency. For example, a receiver is tuned to 20 MHz and can experience interference from a strong 5 MHz signal. Note that 20 MHz represents the 4th harmonic of 5 MHz. In order to eliminate interference of this type, filters can be installed at the receiver antenna terminals to reduce these strong signals to insignificant levels.

Constructing the "Sudden" 160- to 20-Meter Direct Conversion Receiver

The receiver kits described earlier in this chapter make excellent HF receivers for the beginning amateur operator. However, some amateurs prefer to build a receiver from scratch. The "Sudden" Receiver described in this construction project was originally developed and constructed by the Rev. George Dobbs [G3RJV, St. Aidan's Vicarage, 498 Manchester Rd., Rochdale, Lancs, OL11 3HE, Great Britain.] Rev. Dobbs' versatile design of this receiver permits it to be configured for any amateur band between 160 to 20 meters (160, 80, 40, 30, and 20 meters). He documented the construction of the Sudden Receiver in the October 1991 issue of 73 Amateur Radio Today. You can assemble this proven receiver in an evening or two of enjoyable effort.

Figure 9.14 shows a schematic diagram for the Sudden Direct Conversion Receiver. The receiver is a simple design based on two integrated circuits (ICs) and three inductors. A potentiometer, R1, acts as an RF control to prevent signal input overload, particularly on the lower HF bands. The T1/C1/C2/C3/T2 tuned network forms a bandpass filter for the particular amateur band involved. The NE602 Oscillator-Mixer IC provides the RF mixer and local oscillator functions for direct conversion of the SSB or CW signals to an audio signal. The local oscillator section of the NE602 is connected to a tuned Colpitts oscillator circuit tuned by T3, C10, and the Variable Capacitor VC1. Note that the receiver tuning range is determined by VC1. Table 9.1 provides a list of the frequency determining components that permits you to configure the Sudden Receiver for a particular amateur band.

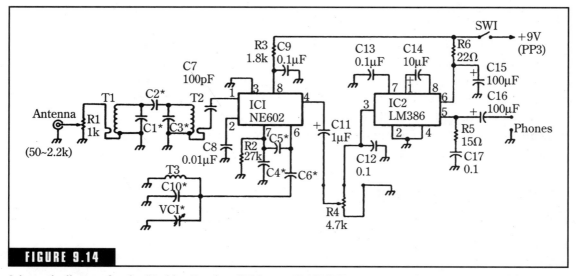

Schematic diagram for the "Sudden Receiver." This two-IC SSB/CW amateur receiver covering the 160- to 20-meter bands was developed by the Rev. George Dobbs, G3RJV, Great Britain. *Rev. George Dobbs, G3RJV, and 73 Amateur Radio Today, October 1991, WGE Center, Forest Road, Hancock, NH 03449, Telephone: (603) 525-4201*

Table 9.1 Frequency Determining Components for the Sudden Receiver

Band	C1	C2	C3	T1	T2
160	220 pF	10 pF	220 pF	BKXN-K3333R	BKXN-K3333R
80	47 pF	3 pF	47 pF	BKXN-K3333R	BKXN-K3333R
40	100 pF	8.2 pF	100 pF	BKXN-K3334R	BKXN-K3334R
30	47 pF	3 pF	47 pF	BKXN-K3334R	BKXN-K3334R
20	100 pF	3 pF	100 pF	BKXN-K3335R	BKXN-K3335R

Band	VC1 + C10		C4	C5	C6	T3
160	All Sections	+ 100 pF	0.001 mF	0.001 mF	560 pF	BKXN-K3333R
80	All Sections	+ 100 pF	0.001 mF	0.001 mF	560 pF	BKXN-K3333R
40	1 Section	+ 47 pF	560 pF	560 pF	270 pF	BKXN-K4173A0
30	1 Section	+ 68 pF	680 pF	680 pF	220 pF	BKXN-K3335R
20	1 Section	+ 68 pF	220 pF	220 pF	68 pF	BKXN-K3335R

Source: *Rev. George Dobbs, G3RJV, and 73 Amateur Radio Today, October 1991*

The detected audio output from the NE602 IC is applied to the LM386 Audio Amplifier via Coupling Capacitor C11 and the Volume Control R4. The LM386 Amplifier IC is configured for a voltage gain of about 200 and provides adequate audio output via Coupling Capacitor C16 for headphone reception. The LM386 Audio Amplifier provides the necessary gain to amplify the detected audio to usable levels for headphone listening.

Parts for the Sudden Receiver are listed in Table 9.2. Most of these parts are readily available in local Radio Shack stores or electronic supply houses. The Variable Tuning Capacitor VC1 can be difficult to locate from local suppliers. You can order a variable tuning capacitor from the source listed in the parts list or salvage a surplus variable capacitor from an abandoned AM radio.

Table 9.2 Parts List for the Sudden receiver

Resistors

R1	1 k	potentiometer
R2	27 k	resistor
R3	1.8 k	resistor
R4	4.7 k	potentiometer
R5	15 ohm	resistor
R6	22 ohm	resistor

All resistors are 1/4 watt

Capacitors

C1–C6	See the Table	
C7	100	pF capacitor
C8	0.01	mF capacitor
C9, C12, C13, C16, C17	0.1	mF capacitor
C10	See the Table	
C11	1.0 mF/35 V tantalum capacitor	
C14	10 mF/16 V tantalum capacitor	
C15	100 mF/25 V electrolytic capacitor	

(continued)

Table 9.2 Parts List for the Sudden receiver (*Continued*)

Coils, ICs and Misc.	
T1, T2, T3	See the Table
IC1	NE602
IC2	LM386
SW1	SPST switch
VC1	Variable tuning capacitor (three sections: 10 pF, 10 pF, and 20 pF); see the table and the note below.

A kit of all parts including the PC board, the TOKO coils and the tuning capacitor is available in the US for $29.95 + $3 shipping from Kanga US, c/o Bill Kelsey N8ET, 3521 Spring Lake Dr., Findlay, OH 45840. Tel. (419) 423-5643, 7–11 P.M. eastern. Kanga US will supply the blank PC board separately for $6 + $3 shipping. The complete kit is also available overseas from Kanga Products, 3 Limes Road, Folkestone, Kent CT19 4AU, Great Britain.

Variable tuning capacitor VC1 is also available as part #2311007 from A.R.E. Surplus, 15272 S.R. 12E, Findlay, OH 45840. Tel. (419) 422-1558.

The TOKO coils are also available from Penstock at (800) 736-7862.

Source: *Rev. George Dobbs, G3RJV, and 73 Amateur Radio Today, October 1991*

The Sudden Receiver can be constructed using a variety of construction techniques. Methods of construction for this receiver's circuitry include predrilled "prototyping" boards (Radio Shack P/N 276-170, 276-147, or 2726-158) or "perfboard" material (Radio Shack 276-1394).

Receiver checkout and alignment is straightforward with this type of receiver. You should check the audio section for proper operation before attempting test and alignment of the RF sections. With power applied to the LM386 IC, you should hear a loud hum in the headphones when you touch the input terminal at PIN 3. The availability of an audio oscillator or function generator will permit a more detailed analysis of the operation of the LM386.

Alignment of the RF and local oscillator sections will require a source of RF signals, either from an antenna or, better still, from an RF signal generator. First, you must get the local oscillator tuned to the desired HF amateur band. Connect a signal from an RF signal generator to C7 or an antenna to the antenna input connector. A QRP CW transmitter makes an excellent alternate signal source for testing and aligning the receiver. Adjust the core of T3 until the signal is heard in the headphones. Now you can align the RF input bandpass filter by peaking the adjustments of T1 and T2. The final adjustment involves the proper band coverage as dictated by the local oscillator tuning range. You can use a calibrating RF signal in the center of the band (adjust the Tuning Capacitor VC1 to the center of its range) and then adjust T3 for maximum reception. An alternate method is to use an RF calibrating signal at the high end (or the low end) of the band, adjust the tuning of the Tuning Capacitor VC1 to minimum (or maximum) capacitance, and adjust T3 for maximum reception. Your "Sudden Receiver" is now ready for operation. With a companion QRP CW transmitter, you are ready to get on the air and make exciting contacts with fellow hams.

FM Receivers

FM receiver design and construction is very similar to that of the CW/AM/SSB receivers described in earlier sections. Today's FM receivers use the superheterodyne concept. Thus the RF, IF, and audio amplifier and the local oscillator stages for each type of receiver are virtually identical. To build an FM receiver, you need three additional stages: a limiter IF stage, an FM detector or demodulator, and a deemphasis stage. By now, you have noticed that most of the newer amateur radios have an "all-mode" capability. Thus CW, AM, SSB, and FM modes of communications can be handled by one amateur radio.

Figure 9.15 shows a block diagram of an FM receiver. RF signals collected by the antenna are applied to the RF amplifier stage. This stage performs two important functions. First, the desired RF signal, sometimes only a few microvolts in amplitude, is amplified before being fed to the mixer stage. The second function is that the RF amplifier rejects or suppresses unwanted signals near the operating frequency.

The mixer stage heterodynes or mixes the incoming operating RF frequency signal with the local oscillator signal. This action produces the two original signals and the sum and difference frequencies. For example, let the incoming RF FM signal frequency be 146.00 MHz and the local oscillator signal frequency be tuned to 135.3 MHz. The mixer out signals will have four combinations of the above frequencies: 146.0 MHz, 135.3 MHz, 281.3 MHz (from 146.0 + 135.3 MHz), and 10.7 MHz (from 146.0 − 135.3 MHz). The new 10.7-MHz signal is the only one accepted by the tuned 10.7-MHz IF amplifier. Note that both the sum and difference frequency signals possess the original FM modulation characteristics. This process converts the FM signal to a

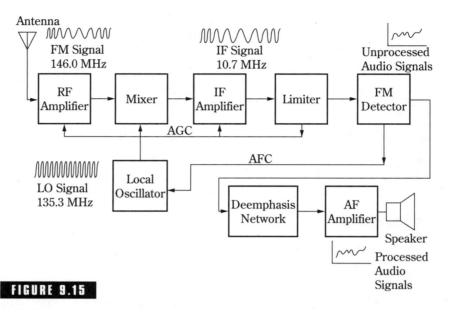

FIGURE 9.15

Block diagram for a typical FM receiver using the superheterodyne concept. This design is applicable for most FM receiver requirements: broadcast band (88–108 MHz), commercial radiotelephone, and amateur operation on authorized FM emissions on the amateur bands.

new FM signal frequency. Thus, tuning the FM receiver to any frequency in the design coverage (for example, the 2-meter band of 144.0 to 148 MHz) always produces an IF signal of 10.7 MHz.

Now check the operating frequency of the IF amplifier. It is 10.7 MHz! Since the IF amplifier can operate at a single frequency (plus and minus the FM modulation sidebands), it can be designed for very high gain and sharp selectivity. 10.7 MHz is a standard IF frequency used in most FM receivers. Some FM receivers will have one IF amplifier stage while others will have two or more stages.

The limiter stage, tuned to the 10.7 MHz IF frequency, removes all amplitude variations of the received signal as well as electrical noise. This is accomplished by adjusting the limiter amplifier to operate in the saturated mode. A parallel LC circuit, tuned to 10.7 MHz, restores the distorted signal to a perfect sinewave with no amplitude variations. The limiter stage is responsible for producing sharp, clear voice communications!

Unlike the AM detector, the FM detector detects variations in the frequency of the incoming FM signal. There are many types of FM detector, each being developed from the electronics state of the art—from vacuum tubes to transistors and finally to integrated circuits. Most modern FM communications receivers use a phase locked loop (PPL), a veritable workhorse in current electronics technology. The PLL is also used in local oscillator stages of superheterodyne receivers to provide precise frequencies and in FM modulator stages.

A deemphasis network is required to restore the original frequency response of the audio (or data) modulating signals. For improved high-frequency response and reducing noise, the higher frequencies of the audio modulating signal are boosted while the lower frequencies are attenuated prior to transmission. A frequency-sensitive circuit is needed in the receiver to restore the audio signal to its original condition. Thus a "copy" of the original audio signal is applied to the AF amplifier. Virtually all amateur receivers (or the receive section of transceivers) provide an audio jack for headphone use.

FM receivers usually include two important features to provide clear and constant audio output signals. First, an automatic gain control (AGC) controls the gain of the RF and IF amplifiers to compensate for varying signal levels from the antenna. The automatic frequency control (AFC) control signal, generated by the FM detector stage, keeps the local oscillator frequency at the proper frequency for optimum reception.

Although not indicated by this general block diagram for FM receivers, some communications receivers employ a second conversion stage to down-convert the 10.7-MHz IF signal to a frequency of 455 kHz. Then the 455 kHz IF amplifiers provide for narrow frequency response, which eliminates any remaining image frequency problems and adjacent interfering signals.

All about Transmission Lines and Antennas

In the previous two chapters we covered transmitters and receivers Now let us connect this equipment to a transmission line and antenna. We will consider a systems approach because all of these components must be properly matched for successful operation.

The transmission line is simply a device for conducting radio-frequency energy between the amateur transmitter or receiver and the antenna. The distances between these two locations will vary depending on where the amateur operator can install the transmitter and receiver and the antenna. At HF frequencies, losses in the transmission line will be negligible at distances between 50 to 100 feet (about 150 to 300 meters). However, if we increase distance or operating frequency, losses in the transmission line might degrade system operation.

In the transmit mode, the transmission line must be capable of delivering many watts of RF energy from the transmitter to the antenna with negligible loss. In most cases, the same antenna and transmission line is used for both transmitting and receiving. During the receiving mode, the transmission line must be able to deliver low-level RF signals of a few micromicrowatts with minimum attenuation. A loss in the transmission line can push these small signals into the noise level, making reception impossible.

Definitions

Antenna An antenna is a conductor or group of conductors used for radiating RF energy into space or collecting radiated RF energy from space. In general, any antenna used for transmitting can also be used for receiving purposes.

Antenna input resistance The ideal antenna acts as a resonant circuit and presents a pure resistance to the transmission line. In practice, antennas usually exhibit inductive or capacitive reactance, resulting in a complex impedance. One definition of this input impedance is that it is equal to the voltage across the antenna terminals divided by the current flowing into the antenna. A related term is radiation resistance. The ideal antenna accepts RF energy as if it were a resistor connected across the end of the transmission line. For example, a dipole antenna in free space has a radiation resistance of 73 ohms. To the transmission line, this appears as a 73-ohm resistor.

Antenna tuner Sometimes called a transmatch or antenna matching unit, this device provides for matching the impedance of the transmission line to that of the transmitter or antenna. For example, an antenna tuner can be used to match a 50-ohm transmitter to a 300-ohm transmission line.

Balun A balun is a device for matching or connecting balanced and unbalanced circuits. For example, a balun can be used to match a coaxial cable (an unbalanced transmission line) to a dipole antenna (a balanced electrical device).

Characteristic impedance As applied to transmission lines, this term defines the input impedance characteristics. Characteristic impedance is a function of the lumped inductance and capacitance contained in the transmission line and is approximately equal to L/C. It appears as a resistive element when the transmission line is properly matched to the transmitter and antenna. The popular RG-8/U coaxial cable has a characteristic impedance of 52 ohms.

Coaxial-cable transmission lines This type of transmission line employs an outer conductor that surrounds an inner conductor. An insulating material such as polyethylene separates the two conductors. The outer conductor is normally held at ground potential while the inner conductor is considered as the "hot" lead. A coaxial cable is referred to as an unbalanced line. Typical coaxial cables used in amateur radio stations have a characteristic impedance of 50 to 75 ohms.

Dipole antenna The dipole antenna consists of two "poles" or conductors arranged in a straight line and separated at the center by an insulator. The transmission line is connected to the center of the dipole. A dipole antenna is normally one-half wavelength long from end to end. Sometimes the dipole is called a *center-fed Hertz antenna* or *half-wave doublet antenna*.

Electrical length The electrical length of an antenna is defined in terms of wavelengths. A vertical antenna, for example, may have an electrical length of one-quarter wavelength.

Harmonic operation Some antennas, such as the dipole, can be used at an odd multiple of its operating frequency. For example, a 40-meter dipole antenna will also perform as an antenna in the 15-meter band. Note that 21 MHz is an odd (3rd) harmonic of 7 MHz. The use of a 7-MHz doublet antenna at 21 MHz is called *harmonic operation*.

Multiband antenna A multiband antenna is one designed to operate on two or more bands of frequencies. For example, vertical antennas designed for operation on 80, 40, 20, 15, and 10 meters can be purchased from commercial sources or constructed from antenna designs available in amateur handbooks.

Single-wire transmission lines Used primarily to connect a long-wire antenna to a receiver, this is simply a single-conductor transmission line. It is not used with transmitters because it tends to act as a radiator and presents a varying impedance if allowed to swing with the wind.

Standing waves These are essentially stationary waves of voltage and current that might exist on a transmission line or antenna. Standing waves are useful in antennas because they produce maximum radiation of the RF signal. However, standing waves in a transmission line are undesirable because they represent a mismatch of impedances. This results in a portion of the RF energy delivered to the antenna being reflected back to the transmitter.

Standing-wave ratio (SWR) This defines the ratio of the maximum to minimum voltage or current levels in a transmission line. It is also a function of the power delivered to the antenna and the power being reflected back to the transmitter. An SWR bridge is a device for measuring the SWR of a transmission line connected between a transmitter and an antenna.

Vertical antenna In general, a vertical antenna consists of a vertically positioned conductor, rod, or tower. A vertical quarter-wave antenna that employs a ground plane is called a *Marconi antenna*. The ground plane serves as a mirror, reflecting the radiated waves into space.

Transmission Line Basics

Let us consider the important role that the transmission line plays in the amateur radio station. The function of the transmission line is to transfer RF energy between the transmitting/receiving equipment and the antenna. Figure 10.1 illustrates a typical Technician amateur installation employing an HF transceiver, a 50-ohm coaxial transmission line, and a half-wave doublet antenna. Antennas designed for operation on the 80- to 10-meter bands can be used in this basic configuration.

We have a properly matched RF system in Fig. 10.1. The transmitter output impedance is adjusted to 50 ohms and matches into the 50-ohm coaxial cable. At the antenna end, the 50-ohm cable matches the 50 ohms presented by the antenna. These conditions allow for maximum transfer of RF energy, both in transmit and receive modes of operation. The half-wave doublet antenna shown in Fig. 10.1 is cut for 40-meter operation. However, antennas for any HF band can be substituted. Also, an arbitrary length of 100 feet was selected for the trans-

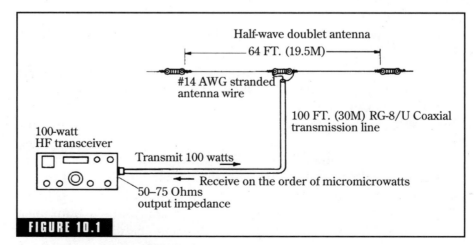

FIGURE 10.1

A typical Novice antenna installation for 40 meters.

mission line. In practice, this length should be as short as possible to reduce losses.

Types of Transmission Lines

Transmission lines come in a variety of types—single wire, parallel two-wire open line, twin lead, twisted pair, and coaxial cable. This discussion is limited to the more common types of transmission lines used by most hams, twin-lead and coaxial lines. Figure 10.2 illustrates these types of transmission lines. From top to bottom, Fig. 10.2 shows 300-ohm heavy-duty twin lead used for TV as well as amateur radio, 72-ohm parallel transmission line, RG-58A/U 50-ohm coaxial cable with UHF-type connector, and RG-8/U 50-ohm foam-type coaxial cable.

Twin-lead lines

Twin-lead line, available with a characteristic impedance of 72 to 300 ohms, was originally developed for television receivers. This line consists of two parallel conductors embedded in a thin ribbon of plastic or polyethylene material. Sometimes called ribbon-type TV cable, it generally was not suitable for transmitting purposes. However, heavy-duty versions with larger sizes of conductors are available for the amateur market.

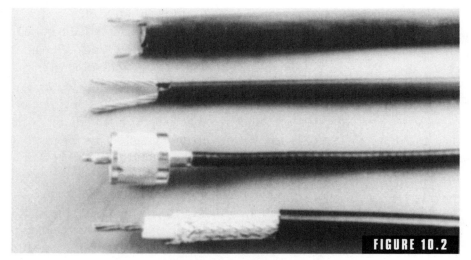

FIGURE 10.2

Typical transmission lines used in amateur radio applications.

The twin-lead cable, a balanced form of transmission line, is used with some special forms of antennas. However, twin-lead lines have several limitations. Deposits of moisture and dirt on the line can change the characteristic impedance, causing a mismatch and loss of efficiency. The twin-lead line cannot be installed close to a metal surface, such as being taped to the side of a tower. Again, this causes a change in the characteristic impedance. The heavy-duty 72-ohm versions are less affected by these limitations because the conductors are spaced closer together to confine the electric field within the dielectric material.

Coaxial transmission lines

A coaxial line, or "coax" cable, is the most common type of transmission line used in amateur radio operations from HF to UHF frequencies. Coax, an unbalanced form of transmission line, consists of an outer conductor surrounding an inner conductor. An insulating material, usually polyethylene or polyfoam, is placed between the conductors. The outer conductor, or braid, serves as a shield and is normally held at ground potential. Figure 10.3 shows the construction used in a typical coaxial cable.

The major advantage of the coaxial line is that the RF field is contained within the outer conductor. Also, the outer insulating jacket is virtually waterproof, permitting outdoor installation. These factors allow coax transmission lines to be installed almost anywhere—taped to towers, placed in metal pipes, or even run underground.

Coax cables are available in many sizes and characteristic impedances. Table 10.1 shows a group of the more common types of coaxial cables used by hams. Attenuation factors have been converted to transmission efficiencies to illustrate how cable length affects line losses. For example, if the transmission line and antenna in Fig. 10.1 is operated on 40 meters, the transmission efficiency is approximately 90 percent. This means that 90 percent of the 100 watts of power delivered to the transmission line, or 90 watts, will be made available to the antenna for radiation into space. Thus, 10 watts of RF power will be dissipated in the transmission line. (This example assumes that the antenna length is adjusted to the proper

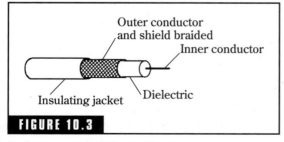

Outer conductor and shield braided

Inner conductor

Insulating jacket

Dielectric

FIGURE 10.3

Basic construction of coaxial cable.

length and that the transmission line is matched to the antenna. For example, an antenna too long or too short can cause additional loss of power actually being radiated into space.) Reducing the line length to 50 feet (or 15 meters) will cut the power loss in the line to about 5 watts.

Both RG-8/U (52 ohms) and RG-11/U (75 ohms) are heavy-duty, low-loss types of coaxial cable suitable for permanent, long-run installations. Figure 10.4 shows RG-8/U foam cable that is available in standard lengths up to 100 feet or more, with RF connectors installed. This type of cable is highly resistant to all weather conditions and should provide many years of service.

Lightweight, smaller coaxial cables, such as RG-58/U (52 ohms) and RG-59/U (73 ohms), are more suitable for short cable runs (such as mobile installations and portable operation) where the higher cable losses are not important. Figure 10.4B shows RG-58/U, which is available in standard lengths with RF connectors installed.

An intermediate size cable, the RG-6, 75-ohm impedance, is adequate for many installations where small size and weight is required.

Standing Waves

The transmission line is said to be matched to the antenna when each impedance is resistive and equal in value. For example, a 50-ohm transmission line connected to a 50-ohm antenna results in a maximum transfer of RF power from the transmitter to the antenna. Thus properly matched to the antenna, the transmission line exhibits a constant impedance at any point along the line. This means that voltage and current amplitudes are constant throughout the line. However, if the antenna is not constructed or adjusted for the required frequency of operation, the antenna will not match properly to the transmission. Extreme cases of mismatch occur when the antenna appears to be either a short or an open circuit. Figure 10.5 shows voltage and current relationships (standing waves along the transmission line) for these two extreme conditions. Note that the transmission line is approximately one wavelength long.

Standing-wave ratios

An antenna that is not matched to the transmission line will cause standing waves to appear on the line. The ratio of maximum to minimum

A

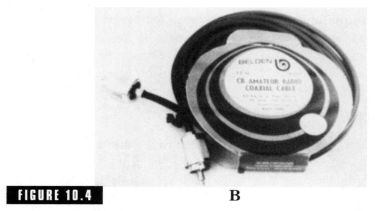

FIGURE 10.4   **B**

Coaxial cables available in standard lengths with RF connectors installed. (A) Heavy-duty RG-8/U is suitable for long runs of transmission lines. (B) The smaller RG-58/U is handy for short cable runs such as modible installations.

levels of voltage (or current) is called the standing-wave ratio (SWR). This can also be defined as the ratio of the impedances involved, or:

$$\text{SWR} = \frac{Z_L}{Z_O} \text{ or } \frac{Z_O}{Z_L} \text{ (whichever is larger)} \qquad \textbf{(Eq. 10.1)}$$

where
    SWR is the standing-wave ratio
    Z_L is the impedance of the load in ohms
    Z_O is the impedance of the transmission line in ohms

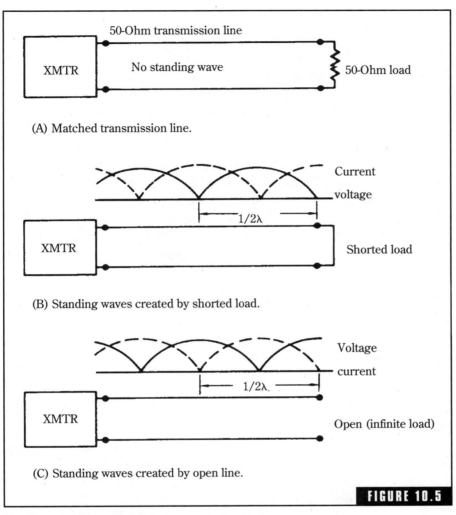

(A) Matched transmission line.

(B) Standing waves created by shorted load.

(C) Standing waves created by open line.

FIGURE 10.5

Standing-wave characteristics of transmission lines.

Example If a 50-ohm coaxial cable is connected to a 75-ohm antenna, what is the resulting SWR?

Solution

$$\text{SWR} = \frac{Z_L}{Z_O} = \frac{75}{50} = \quad 1.5$$

It is customary for hams to express an SWR such as this as "1.5 to 1."

THE MEANING OF SWR

Now that SWR has been defined, let's see how it affects the performance of our radiating system.

Table 10.1 shows us that any transmission line, even when properly matched, has some attenuation or loss. This loss is due mostly to resistance in the conductors and power dissipated in the dielectric.

The ideal SWR of a transmission line is 1 to 1. Any increase in SWR results in additional power loss. For example, a mismatched antenna causes power to be reflected back toward the transmitter. For high values of SWR and large transmission-line losses, power lost in the system can be greater than power radiated by the antenna.

In general, an SWR of 2 to 1 or less results in minimal loss in the radiating system. When receiving a transmitted signal, you can barely notice the difference between an SWR of 2 to 1 and that of 1 to 1.

STANDING-WAVE MEASUREMENTS

Standing-wave ratios can be measured with devices called SWR bridges or SWR meters. You can purchase an SWR meter from most amateur supply houses or build one from plans in ham publications such as the *ARRL Radio Amateur's Handbook*. Figure 10.6 shows a commercial SWR meter that also measures output power levels. Many SWR meters cover a frequency range of about 3 to 150 MHz and work with either 50- or 75-ohm transmission lines.

These devices are particularly useful when installing and adjusting antennas. To measure SWR, simply connect the SWR meter between the transmitter and the transmission line as shown in Fig. 10.7. Because SWR measurements require only a minimum amount of power, you can make SWR measurements without causing interference to other ham operators. After an initial sensitivity adjustment, the SWR can be read directly from the meter scale.

High-Frequency Antennas

The antenna is one of the most important elements in amateur operation. When transmitting, the antenna converts radio-frequency energy into electromagnetic waves for radiation into space. For receiving, the antenna collects electromagnetic waves and converts these waves into electrical energy for input into the receiver. In general, the gain and

TABLE 10.1. Typical coaxial cabletransmission lines used in Amateur Radio installations.

Charst. ype of cable	O.D. imped. ZO ohms	inches (mm)	Attenuation in db/100 feet (30 meters) (also expressed as transmission efficiency in percent)					Notes
			3.5 MHz	7 MHz	14 MHz	21 MHz	28 MHz	
RG-8/U	50	0.405	0.30	0.45	0.66	0.83	0.98	1, 2, 3
		(10.3)	(93)	(90)	(86)	(83)	(80)	
RG-11/U	75	0.405	0.38	0.55	0.80	1.0	1.15	1, 2, 3
		(10.3)	(92)	(88)	(83)	(79)	(77)	
RG-8X	50	0.242	0.56	0.78	1.1	1.4	1.5	4, 5
		(6.1)	(87)	(84)	(78)	(72)	(71)	
RG-58/U	50	0.195	0.68	1.0	1.5	1.9	2.2	2, 3
		(5.0)	(86)	(79)	(71)	(65)	(69)	
RG-59/U	73	0.242	0.64	0.90	1.3	1.6	1.8	1, 2, 3
		(6.1)	(86)	(81)	(74)	(69)	(66)	

Notes:
1. "A" versions (for example, RG-8A/U) have an improved weather resistant which ensures a longer life in outdoor installations.
2. The velocity of propagation factor for these cables is 66 percent.
3. These cables are available with a polyfoam, or "foam," dielectric material to reduce attenuation factors.
4. The velocity of propagation factor for this cable is 78 percent.
5. Attenuation factors and transmission efficiency values for this cable are approximated, based on values available for other frequencies.

direction of maximum radiation (or maximum reception) depends on the shape of the antenna.

Antennas used in amateur radio communications are available in many types, sizes, and shapes. This discussion of amateur antennas is limited to those types most suited for Technician operation, namely the half-wave doublet, quarter-wave vertical, and the beam antenna.

Half-wave doublet (or dipole) antenna

Antenna operation involving the radiation and reception of RF signals is an extremely complex process that engineers and scientists describe

FIGURE 10.6

A direct-reading standing-wave ratio (SWR) meter.

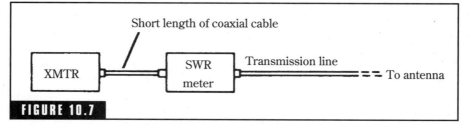

FIGURE 10.7

Test connections for making standing-wave ratio measurements.

in terms of mathematical analysis. However, we can employ some fundamental concepts that will help to explain how antennas work. Because the half-wave dipole antenna is a basic form of an antenna, we will use this as a starting point.

When a radio-frequency current flows in a dipole, electric and magnetic fields are generated along the antenna conductors or elements. If the dipole antenna has a physical length of one-half wavelength, it acts as a resonant circuit at half the wavelength frequency. Figure 10.8A shows a half-wave dipole antenna connected to a transmitter through a transmission line. The relationship between the physical length of a half-wave antenna and the operating frequency is given by:

$$L = 1/2 \; \lambda = \frac{468}{f_{\text{MHz}}} \text{ feet, or} \qquad \textbf{(Eq. 10.2)}$$

$$L = 1/2 \; \lambda = \frac{142.6}{f_{\text{MHz}}} \text{ meters} \qquad \textbf{(Eq. 10.3)}$$

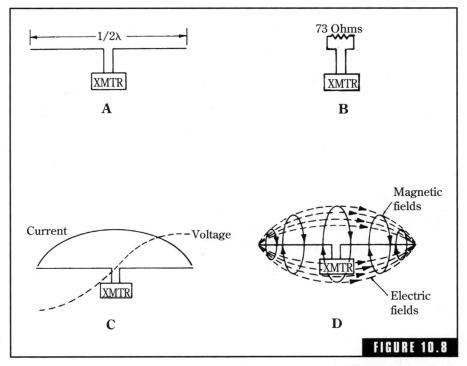

FIGURE 10.8

Characteristics of the half-wave dipole antenna in free space. (A) The half-wave dipole antenna. (B) Half-wave dipole antenna appears as 73-ohm resistor to transmission line. (C) Voltage and current standing waves developed on half-wave dipole antenna. (D) Electric and magnetic fields created by half-wave antenna during one-half RF cycle.

where

L is the length of the half-wave antenna in feet or meters

λ is the symbol for wavelength

f_{MHz} is the frequency in MHz

In theory, the length of a half-wave dipole is about 5 percent longer than indicated by the above equations. However, additional loading effects of insulators and supporting wires result in *end effects*. This correction applies for wire-type antennas up to about 30 MHz.

When installed in free space, the half-wave dipole presents a pure resistance of 73 ohms to the transmission line. Figure 10-8B shows how the antenna appears to the transmission line and the transmitter. This input resistance, sometimes called *radiation resistance*, of the antenna is defined as the voltage applied to the antenna terminals divided by the current flowing into the antenna.

The radio-frequency currents generated by the transmitter produce standing waves of voltage and current along the length of the antenna. This is illustrated in Fig. 10.8C. The voltage standing wave is highest at each end and drops to zero at the center of the antenna. The current standing wave is out of phase with the voltage wave and is a maximum at the center. The dipole is referred to as a center-fed device with current-feed characteristics. As stated earlier, the impedance or resistance at the center is 73 ohms—this value increases to about 2500 ohms at each end of the antenna.

The voltage and current standing waves produce alternating electric and magnetic fields. These fields change direction during each cycle of the RF current. Figure 10.8D shows orientation and direction of the alternating fields during one-half of an RF cycle.

Radiation of electromagnetic waves is created when the alternating electric and magnetic fields attempt to collapse back toward the antenna. However, new fields being created by the next half cycle push the preceding fields into space, which is the mechanics of radiation.

LONG AND SHORT DIPOLES

If the dipole antenna is made longer or shorter than a half-wave length, the input impedance no longer acts as a 73-ohm resistor. Instead, it presents a complex impedance with inductive or capacitive loading effects to the transmission line. Figure 10.9 illustrates the results and corrective action when we cut the dipole too long or too short.

A dipole antenna whose length is greater than one-half wavelength (but less than one wavelength) acts as an inductor in series with a resistor. Capacitance can be added to shorten the electrical length to an equivalent half wavelength. The capacitance cancels out the inductive effect and the antenna appears purely resistive to the transmission line. Of course, we can also correct for the inductive reactance by shortening the antenna.

When the dipole antenna length is less than one-half wavelength, the load presented to the transmission line appears as a capacitor in series with a resistor. Adding inductors in each leg of the dipole cancels the capacitive effect, leaving only the resistive load. This is one way that amateurs "tune" short antennas when space restrictions prevent installation of half-wave dipoles. Vertical antennas can also be made physically short if an inductive *loading coil* is installed in the vertical element.

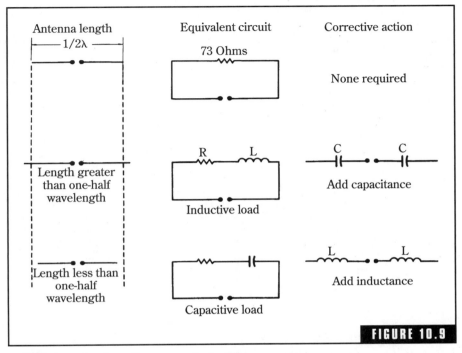

FIGURE 10.9

Effects of antenna length on transmission lines.

RADIATION PATTERNS FROM DIPOLE ANTENNAS

The radiation from a half-wave dipole is strongest on each side of the axis of orientation and decreases to a minimum at the ends. Figure 10.10A shows a plot of the radiation pattern of a dipole from a view looking down on the antenna. For example, if you desire an antenna orientation with maximum radiation in a north-south direction, you must install the antenna in an east-west line.

The radiation pattern of a dipole antenna changes when it is installed above a ground surface. The ground acts as a mirror and waves striking the ground will be reflected into space. To illustrate this, Fig. 10.10B shows an end view of the radiation pattern of a dipole antenna in free space.

If the dipole is installed about one-fourth wavelength above a good conductive ground (damp or moist soil), high-angle radiation will be created due to some of the waves being reflected by the ground. This radiation pattern, shown in Fig. 10.10C, is useful for contacts on 80

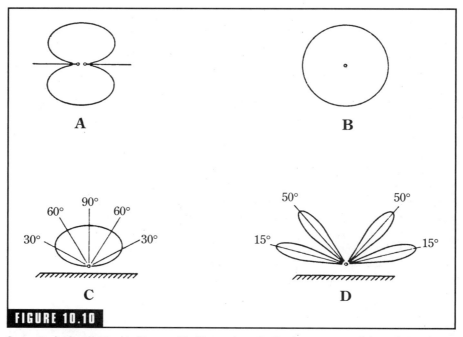

FIGURE 10.10

Some typical radiation patterns of half-wavelength dipole antennas. (A) Horizontal pattern, sometimes called a "figure-eight" pattern. (B) End view of radiation pattern of a dipole in free space. (C) Vertical pattern of a dipole installed one-fourth wavelength above a reflecting surface. (D) Vertical pattern of a dipole installed one wavelength above a reflecting surface.

and 40 meters out of about 200 miles (320 kilometers). However, this high-angle radiation is not effective for long DX contacts.

In general, raising the dipole antenna higher above ground lowers the radiation angle. Figure 10.10D shows the radiation pattern when the antenna height is about one wavelength. This results in four distinct *lobes*, or maximum radiation angles. It should be noted that the type and condition of the ground is important for it to act as a mirror to radio waves. A heavy rain, for example, can raise the effective ground level for an antenna, pushing up the angle of radiation.

RADIATION RESISTANCE CHANGES WITH ANTENNA HEIGHT

We stated earlier that the radiation resistance of a half-wave dipole in free space is 73 ohms. Installing this antenna above a ground surface will cause the radiation resistance to vary from about 95 ohms down to almost zero ohms. Figure 10.11 shows how the radiation resistance for a half-wave dipole will vary with respect to its height above ground. This graph can help you in selecting the type of transmission line to be used with your antennas. In general, 75-ohm coaxial cable is suitable for dipole or doublet antennas installed at heights of one-fourth wavelength or higher. Below heights of one-fourth wavelength, the 50-ohm coaxial cable should be used.

PRACTICAL HALF-WAVE DOUBLET ANTENNAS

The half-wave doublet or dipole antenna is a simple but effective antenna for the Technician HF bands, particularly 80- and 40-meter

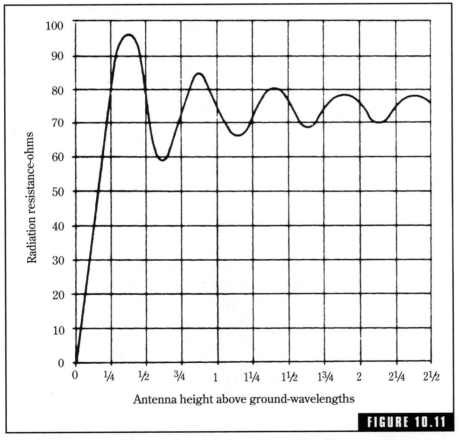

FIGURE 10.11

Radiation resistance of half-wave horizontal dipole antenna installed at varying heights above the ground.

bands. This antenna can be installed with a minimum of time and materials. Figure 10.12 shows some of the practical details for installing the doublet antenna. Also, you can pick up most of the coaxial cable and antenna wire from swap-fests or auction nights at the local ham club. Many of the experienced hams will be glad to help you install the antenna or answer any questions you might have on parts or installation problems.

Safety and other considerations

Prior to any actual installation of antennas, poles, or tower work, you should be aware of all related safety rules. The installation manual for any new equipment, including antenna kits, will have safety instructions concerning installation and grounding procedures. Read these carefully because they might save your life or the life of another person helping you with the installation. Additional suggestions for your safety are as follows:

■ Connect all major electronic equipment (transmitter, receiver, antenna tuner, and other components) with a ground terminal, to a common ground. Chapter 11 provides additional information on this procedure.

■ When installing or working on an antenna mounted on a tower, always use a safety belt to ensure maximum safety. While on the tower, be sure that no power (RF or 60-Hz ac power) is applied to antennas or other equipment on the tower. Battery-operated tools, such as drills or soldering guns, are recommended instead of 120-V ac operated tools.

■ People on the ground, helping with installers on the tower, should wear protective items such as hard hats and safety glasses. This will prevent falling objects from injuring the persons on the ground.

■ Never install an antenna close to an electric power line. During installation, a wire or part of the tower could fall against the power line and cause severe burns or fatal electric shock to people on or near the tower. After installation, there is always the possibility that the power line might be blown or fall across the antenna installation. This could result in dangerously high voltages being conducted inside the ham shack.

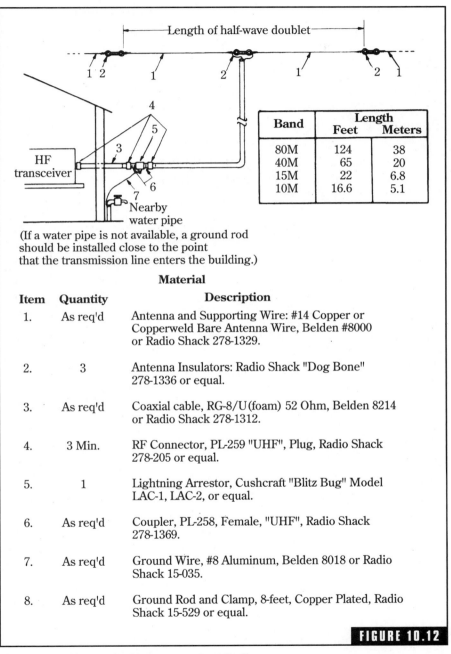

Band	Length	
	Feet	Meters
80M	124	38
40M	65	20
15M	22	6.8
10M	16.6	5.1

(If a water pipe is not available, a ground rod
should be installed close to the point
that the transmission line enters the building.)

Material

Item	Quantity	Description
1.	As req'd	Antenna and Supporting Wire: #14 Copper or Copperweld Bare Antenna Wire, Belden #8000 or Radio Shack 278-1329.
2.	3	Antenna Insulators: Radio Shack "Dog Bone" 278-1336 or equal.
3.	As req'd	Coaxial cable, RG-8/U (foam) 52 Ohm, Belden 8214 or Radio Shack 278-1312.
4.	3 Min.	RF Connector, PL-259 "UHF", Plug, Radio Shack 278-205 or equal.
5.	1	Lightning Arrestor, Cushcraft "Blitz Bug" Model LAC-1, LAC-2, or equal.
6.	As req'd	Coupler, PL-258, Female, "UHF", Radio Shack 278-1369.
7.	As req'd	Ground Wire, #8 Aluminum, Belden 8018 or Radio Shack 15-035.
8.	As req'd	Ground Rod and Clamp, 8-feet, Copper Plated, Radio Shack 15-529 or equal.

FIGURE 10.12

A practical half-wave doublet antenna installation.

- Install the antenna with as much clearance away from nearby objects as possible. Trees, buildings, power or telephone lines, or metal fences will distort the radiation pattern as well as change the radiation resistance.

- Cut the transmission line at the point where it enters the side of the building and install PL-259 (or equivalent) coaxial connectors on each end. Using a PL-258 (or equivalent) coaxial coupler, install a coaxial static discharge unit between the two ends of the coaxial cable. Now connect a No. 12 bare copper ground wire between the coaxial static discharge unit and a water pipe or a ground rod.

The installation of the coaxial static discharge unit serves two important functions. First it allows a discharge path for the buildup of static electricity on the antenna wires. During electrical storms, this electrical potential can build up to thousands of volts, possibly causing electrical shock to the operator or damage to the ham equipment.

The second reason for installing the coaxial static discharge unit is that of protection against direct lightning striking the antenna. Sometimes, however, any protection short of disconnecting the antenna from the ham equipment might not be effective against a direct lightning strike. The coaxial static discharge unit does not guarantee that the lighting will not enter the ham shack—it only reduces the possibility of such an occurrence. During inactive periods or in case of an impending storm, you are strongly urged to disconnect the coaxial cable from the ham equipment. However, don't do this during a storm—a bolt of lightning could strike the antenna the instant you take hold of the coaxial cable. In any event, never operate your ham rig during storm conditions.

In the event you do not have a space for installing a conventional half-wave doublet, an *inverted-vee* antenna will do almost as well. An inverted-vee consists of a half-wave doublet with one or both elements slanted down toward the ground and the center section installed as high as possible. A suggested inverted-vee antenna is shown in Fig. 10.13.

You can use a telescoping TV mast for the center section of the inverted-vee antenna. This will provide a height of about 30 to 45 feet. If the upper section of the mast is attached to the gable of your house, this will reduce guying requirements. To prevent interaction of the guy

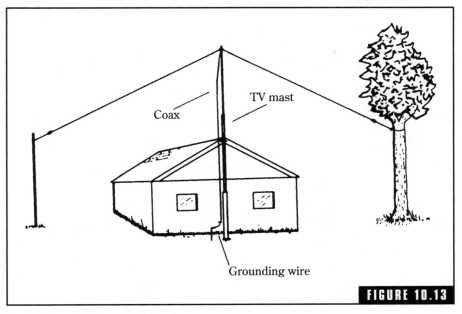

FIGURE 10.13

Inverted-vee half-wave antenna.

wires with the antenna, be sure to break up the guy wires into short sections joined by insulators. Any section of guy wire should not be greater than about one-eighth wavelength of the operating frequency. Also, the angle between the "drooping" legs of the doublet should never be less than 90 degrees. Otherwise, a smaller angle will cause cancellation of the radio waves and less radiating efficiency. Install the antenna with about 2 or 3 percent of additional wire on each end. The added wire can be folded back from the insulator and wrapped around the antenna wire. Having this extra length of wire will allow you to "prune" the antenna to the desired frequency of operation.

To test and adjust the antenna, connect an SWR meter in the transmission line as shown in Fig. 10.7. It is a good idea to measure the SWR over the antenna's range of operation at about 10.kHz steps and plot these readings on a graph. This shows you if the resonant frequency falls within the Technician portion of the band or if the antenna is too long or too short. Figure 10.14 illustrates a typical set of readings for each condition. Using the SWR approach requires that you transmit a test signal for each SWR reading. If you are a Novice

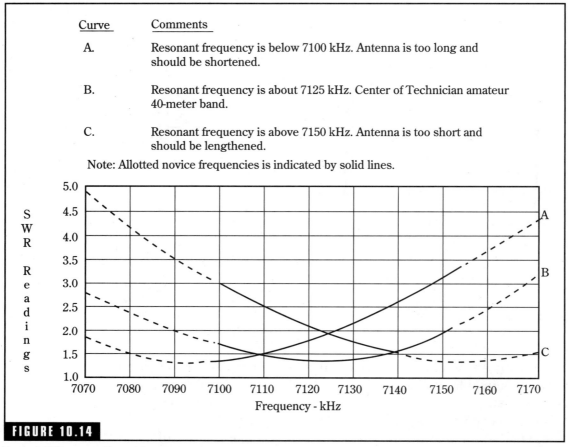

Curve	Comments
A.	Resonant frequency is below 7100 kHz. Antenna is too long and should be shortened.
B.	Resonant frequency is about 7125 kHz. Center of Technician amateur 40-meter band.
C.	Resonant frequency is above 7150 kHz. Antenna is too short and should be lengthened.

Note: Allotted novice frequencies is indicated by solid lines.

FIGURE 10.14

Typical SWR curves for a 40-meter antenna.

operator, you are limited to emitting test signals within the Technician bands. Also, remember that your test signal should be adjusted to the lowest level possible so that you will not interfere with other amateurs operating on the same frequencies. By plotting the SWR readings on a graph you can get a rough indication of the antenna's resonant frequency even though it may fall outside of the Technician band. This is shown by the dotted lines on the graphs in Fig. 10.14.

Newer forms of SWR meters, designated as *antenna analyzers*, are available that will let you check the performance of an antenna at any frequency within the range of the antenna analyzer. This instrument has a low-level internal oscillator along with the SWR circuitry, eliminating the need for using your transmitter in SWR measurements. Figure 10.15 shows antenna analyzers marketed by MFJ Enterprises Inc. You merely connect the antenna analyzer to the antenna being tested, tune the analyzer's internal oscillator to the desired frequency, and read the SWR on the analyzer's meter.

If the resonant frequency does not fall within the desired band of frequencies, the slope of the SWR plot indicates whether the antenna

FIGURE 10.15

Antenna SWR antenna analyzers. These innovative, portable devices permit SWR measurements of amateur antennas over an entire band without the need for a transmitter, SWR meter, or other equipment. (A) The MFJ HF/VHF SWR Antenna Analyzer covers a frequency range of 1.8 to 170 MHz. This provides a capability to measure SWR in all the HF bands as well as the 6-meter and 2-meter bands. The antenna analyzer has a built-in frequency counter for precise frequency measurements and also measures RF impedance (Z), resistance (R), and phase angles. (B) The MFJ-219 UHF SWR Analyzer covers 420 to 470 MHz for the 70-cm band and provides an output jack for a frequency meter for accurate frequency measurements. *Courtesy of MFJ Enterprises Inc., P.O. Box 494, Mississippi State, MS 39762; Telephone: (601) 323-5869* is too long or too short. Vary the antenna length about 1 inch (2.5 centimeters) in the required direction and recheck the SWR readings. In this manner you will be able to bring the resonant frequency of the antenna to the desired frequency.

Some amateurs prefer to use an antenna-impedance bridge or noise bridge for measuring antenna impedances and the resonant frequency. Figure 10.16 shows a typical noise bridge that allows you to measure the resistance and reactance (inductive or capacitive) characteristics of an antenna at any frequency. This and similar noise bridge test instruments are powered by an attached battery. Because this type of device does not require the use of a transmitter test signal, it affords portable operation, even testing at the antenna input terminals. Furthermore, the antenna noise bridge can be used by the Technician operator at frequencies outside of the Technician bands.

The antenna noise bridge shown in Fig. 10.16, called the R-X Noise Bridge, is manufactured by Palomar Engineers. It consists of a noise generator and a bridge circuit for measuring resistive, capacitive, and inductive characteristics. An HF receiver is required as a detector for the bridge circuit. When the R and X_C-X_L controls are adjusted for a mull (minimum noise out of the receiver), their dial settings are read directly in resistance and reactance.

FIGURE 10.16

The Palomar Model RX-10 R-X Noise Bridge. This handy, portable test instrument is useful in measuring the electrical characteristics (resistance and reactance) of RF devices such as antennas. *Palomar Engineers, P.O. Box 462222, Escondido, CA; Telephone: (619) 747-3343* compromise and is used primarily when only limited space is available for installing antennas. One disadvantage of this type of installation is that one end of the antenna is brought into the ham shack causing strong RF fields. These fields can interfere with the operation of the ham equipment or nearby TV, AM radio, or hi-fi sets. Another disadvantage is that long-wire antennas with lengths of one wavelength or more produce highly directional radiation patterns.

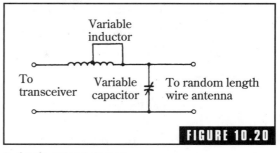

FIGURE 10.20

A simple antenna tuner.

The antenna tuner is used to match the transmitter and transmission line to the varying impedances of the long-wire antenna. As we switch from one band to another, the impedance presented by the antenna might shift to inductive or capacitive loading effects. The antenna tuner "tunes" out these reactive effects by adding capacitance or inductance to the antenna circuit. One form of an antenna tuner, shown in Fig. 10.20, consists of a variable inductor and capacitor. A commercial antenna tuner is illustrated in Fig. 10.21. The more elaborate models may contain a built-in SWR meter.

FIGURE 10.21

The MFJ Enterprises Model MFJ-971 HF Portable Antenna Tuner. This compact antenna tuner covers the 1.8–30 MHz frequency range with power levels up to 300 watts. The MFJ-971 allows you to match virtually all types of transmission lines to your amateur HF transceiver, such as coaxial cables and even random lengths of wire. This antenna tuner has a unique "Cross-Needle" SWR/watt meter with two switchable power ranges of 30 and 300 watts.

Vertical antennas

The quarter-wavelength vertical antenna, sometimes referred to as a Marconi antenna, provides good low-angle radiation for long-haul DX contacts. It is also useful for installation in limited space. One disadvantage of the vertical antenna is that it does not provide as much gain as the horizontal half-wave doublet. A simple quarter-wavelength vertical antenna is shown in Fig. 10.22.

The operation of the quarter-wavelength vertical antenna is similar to that of the half-wave dipole because the ground plane reflects a mirror image of the vertical section. Voltage and current standing waves for the vertical antenna are illustrated in Fig. 10.23.

For optimum operation, the vertical antenna must be installed on a good conductive ground plane. If the vertical antenna is to be installed on dry rocky soil, radial wires or conductors at least one-quarter wavelength long should be installed beneath the antenna. A minimum of four radials should be used as the ground plane.

The radiation resistance of a quarter-wavelength vertical antenna installed on a flat ground plane is about 30 ohms. An SWR of about 1.7 will be experienced if a 50-ohm coaxial transmission line is used. The radiation resistance can be increased to about 50 ohms if a ground plane consisting of four "drooping" radials is used. Each radial must be

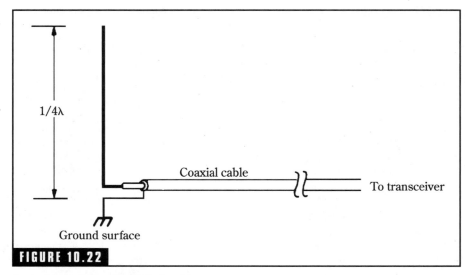

1/4λ

Coaxial cable

To transceiver

Ground surface

FIGURE 10.22

A quarter-wavelength vertical antenna.

high gain and directivity. By arranging a series of elements in a plane as shown in Fig. 10.27, an antenna array with high gain and narrow beamwidth can be obtained. An important advantage during reception is that antenna gain is reduced off to the sides and the back of the antenna. This minimizes interference from signals in other directions.

The beam or Yagi antenna provides considerable gain over that of dipoles or vertical antennas and is an excellent antenna for DX contacts. The spacing between elements is on the order of 0.15 to 0.2 wavelengths for maximum gain. A simple transmission line is shown to illustrate how this antenna is fed. In actual practice, some form of impedance matching device must be used to couple the antenna to the transmission lines. The radiation resistance of a three-element beam is on the order of 10 ohms. If a second director element is added to increase the gain, the radiation resistance drops to the vicinity of 4 to 6 ohms. Manufacturers of amateur beam antennas use various techniques for matching the low-impedance antenna feedpoints to 50-ohm coaxial cable.

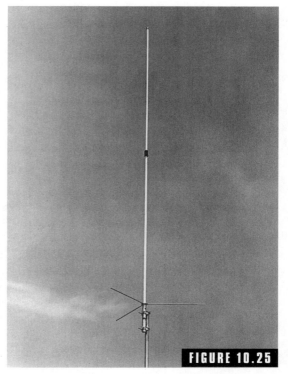

FIGURE 10.25

The cushcraft Model R-7 HF Vertical Antenna. This high-performance, half-wave vertical antenna covers all amateur bands from 10 meters to 40 meters. The R-7 features low SWR over specified portions of each amateur band, a power rating of 1800 watts pep, and low radiation angle of 16 degrees for long distance DX contacts. Topping only 22.5 feet (or 7 meters) in height and requiring no ground radials, this antenna can easily be installed almost anywhere.

A typical four-element beam antenna is shown in Fig. 10.28. This type of antenna provides excellent gain and directivity characteristics on 20, 15, and 10 meters. Note that this antenna uses resonant traps to obtain multiband operation. Antennas of this type can easily be installed on towers and rotated with antenna rotating systems. The antenna shown in Fig. 10.28 provides a forward gain of 8.9 dB on all specified bands and will handle up to 2 kilowatts of RF power. The nominal input impedance is 50, ohms and cable such as RG-8/U can be used to feed the antenna.

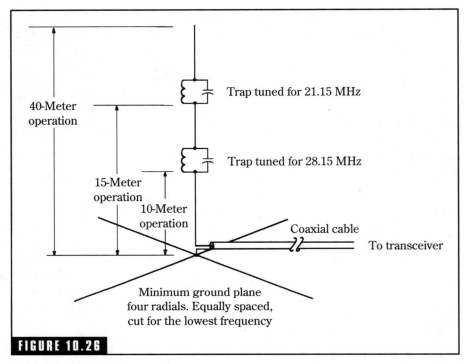

40-Meter operation

15-Meter operation

10-Meter operation

Trap tuned for 21.15 MHz

Trap tuned for 28.15 MHz

Coaxial cable

To transceiver

Minimum ground plane four radials. Equally spaced, cut for the lowest frequency

FIGURE 10.26

A multiband vertical antenna for 40, 15, and 10 meters.

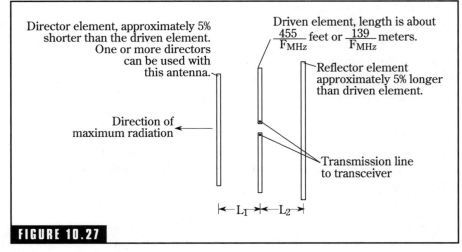

Director element, approximately 5% shorter than the driven element. One or more directors can be used with this antenna.

Driven element, length is about $\dfrac{455}{F_{MHz}}$ feet or $\dfrac{139}{F_{MHz}}$ meters.

Reflector element approximately 5% longer than driven element.

Direction of maximum radiation

Transmission line to transceiver

$\leftarrow L_1 \rightarrow \leftarrow L_2 \rightarrow$

FIGURE 10.27

The beam or Yagi antenna.

The Cushcraft Model A4S 20-15-10 Meter Beam Antenna. This four-element beam antenna features high gain and excellent directional characteristics.

Station wiring diagrams

We have now covered transmitters, receivers, transmission lines, and antennas. Next we examine station wiring diagrams. The FCC has a number of questions that relate to station configurations. Every possible combination is not shown. Develop some of the required station configurations yourself to help in the learning process!

Figure 10.29A shows a station consisting of a transmitter, receiver, transmit-receive switch, transmission line, and an antenna. This is one of the simplest configurations. If a transceiver is substituted for the transmitter and receiver, then the transmit-receive switch is not required as this function is built into the transceiver.

The addition of an SWR meter and an antenna tuner to the first configuration provides the operator with the capability to present a low SWR to the transmitter. This equipment configuration is illustrated in Fig. 10.29B.

The last configuration shows an amateur radio station with a transmitter, receiver, transmission line, antenna, transmit-receive switch, grounding provisions, and a telegraph key. This is shown in Fig. 10.29C.

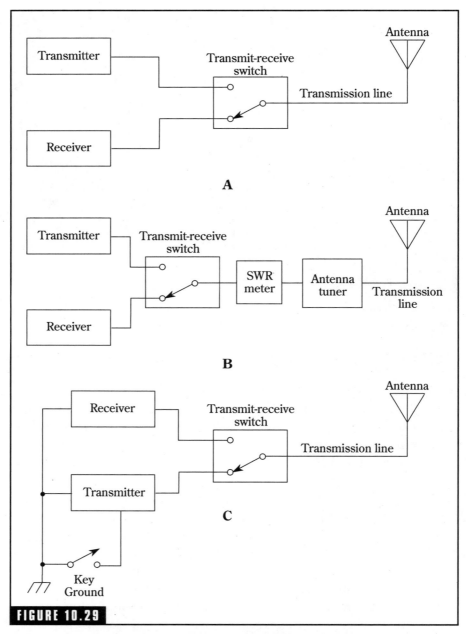

FIGURE 10.29

Typical amateur station wiring for transmitter-receiver combinations.

RF Radiation Safety and Radio Communications Practices and Procedures

This chapter provides a review of the required amateur radio communications practices and operating procedures. First, we will cover the recent FCC Rules on RF Environmental Safety Practices, which apply to three questions on the Technician Class Element 2 Examination. The next sections will provide information and recommendations for the design, installation, and operation of your amateur radio station. The FCC expects you to know this material for the Technician exam and subsequent operation of your amateur radio station (the exam contains five questions on operating procedures). Understanding the material and related questions in the Technician Question Pool (see Appendix A) is crucial to passing the Technician Examination, Element 2!

Definitions

Ammeter   The ammeter is an instrument (i.e., test equipment) for measuring current flow (amperes) in an electrical circuit. It can be designed to measure either ac or dc current.

Average power   The time-averaged rate of energy transfer.

Controlled environment   For FCC purposes, this applies to human exposure to RF radiation when people are exposed as a consequence of their employment and after those who are exposed have been made fully aware of the potential for exposure and can exercise control over their exposure. This includes amateur radio operators and their immediate families.

Decibels (dB)   A mathematical expression, decibels are equal to ten times the logarithm to the base ten of the ratio of two power levels. A complete description of decibels and how to calculate them is given in Chap. 7. Decibel calculation is a convenient way to express the gain (or loss) of amplifiers and small and large power, voltage, and current levels.

Duty factor   Applied to RF safety requirements, duty factor is defined as a measure (or ratio) of the total time that an RF transmitter is turned "on" (say 20 minutes) to the total observation time, usually 30 minutes. The duty factor in this observation period would be 20 minutes divided by 30 minutes, or 0.667. Note that a duty factor of 1.0 corresponds to continuous "on" operation of the transmitter.

Effective radiated power (ERP)   Transmitted power being radiated in a given direction. The ERP is equal to the RF power supplied to the antenna multiplied by the antenna gain. Thus an RF transmitter providing 100 watts of power to an antenna with a gain of 5 will produce an ERP of 500 watts in a given direction.

Exposure   Exposure occurs whenever and wherever a person is subjected to electric, magnetic, or electromagnetic fields produced by a RF transmitter.

Far field   That region of the antenna's electromagnetic field, a great distance from the antenna in terms of wavelengths, where the wave front has a plane wave with uniform characteristics.

Gain (of an antenna)   Usually expressed in decibels, the gain of the subject antenna is determined by comparison with a "standard antenna," such as a half-wavelength dipole or an isotropic radiator.

General population/uncontrolled exposure   For FCC purposes, this applies to human exposure to RF fields when the general public is exposed or when persons who are exposed as a consequence of their employment may not be fully aware of the potential for exposure or cannot exercise control over their exposure. Therefore, members of the general public always fall under this category when exposure is employment related.

Ground potential   This is described as a zero voltage potential with respect to ground or earth.

Ground rod   A ground rod may be either a metal pipe or a rod driven into the ground for an electrical connection with the earth. A buried cold-water (metallic only) pipe usually makes an excellent ground connection.

Harmonic radiation   This is defined as unintentional radiation of signals whose frequencies are harmonics or multiples of the fundamental operating frequency.

Interference   In terms of radio communications, interference is the reception of undesired signals that interferes with the reception of desired signals. Radio signals may also interfere with the performance of other electronic equipment, such as audio systems.

Interfering signals present in the broadcast bands are called broadcast interference (BCI).

Interfering signals in the television bands are called television interference (TVI).

Low-pass filter   A low-pass filter passes all signal frequencies below a critical frequency and reflects higher frequency signals. Some amateurs use a low-pass filter between the transmitter and transmission line to prevent radiation of harmonic or spurious signals above 10 meters. This is an effective way of eliminating TVI.

Maximum permissible exposure (MPE)   The maximum permissible electromagnetic field adjacent to the antenna that a person may be exposed to without harmful effects and with an acceptable safety factor. It is to be noted that all amateur transmitters having a maximum power output of 50 watts or less require only a routine evaluation for RF safety. This means that as a rule, a transmitter output of 50 watts or less will pose no threat to the amateur or nearby persons.

Near field   The region adjacent to the antenna and nearby structures where the electromagnetic fields are the most intense and nonuniform. For most antennas, the near-field boundary is estimated to be about one-half wavelength from the antenna.

Ohmmeter   The ohmmeter is an instrument for measuring resistance (in ohms). *The electrical power to a circuit must be turned off during all resistance measurements.*

Peak envelope power (PEP)   This is the average power supplied to the antenna transmission line by a radio transmitter during one radiofrequency cycle at the crest of the modulation envelope taken under normal operating conditions.

Plate circuit   This term is normally used to describe the external circuit that is connected to the plate and cathode terminals of a vacuum tube. The FCC refers to the "amplifier plate circuit" when describing the plate and cathode connections to the final RF amplifier stage of a tube-type transmitter or external tube-type RF power amplifier.

Power density (S)   The instantaneous power density integrated over a source repetition period.

Public service operation   Amateurs provide a valuable public service in terms of a voluntary noncommercial communications service, particularly with respect to emergency communications.

Q signals   This consists of a set of international abbreviated signals used primarily in Morse code communications. Each Q signal is a three-letter group starting with the letter Q. For example, QTH means "my location is —." If followed by a question mark, the Q signal is a question. For example, QTH? would be translated as "what is your location?"

RST reporting system   This is a signal-quality reporting system used by amateurs and other communications activities. RST stands for readability, strength, and tone. Numerals are used to indicate the quality of each characteristic, with scales of 0 to 5 for readability and 0 to 9 for both strength and tone. For example, a signal report of 5-9-9 means the received signal is "perfectly readable," an "extremely strong signal," and a "perfect tone."

Specific absorption rate (SAR)   This is a measurement of the rate of energy being absorbed by (or dissipated in) an incremental mass contained in a volume element of dielectric material, such as biological tissues. SAR is usually expressed in watts per kilogram (W/kg) or milliwatts per gram.

Thermal effects   As with microwave ovens, RF radiation at the lower frequencies can produce heating effects within biological tissue. If severe, this heating could have adverse effects.

Universal coordinated time (UTC)   This is a worldwide system of keeping time. The National Bureau of Standards standard time and frequency broadcast stations, WWV and WWVH, broadcast time signals based on the UTC time system. The specific hour and minute stated is actually the time zone centered around Greenwich, England. This time system is based on a 24-hour system—the hours are numbered beginning with 00 hours at midnight to 23 hours, 59 minutes just before the next midnight. For example, 12 hours would be noon and 18 hours would be 6:00 pm.

Voltmeter   The voltmeter is an instrument for measuring voltage (volts) or differences of potential. Voltmeters may be designed to measure ac or dc voltages.

Wattmeter   This is an instrument for measuring the electrical power being delivered to a circuit and is usually calibrated in units of watts or kilowatts. Wattmeters are available for measuring ac or dc power. For example, an RF wattmeter can be connected in the transmission line between the transmitter and the antenna in order to measure the amount of RF power being delivered to the antenna.

Radio Frequency Environmental Safely Practices

FCC rules and regulations require every amateur radio operator to evaluate the RF environment surrounding his or her amateur station and antenna installations. You must insure that all persons in the vicinity of your station are not exposed to excessive RF radiation. Here the FCC identifies two categories. The first or immediate group, including the amateur operator, the amateur's family, and other persons who are aware of the RF radiation, are in a Controlled Environment. An Uncontrolled Environment is the area around the amateur station where persons, unaware of the RF radiation, may be present. Persons in this category include neighbors and the general public.

Excessive RF radiation is defined by the FCC in terms of Maximum Permissible Exposure (MPE). If a preliminary evaluation indicates that your antennas are producing RF electromagnetic fields in excess of the FCC limits, you must take immediate action to correct this potential hazard. Simple steps such as reducing the RF power output of your transmitter(s) to recommended levels or elevating the antennas and installing them away from buildings may be the best solution.

In general, amateur radio is a relatively safe hobby in terms of RF radiation hazards. The typical amateur radio transmitter power levels rarely exceed the FCC radiated power limits. Again, sometimes all that may be needed is to reduce the power level. It is to be noted here that the FCC requires the amateur radio operator to operate with the minimum power required to successfully complete a contact. Why use 100 watts of RF power when 5 watts may be sufficient when contacting a nearby amateur on a VHF repeater?

Considerable research by the FCC and other organizations has been conducted on human exposure to RF electromagnetic radiation. Two important findings of this research are as follows:

1. High levels of thermal or heating effects of RF radiation can cause serious heat damage to biological tissue. The magnitude of these heating effects will vary with the frequency (and wavelength), power density, and propagation factors of the radiation field. Also, the physical size or height of a person can create a "natural resonance" which causes maximum heating effects. These resonant effects are most dominant in the 30- to 300-MHz frequency range. Individual body parts such as the head will resonate at frequencies about 400 MHz, and eyes will resonate at certain microwave frequencies. For example, excessive RF heating effects in the human eye may cause cataracts to form.

2. Nonthermal or nonheating RF radiation, usually associated with low-level RF and low-frequency energy fields, may have some biological effect on living beings, such as humans and animals. This includes ac (60 Hz) power-line radiation. Long-term statistical studies have produced no definite link between this radiation and any health problems. We can expect considerable research in the future concerning the effects of RF radiation on humans and animals.

The applicable FCC documents concerning RF safety are the FCC Part 1 and Office of Engineering and Technology (ORT) Bulletin 65. You may want to obtain a copy of ORT Bulletin 65, "Evaluating Compliance with FCC Guidelines for Human Exposure to Radiofrequency Electromagnetic Fields." This bulletin provides a summary of the FCC limits for maximum permissible exposure (MPE) in reference to specific frequencies. Table 11.1, provided in OET Bulletin 65, shows these MPE values for two conditions, Occupational/Controlled Exposure and General Population/Uncontrolled Exposure. The key to this

TABLE 11.1 The FCC Limits for Maximum Permissible Exposure (MPE) to RF radiation in Reference to Frequency Bands. (Covers both Occupational/Controlled and General Population/Uncontrolled Exposure Limits)

A. Limits for Occupational/Controlled Exposure

Frequency range (MHz)	Electric field Strength (E) (V/m)	Magnetic field strength (H) (A/m)	Power density (S) (mW/cm2)	Averaging time $|E|^2$, $|H|^2$ or S (minutes)
0.3–3.0	614	1.63	(100)*	6
3.0–30	18–12/F	4.89/F	(900/F)°	6
30–300	61.4	0.165	1.0	6
300–1500	—	—	F/300	6
1500–100,000	—	—	5	6

B. Limits for General Population/Uncontrolled Exposure

Frequency range (MHz)	Electric field Strength (E) (V/m)	Magnetic field strength (H) (A/m)	Power density (S) (mW/cm2)	Averaging time $|E|^2$, $|H|^2$ or S (minutes)
0.3–1.34	614	1.63	(100)*	30
1.34–30	124/F	2.19/F	(180/F)°	30
30–300	27.5	0.073	0.2	30
300–1500	—	—	F/1500	30
1500–100,000	—	—	1.0	30

F = frequency in MHz

*Plane-wave equivalent power density

Note 1: *Occupational/controlled* limits apply in situations in which persons are exposed as a consequence of their employment provided those persons are fully aware of the potential for exposure and can exercise control over their exposure. Limits for occupational/controlled exposure also apply in situations when an individual is transient through a location where occupational/controlled limits apply provided he or she is made aware of the potential for exposure.

Note 2: *General population/uncontrolled* exposures apply in situations in which the general public may be exposed, or in which persons that are exposed as a consequence of their employment may not be fully aware of the potential for exposure or can not exercise control over their exposure.

table contains the maximum power density in mW/cm² (milliwatts per square centimeter). Prior research determined the whole body's specific absorption rate (SAR), the value on which the MPEs are based.

Figure 11.1 incorporates the data given in Table 11.1 to provide a graphical representation of the MPE limits in terms of a plane-wave equivalent power density. Both types of environments are given, i.e., the Occupational/Controlled Exposure and the General Population/ Uncontrolled Exposure. Both Table 11.1 and Fig. 11.1 show the following conclusions:

1. The human body absorbs energy differently at various frequencies. The most stringent frequency band for RF radiation exposure is that of the 30 to 300 MHz (VHF). Human body resonance effects occur in this region.

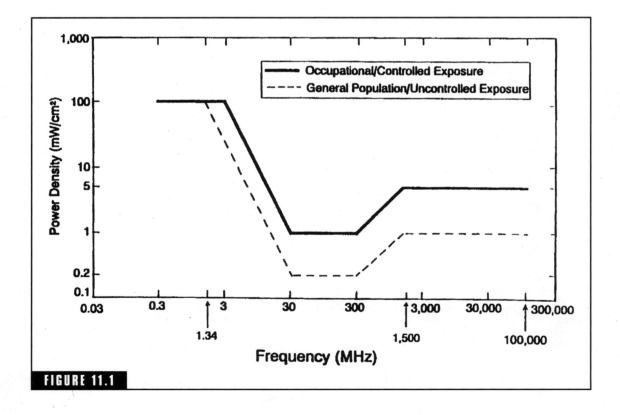

FIGURE 11.1

2. Regarding the microwave region (about 1500 MHz and above), intense RF radiation in these frequencies can cause severe heating in human biological tissues. In particular, the eye's resonant frequency is highly susceptible to microwaves. The heating effect in the ocular fluids can cause cataracts. This is one reason that you should never look into the open end of a microwave feed horn antenna when the transmitter is operating!

3. The FCC's limits for maximum permissible exposure show that any radiation tests must cover specified periods of time. In order to determine compliance with the MPE levels, safe exposure levels for RF energy are averaged as follows:

RF test environment	Time period
Uncontrolled RF Environment	30 minutes
Controlled RF Environment	6 minutes

Determining Compliance with FCC RF Safety Rules and Regulations

Every amateur radio operator must determine if his or her amateur radio station is in compliance with the FCC RF safety requirements. Even the applicant for an amateur license must indicate his or her understanding of these requirements on an amateur radio license application form at the time of application. As stated earlier, there will be three questions concerning this on your Technician license examination. By learning this RF safety material now, you will be prepared both for the Technician exam and subsequent operation of your ham station.

To evaluate compliance with the FCC rules, the first step is to determine the RF power output of your transmitter(s) or transceiver(s). Part 97.13 states that you must perform the routine RF evaluation if the RF power output of your station exceeds the limits given in Table 11.1. That's right! You don't need to perform any further evaluation as long as you don't exceed these RF power limits. As an example, you have the 5 WPM certification and want to operate on the 80-meter and 40-meter CW bands. If you have a typical HF transceiver that provides

TABLE 11.2 FCC Power Thresholds for
Routine Evaluation of Amateur Radio Stations

Wavelength band (meters)	Transmitter power (watts)
160 m	500
80 m	500
75 m	500
40 m	500
30 m	425
20 m	225
17 m	125
15 m	100
12 m	75
10 m	50
VHF (all bands)	50
UHF	
70 cm	70
33 cm	150
23 cm	100
13cm	250
SHF (all bands)	250
EHF (al bands)	25

Note: This table covered in Part 97.13, shows the maximum RF opersting power limit of an amateur radio station unless the amateur operator performs a routine RF environmental evaluation as specified by the FCC.

100 watts output, Table 11.2 shows that you are well within the RF output limits for these two bands.

If your station RF power output exceeds the power limits in Table 11.1, you must perform a routine RF environmental evaluation in accordance with FCC rules and regulations. Remember, the purpose of this evaluation is to ensure a safe operating environment for amateurs, their families, and their neighbors.

This evaluation can be performed using one of three methods: direct power density measurements, mathematical calculations, or using reference tables. The first two of these techniques are generally beyond the capability of the average amateur operator. For example, the complex test equipment is too expensive for most amateurs. Also, considerable skill and technical knowledge are required for making and interpreting power density measurements. The second technique involves predicting the theoretical power density by applying mathematical equations for "near-field" and "far-field" electromagnetic field power intensities. These equations take into account the antenna gain, which varies with the type of antenna being used.

The far-field equation predicts the value of the power density at a distant point given the power input to the antenna and the gain of the antenna:

$$S = \frac{P\,G}{4R^2}\ \text{mW/cm} \qquad \textbf{(Eq. 11.1)}$$

where

S is the power density in milliwatts per square centimeter

P is the input power to the antenna in watts

G is the antenna gain

R is the distance in centimeters

Stated in another way, "In the far field, as the distances from the source increases, the power density is proportional to the inverse square of the distance."

The calculation of power density in the near field is complex due to the particular antenna being used. Equation 11.1 will provide an "overprediction" but could be used in a "worst-case" or conservative prediction. Stated in terms of field strength, "In the near field, the field strength depends on the type of antenna used."

The W5YI RF Safety Tables, given in Appendix B, provide a convenient and fast means for determining if your station is within the MPE guidelines established by the FCC. These tables provide safe distances from your antenna for both controlled and uncontrolled environments in terms of frequency and power levels.

General RF safety recommendations

The material up to this point covers most of the RF safety questions in the Technician Examination Pool in Appendix A. Some additional items are provided to cover the remaining examination questions as well as important aspects of RF safety guidelines.

ADDING A NEW ANTENNA TO AN EXISTING ANTENNA "FARM"

When a new antenna is added to a group of antennas, such as multiple repeated antennas on top of a building, the new repeater signal must be considered as part of the total radiation from the site when determining RF radiation exposure levels.

RF SAFETY CONSIDERATIONS FOR USING A HAND-HELD TRANSCEIVER

You should make sure that the antenna is not held too close during transmission. The antenna should be positioned away from your head and away from others.

PRECAUTIONS FOR INSTALLING TRANSMITTING ANTENNAS

Highly directional antennas such as VHF and UHF Yagi beam antennas should be installed in a location where no one can get close to them while you are transmitting. Any transmitting antenna, including wire antennas, should be located so that no one can touch them while you are transmitting. This touching may cause RF burns.

PRECAUTIONS FOR INSTALLING MOBILE ANTENNAS ON VEHICLES

For the lowest RF radiation exposure to occupants inside the vehicle, mobile antennas should be installed in the roof of the vehicle. When this type of antenna is mounted on the metallic vehicle roof, the mobile transceiver will generally produce less RF radiation exposure than hand-held transceivers used inside the vehicle.

PRECAUTIONS FOR REMOVING THE SHIELDING COVER OF A UHF AMPLIFIER

If the shielding cover of a UHF power amplifier is removed, make sure the amplifier cannot be turned on. In general, all RF power amplifiers should never be operated unless the covers are in place. This precaution will help reduce the risk of shock from high voltages and reduce RF radiation exposure.

Radio Communications Practices

Good communications practices for an amateur radio station include the following related items: properly installed equipment and antennas with adequate grounding, use of filters and shielding when required to eliminate interference, proper use of test equipment, proper tuning and adjustment of transmitters, protection against lightning, good safety practices, and good operating procedures. Note that actions in one area might affect other areas. For example, safety is directly related to all other areas.

You can develop good communications practices for your amateur station only by diligent study and careful planning. A job well done results in a sense of achievement and self-satisfaction.

Station installation

Planning an amateur station involves a variety of actions. By this time you might have decided what type of equipment and antennas are best suited for your needs. Now you have to find a space for the ham shack and the antennas. To get started, develop an installation plan that identifies the general layout and the required installation material. By the time you receive your license, you will be ready to get on the air with a minimum of effort.

Station layout

Select a location for the ham shack that will provide space for a comfortable operation position. A large table is ideal for locating the receiver, transmitter, key, and other equipment you might have decided upon. Be sure to leave plenty of room for operating the key, writing traffic on a pad of paper, and storage of log books.

If you have to locate the ham station in a small space, you can install a shelf over a small table to hold most of the equipment. Figure 11.2 shows a typical layout with the receiver located on the

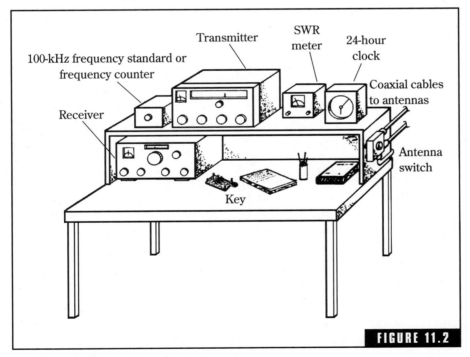

FIGURE 11.2

Typical layout for a Novice radio station.

table and the remaining equipment installed on the shelf. Because most operation involves tuning the receiver, it is convenient to have the receiver within easy reach. Many variations of this type of installation can be developed to meet your particular needs. Some amateurs have designed elaborate operating consoles for their ham stations.

Keep in mind that personal safety is a major factor is designing the station layout. This applies to the operator as well as any visitors to the ham shack. Being able to lock the entrance to the ham shack is a good idea, especially to keep out small children when you are away.

Station wiring

Make sure that you have adequate wiring to handle the required ac power for the ham shack. All power lines should be properly fused to protect against short circuits in the equipment. In addition, each piece of equipment should be individually fused to prevent internal shorts from causing a fire hazard or damage to the equipment. The use of three-conductor power cords with a ground connection will help to ensure safety in the ham shack.

It is a good idea to install all station wiring in hidden or concealed locations to prevent physical contact with the wiring or possible damage to the wiring. This also results in a neat and professional appearance to your ham shack.

A station ground is a must for both personal safety and proper operation of the amateur equipment. Be sure that all equipment in the ham shack is connected to a common ground with a heavy conductor such as No. 10 or 12 AWG wire. A buried water pipe makes an excellent earth ground. In the absence of a water pipe, drive a copper ground rod about 6 to 8 feet into the ground. If you have a tower installed next to the ham shack, you can also use the ground rod to ground the tower as well as the station equipment. Grounding wire, rods, and special connectors can be obtained at a local electrical supply house, electronics parts supplier, or hardware store.

One final recommendation—design your station wiring so that all equipment can be disconnected from the ac power lines and antennas with a minimum of effort. This helps to eliminate safety hazards or unauthorized operation during your absence. In addition, this capability offers excellent protection against lightning damage.

Radio-Frequency Interference (RFI)

Many radio amateurs are blamed for causing interference to nearby radios, television sets, or audio systems. In some cases, the amateur's transmitter or transceiver can be radiating harmonic or spurious signals that interfere with the operation of these home-entertainment systems. However, studies to date have shown that most cases of interference can be traced to design or construction deficiencies inherent in these home-entertainment systems. For example, a television set with poor selectivity in the tuner section can allow amateur signals to enter the set and cause interference. Unfortunately, this type of situation does not relieve the amateur of some responsibility in trying to help resolve the problem.

When confronted with a radio-frequency interference (RFI) complaint, there are a number of actions that you can take to solve the problem. If you are a newcomer to amateur radio with no experience in RFI problems, you can often contact a local ham radio club for advice and assistance. Almost every ham club has a "TVI committee" that offers technical assistance in tracking down RFI problems by checking your rig and the neighbor's radio, TV set, or audio system. In addition, most radio clubs are affiliated with the American Radio Relay League, Newington, Connecticut 06111. This organization maintains extensive information and technical expertise on amateur matters such as RFI and provides this service upon request. The *ARRL Radio Amateur's Handbook* contains detailed information on RFI problems and recommended solutions.

The first step in clearing up an RFI complaint is to check your transmitter or transceiver for RFI emissions. For BCI or TVI problems, you can check for interference with any available AM radio or TV set. If you find evidence of interference radiations, take immediate action to clear up the problem. If no interference exists, you might want to demonstrate this finding to your neighbor.

Elimination of RFI

Most modern amateur transmitters and transceivers are designed to suppress harmonic or spurious radiation. However, improper adjustment of the operating controls can result in RFI emissions. In other cases, poor grounds or lack of grounding can cause RFI problems. These types of problems can be quickly cleared up with a minimum of effort.

Some older or home-built amateur transmitters can radiate harmonic or spurious signals, thereby causing interference problems. Internal modifications such as shielding the high-power stages or installing bypass capacitors can clear up these problems. This type of effort requires considerable patience and a technical expertise. An alternate solution is to determine if a low-pass RF filter will eliminate the interfering signals. A number of companies offer high-quality RF filters for amateur operation. For example, the Barker & Williamson Company, 10 Canal Street, Bristol, PA 19007 [Telephone: (215) 788-5581], offers a complete line of radio-frequency filters for the amateur station. The B&W Series FL10 Low Pass TVI Filters for 160–10 meters and FL6 Filters for 6 meters are available in either 100 or 1000 watt power capacity ratings. Most of these filters are designed for 52-ohm transmission lines; however, the FL10/1500/70 works with 72-ohm transmission lines. Figure 11.3 shows how these filters can be installed.

Sometimes a TVI problem can be cleared up by installing a high-pass RF filter at the antenna terminals of the TV set. This type of filter allows the higher frequency TV signals to pass unattenuated into the TV set while rejecting the lower frequency amateur signals. The Radio Shack TV Interference Filter, Catalog No. 15-582, is typical of these types of filters.

In general, interference to audio systems can only be corrected by installing filter devices on the audio equipment. In some cases, the interconnecting audio cables, particularly long unshielded speaker wires, act as antennas. As such, the audio or hi-fi system simply operates as an untuned shortwave receiver, picking up any strong radiated signals in the vicinity. Some manufacturers of these systems will pro-

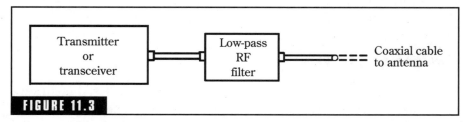

FIGURE 11.3

A low-pass RF filter installed in the transmission line at the transmitter or transceiver end.

vide modification kits to correct this type of problem. Installation of RF bypass capacitors (about 0.001 µF) on input and output terminals will usually clear up this type of interference. In some cases, the speaker wires might require shielding with the shield connected to the equipment ground.

How to Use Test Equipment

An important part of amateur radio is knowing when and how to use test equipment. Technical manuals furnished with amateur radio equipment usually include dc operating voltage and resistance measurements. This information is helpful in locating defective components or in making adjustments for proper operation. With simple test equipment such as voltmeters or ohmmeters, you can quickly learn to isolate defective components and repair your ham rigs.

This discussion of test equipment concentrates on the four basic instruments: the ammeter, voltmeter, ohmmeter, and wattmeter. The FCC expects you to be familiar with these basic instruments for both the Technician examination and subsequent operation of your ham station. As you progress in your career of amateur radio, you will come into contact with more complex forms of test equipment, such as signal generators, frequency counters, and oscilloscopes. By developing a good knowledge of the four basic test instruments listed above, you will be prepared to understand and use the more advanced types of test equipment. Many amateur publications contain descriptions of this test equipment and how to use it.

Meter movements

The D'Arsonval meter movement is used in virtually all test equipment employing meters. This meter movement consists of a horseshoe-shaped magnet with a moving coil placed between the open ends of the magnet. A pointer and spring is attached to the moving coil to establish a zero reference point. When a dc current flows through the coil, the magnetic field produced by this current interacts with the fixed magnetic field produced by the magnet. This causes the moving coil to rotate, and the amount of rotation is proportional to the current flowing in the coil. Thus the D'Arsonval meter movement is used to measure the amount of current flow. In the proper type of circuit, the

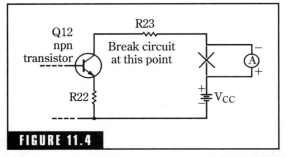

Using an ammeter to measure current in a closed circuit.

meter movement can be calibrated in terms of amperes, volts, watts, or even gallons of gasoline in an automobile gas tank.

Ammeters

The ammeter in its simplest form is a current meter. When connected in series with an electrical circuit, it allows you to measure the current flowing in the circuit. Figure 11.4 shows how the ammeter is connected in a circuit. Note that the circuit must be broken and the ammeter connected in series to form a completed electrical path. As connected, the ammeter measures current flowing in the collector circuit through Q12, R22, and R23. Never connect an ammeter across a source of voltage or potential difference, such as a battery or resistor. This can damage or burn out the ammeter.

Ammeters are available in many types, sizes, and current ranges. Typical dc ammeters used with amateur equipment include current ranges of 0–50 μA, 0–1 mA, 0–200 mA, and 0–1 ampere. Most ac ammeters are rated in terms of frequency response or frequency of operation. For example, a 0–100-mA ac ammeter designed for 60-Hz power line measurements cannot be used to measure current at RF frequencies.

Voltmeters

Voltmeters are used to measure voltage or potential differences in electrical circuits. Unlike the ammeter, the voltmeter must be placed across the component or source of voltage to make the required measurement. Voltage measurements are illustrated in Fig. 11.5.

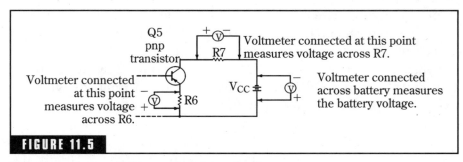

Voltage measurements in a pnp transistor circuit.

FIGURE 11.6

The Icom IC-207H Dual Band FM Mobile Transceiver. The IC-207H provides dual band coverage of the 2–meter and 440-cm amateur bands for FM operation. A "receive only" capability for 118–174 MHz AM is also provided. The mobile transceiver output power is 5–50 watts for VHF and 5–35 watts for UHF operation. The many operating features include FM repeater operation and 9600 bps (bits per second) packet radio with direct connections to an external packet modem.

A simple dc voltmeter consists of a dc ammeter in series with a resistor, as shown in Fig. 11.7. The sensitivity of the meter (current range) and the value of the series resistor determine the range of voltage that can be measured. For operation as a voltmeter, the meter scale is calibrated in volts instead of current.

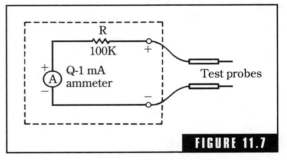

FIGURE 11.7

A simple 0–100 Vdc voltmeter circuit using a 0–1 mA meter.

The value of the series resistor for a given voltage range can be calculated as follows:

$$R = \frac{V_R}{I_M} \qquad \textbf{(Eq. 11.2)}$$

where

R is the resistance of the series resistor in ohms

V_R is the desired voltage range of the voltmeter in volts

I_M is the ammeter sensitivity in amperes

Ohmmeters

As the name implies, an ohmmeter measures ohms or resistance. Figure 11.8 illustrates resistance measurements in an electrical circuit.

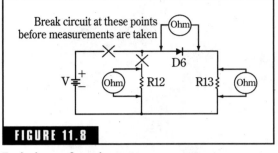

FIGURE 11.8

Typical use of an ohmmeter.

When making resistance measurements, the following conditions should be observed:

- Always turn off the power when measuring the resistance of a component wired into a circuit. Any voltage applied across an ohmmeter can damage the instrument. Also, the resistance measurement will probably be in error because an external voltage will affect the operation of the ohmmeter.

- It might be necessary to remove a component from a circuit before measuring its resistance. Other components in parallel with it might lower the resistance "seen" by the ohmmeter. In most cases, you can simply disconnect one side of the component and take the resistance reading.

- Semiconductor devices such as diodes and transistors will show a greater resistance in one direction than in the opposite direction. Therefore, by reversing the leads of the ohmmeter, you can often determine if the device is defective.

In its simplest form, the ohmmeter consists of an ammeter in series with a resistor and a battery. Figure 11.9 shows a simple ohmmeter. With

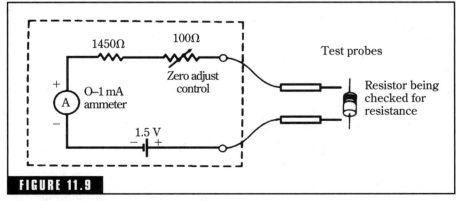

FIGURE 11.9

A simple ohmmeter circuit.

a short circuit connected across the test probes, the zero adjust control is adjusted for a full-scale reading on the ammeter. Thus, when a resistance is connected across the test probes, the ammeter reading will correspond to a known value of resistance. If the ammeter scale is calibrated in ohms, the value of the unknown resistor can be read directly.

The multimeter

One of the most versatile pieces of test equipment, the multitester (or multimeter) can be used to measure voltages, current, resistance, and in some instances, other electrical quantities. For example, the Radio Shack Model 22-220 10-megohm FET-Input Multitester in Figure 11.10A can be used to measure dc and ac voltage levels to 1000 volts in seven ranges with a 10-megohm input resistance. This multitester, also known as a volt-ohmmeter, measures dc current up to 10 amperes, resistance up to 100 megohms, and five ranges of decibels from −20 to +62 dB voltage levels.

A more recent form of multitester, the digital multimeter, is rapidly becoming a popular test instrument. The Radio Shack Model 22-174 Handheld True RMS Autoranging/Manual LCD Digital Multimeter, shown in Figure 11.10B is an example of this new generation of multi-testers. This test instrument measures dc and ac voltages, dc and ac current, frequency up to 4 MHz, dBm levels, capacitance from 4 nF to 400 μF, transistor gain, audible continuity buzzer, and even a temperature probe with C/F scales.

Wattmeters

Wattmeters are used to measure power delivered to a load. Amateurs use RF wattmeters to measure the RF power delivered from a transmitter to the transmission line. Some RF wattmeters contain an SWR (standing-wave ratio) meter for measuring both forward and reflected power. Figure 8.15 in Chap. 8 shows a typical RF wattmeter suitable for amateur use.

Operating Procedures

Finally, after passing the Technician examination and completing the installation of your amateur station, you are ready to begin operation. But first, you need to learn the basic principles of good operating procedures for successful and rewarding operation of that new ham rig. To avoid mass

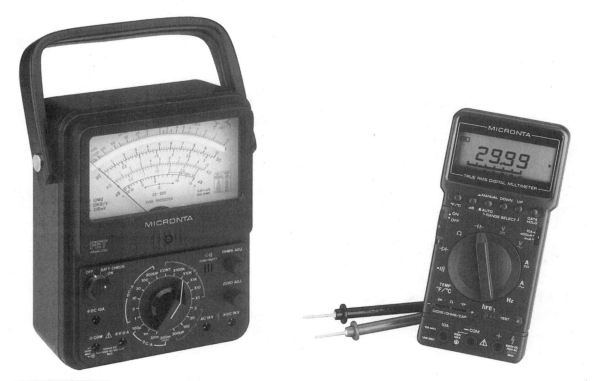

FIGURE 11.10

Analog and digital multimeters. These versatile test instruments provide an accurate means for measuring electrical qualities associated with amateur radio equipment as well as virtually all other types of electrical and electronic devices and equipment. (A) Radio Shack Model 22-220 Multitester employing a sensitive meter movement for measuring a wide range of voltage, current, and resistance. (B) Radio Shack Model 22-174 Handheld True RMS Autoranging/Manual LCD Digital Multimeter. This digital multimeter is capable of precise measurements of voltage, current, resistance, frequency, capacitance and temperature. *Tandy/Radio Shack*

confusion on the amateur bands, all hams use a common and efficient set of operating procedures that have been developed over the years. Each FCC Technician examination includes several questions covering this subject.

Operating courtesy

Before attempting to establish a QSO (communication) with another amateur, be sure you have tuned the transmitter for proper operation.

To avoid excessive tune-up time on the air (you might cause interference to an ongoing communication) use a dummy load for initial tune-up procedures. Check for activity on the desired frequency before tuning up the transmitter on-the-air. If necessary, shift up or down a few kHz to avoid causing interference to other hams.

In any communication with another amateur radio station, always zero beat your signal with the incoming signal from the other station. This ensures a more reliable contact; equally important, you and the other amateur radio station will be occupying less frequency spectrum. In today's crowded bands, this can make a difference in a successful contact.

Before sending out your first CQ, listen on the desired operating frequency to determine if it is already in use. If you attempt to use a frequency with a QSO in progress, the odds are that you are wasting your time as well as interfering with the other parties in communication. Remember, you have to share the amateur bands with many other hams.

Sometimes it is better to listen for someone calling CQ. Using this approach, you avoid interference to other hams and gain valuable experience in how to establish contacts. With a few QSOs under your belt, you will feel more confident in calling CQ.

One last item in operating courtesy—don't try to show off your code proficiency by sending too fast. If you want to make solid contacts, send your transmissions at a rate that can be copied by the distant ham. If he or she is transmitting at a rate too fast for good copy, send a QRS on your next transmission. This tells the person to slow down.

CW or telegraph procedures

Hams use an abbreviated language for CW operation to expedite communications. This universal language allows you to communicate with other hams regardless of country or nationality.

There are three parts to this language: Q signals that state or request specific information; the RST signal reporting system for exchanging information on signal quality; and standard abbreviations. With a little effort, you can quickly learn enough of this new language to pass the Technician examination. After a few months of operating experience, you will pick up the remaining parts.

Q signals

Q signals are always sent as a group of three letters that begin with the letter Q. Each Q signal has a particular meaning that is recognized by hams across the world. For example, QRN means "I am being troubled by static." If the Q signal is followed by a question mark (?), it is a query to the distant operator. QRN? asks "Are you being troubled by static?"

Table 11.3 lists most of the more common Q signals used in amateur radio communications and in the FCC Technician examination. A complete listing of all international Q signals as well as other operating aids can be found in the *ARRL Handbook For Radio Amateurs*.

The RST signal-reporting system

Amateurs use this signal-reporting system to describe the signal readability, strength, and tone characteristics of the received signal. In many instances, you will want to know how your signal is being received by a distant ham. By sending only three numbers, he or she can give you a complete signal report. The RST system is given in Table 11.4.

Prosigns and standard abbreviations

The final part of this new and exciting language used by CW operators is that of prosigns (procedural signs) and standard abbreviations. Most of these CW terms, required by regulations, help you to minimize the keying required to communicate with distant hams. Table 11.5 lists the more common prosigns and standard abbreviations you should know for the FCC examination and subsequent operation on the ham bands. The lines above the prosigns mean that there is no spacing between the characters.

A Typical Contact on the Novice Bands

The following series of messages illustrate how you can use CW transmissions to converse with other hams. In actual CW communications you would probably stay on the air for a much longer period of time, exchanging much information with the other operator. Also, you might want to use more or fewer abbreviations with longer or shorter sentences. Refer to Tables 11.3, 11.4, and 11.5 to determine the full meaning of this short series of messages.

Your initial transmission:
CQ CQ CQ DE WC4HOI WC4HOI WC4HOI K

The first reply:
WC4HOI DE WD6LZZ R K

TABLE 11.3 Some of the More Common Q Signals Used in Amateur Radio

Q signal	Meaning as statement	Meaning as question
QRM	I am experiencing interference.	Are you experiencing interference?
QRN	I am troubled with atmospherics and static.	Are you being troubled with atmospherics and static?
QRP	I am reducing power. (Alternate meanings: I am transmitting on low power or Please reduce power.)	Shall I reduce power?
QRS	Please send slower.(You are sending too fast, I am having trouble copying you.)	Shall I send slower?
QRT	I am going to stop sending.	Are you going to stop sending?
QRX	Wait (standby), I will call again at (time)	When will you call again? (Or: Please call me if convenient.)
QRZ	You are being called by (a third station)	Who is calling me?
QSA	Your signal strength is _____.	What is my signal strength?
	(A scale of 1 to 5: 1 - Barely perceptible 2 - Weak 3 - Fair 4 - Good 5 - Excellent)	
QSB	Your signals are fading.	Are my signals fading?
QSL	I (will) acknowledge receipt (or I will send you a QSL card acknowledging this contact).	Please acknowledge receipt (or Please send me a QSL card acknowledging this contact).
QSO	I can communicate with _____ directly.	Can you communicate with _____ directly?
QSY	I am changing operating frequency to _____ kHz.	Are you going to change operating frequency (or Can you change operating frequency to _____ kHz?).
QTH	My location is _____.	What is your location _____?

TABLE 11.4 The RST Signal Reporting System

Readability

1 - Unreadable

2 - Barely readable, some words understandable

3 - Readable with considerable difficulty

4 - Readable with practically no difficulty

5 - Perfectly readable

Signal strength

1 - Faint signals, barely perceptible

2 - Very weak signals

3 - Weak signals

4 - Fair signals

5 - Fairly good signals

6 - Good signals

7 - Moderately strong signals

8 - Strong signals

9 - Extremely strong signals

Tone

1 - Sixty cycle ac hum or less, very rough or broad

2 - Very rough ac, very harsh and broad

3 - Rough ac tone, rectified but not filtered

4 - Rough note, some trace of filtering

5 - Filtered rectified ac but strongly ripple modulated

6 - Filtered tone, definite trace of ripple modulation

7 - Near pure tone, trace of modulation

8 - Near perfect tone, trace of modulation

9 - Perfect tone, no trace of ripple or modulation of any kind (Add an X for crystal-clear tones)

TABLE 11.5 Prosigns and Standard Abbreviations Used in Amateur Radio.

Prosign	Meaning
AR	End of transmission
AS	Standby (or "Wait until I call you")
BK	Break (or "Please allow me to break into this conversation")
BT	Separation between address and message text, and between message text and signature
CQ	General call to any station
DE	From (or "This is _____ ")
K	Go ahead
R	Roger (or affirmative)
SK	Out (or end of communication and no more communications expected)
V	Test (on-the-air test signal) Normally a series of Vs followed by the operator's call sign.
Popular abbreviations used by many amateurs	
CUL	See you later
DX	Distant or far-away station
GA	Go ahead
GM	Good morning
OM	Old man
Rig	Ham equipment, normally the transmitter or transceiver
TNX	Thanks
UR	Your
WX	Weather
YL	Your lady
XYL	Wife
73	Best regards
88	Love and kisses

Your second transmission:

QRZ? QRZ? WD6? DE WC4HOI AR K

The second reply:

WC4HOI DE WD6LZZ WD6LZZ WD6LZZ AR K

Your third transmission:

WD6LZZ DE WC4HOI R UR CALL BT HAVE QRN HERE. NAME IS CHARLIE. QSA? UR SIG 5-7-9. QTH IS WINSTON SALEM, NC. PLS GIVE UR NAME AND
QTH? BT WD6LZZ DE WC4HOI AR K

The third reply:

WC4HOI DE WD6LZZ R BT NAME IS JAMES. QTH IS LOS ANGE-LES, CA. UR
SIG 5-9-9. MANY TNX FOR QSO. HAVE TO LEAVE FOR SCHOOL. CUL OM BT
WC4HOI DE WD6LZZ SK K

Your final transmission:

WD6LZZ DE WC4HOI R BT TNX FOR QSO. HOPE TO HEAR U ON TOMOR-
ROW AT SAME TIME. CUL BT WD6LZZ DE WC4HOI SK

Element 2—Technician Class Examination Question Pool*

Subelement T1: Commission's Rules (9 Exam Questions—9 Groups)

T1A Basis and purpose of amateur service and definitions; Station/Operator license; classes of US amateur licenses, including basic differences; privileges of the various license classes; term of licenses; grace periods; modifications of licenses; current mailing address on file with FCC

T1A01 Who makes and enforces the rules and regulations of the amateur service in the US?

A. The Congress of the United States

B. The Federal Communications Commission (FCC)

C. The Volunteer Examiner Coordinators (VECs)

D. The Federal Bureau of Investigation (FBI)

*As released by the Question Pool Committee National Conference of Volunteer Examiner Coordinators February 1, 2000

T1A02   What are two of the five purposes for the amateur service?

A. To protect historical radio data, and help the public understand radio history

B. To help foreign countries improve communication and technical skills, and encourage visits from foreign hams

C. To modernize radio schematic drawings, and increase the pool of electrical drafting people

D. To increase the number of trained radio operators and electronics experts, and improve international goodwill

T1A03   What is the definition of an amateur station?

A. A station in a public radio service used for radiocommunications

B. A station using radiocommunications for a commercial purpose

C. A station using equipment for training new broadcast operators and technicians

D. A station in the Amateur Radio service used for radiocommunications

T1A04   What is the definition of a control operator of an amateur station?

A. Anyone who operates the controls of the station

B. Anyone who is responsible for the station's equipment

C. Any licensed amateur operator who is responsible for the station's transmissions

D. The amateur operator with the highest class of license who is near the controls of the station

T1A05   Which of the following is required before you can operate an amateur station in the US?

A. You must hold an FCC operator's training permit for a licensed radio station

B. You must submit an FCC Form 605 together with a license examination fee

C. The FCC must grant you an amateur operator/primary station license

D. The FCC must issue you a Certificate of Successful Completion of Amateur Training

T1A06 What must happen before you are allowed to operate an amateur station?

A. The FCC database must show that you have been granted an amateur license

B. You must have written authorization from the FCC

C. You must have written authorization from a Volunteer Examiner Coordinator

D. You must have a copy of the FCC Rules, Part 97, at your station location

T1A07 What are the US amateur operator licenses that a new amateur might earn?

A. Novice, Technician, General, Advanced

B. Technician, Technician Plus, General, Advanced

C. Novice, Technician, General, Advanced

D. Technician, Technician with Morse code, General, Amateur Extra

T1A08 How soon after you pass the elements required for your first Amateur Radio license may you transmit?

A. Immediately

B. 30 days after the test date

C. As soon as the FCC grants you a license

D. As soon as you receive your license from the FCC

T1A09 How soon before the expiration date of your license should you send the FCC a completed Form 605 or file with the Universal Licensing System on the World Wide Web for a renewal?

A. No more than 90 days

B. No more than 30 days

C. Within 6 to 9 months

D. Within 6 months to a year

T1A10   What is the normal term for which a new amateur station license is granted?

A. 5 years

B. 7 years

C. 10 years

D. For the lifetime of the licensee

T1A11   What is the "grace period" during which the FCC will renew an expired 10-year license?

A. 2 years

B. 5 years

C. 10 years

D. There is no grace period

T1A12   What is one way you may notify the FCC if your mailing address changes?

A. Fill out an FCC Form 605 using your new address, attach a copy of your license, and mail it to your local FCC Field Office

B. Fill out an FCC Form 605 using your new address, attach a copy of your license, and mail it to the FCC office in Gettysburg, PA

C. Call your local FCC Field Office and give them your new address over the phone or e-mail this information to the local Field Office

D. Call the FCC office in Gettysburg, PA, and give them your new address over the phone or e-mail this information the FCC

T1B   Frequency privileges authorized to the Technician control operator (VHF/UHF and HF)

T1B01   What are the frequency limits of the 6-meter band in ITU Region 2?

A. 52.0–54.5 MHz

B. 50.0–54.0 MHz

C. 50.1–52.1 MHz

D. 50.0–56.0 MHz

T1B02   What are the frequency limits of the 2-meter band in ITU Region 2?

A. 145.0–150.5 MHz

B. 144.0–148.0 MHz

C. 144.1–146.5 MHz

D. 144.0–146.0 MHz

T1B03   What are the frequency limits of the 1.25-meter band in ITU Region 2?

A. 225.0–230.5 MHz

B. 222.0–225.0 MHz

C. 224.1–225.1 MHz

D. 220.0–226.0 MHz

T1B04   What are the frequency limits of the 70-centimeter band in ITU Region 2?

A. 430.0–440.0 MHz

B. 430.0–450.0 MHz

C. 420.0–450.0 MHz

D. 432.0–435.0 MHz

T1B05   What are the frequency limits of the 33-centimeter band in ITU Region 2?

A. 903–927 MHz

B. 05–925 MHz

C. 900–930 MHz

D. 902–928 MHz

T1B06   What are the frequency limits of the 23-centimeter band?

A. 1260–1270 MHz

B. 1240–1300 MHz

C. 1270–1295 MHz

D. 1240–1246 MHz

T1B07   What are the frequency limits of the 13-centimeter band in ITU Region 2?

A. 2300–2310 MHz and 2390–2450 MHz

B. 2300–2350 MHz and 2400–2450 MHz

C. 2350–2380 MHz and 2390–2450 MHz

D. 2300–2350 MHz and 2380–2450 MHz

T1B08   What are the frequency limits of the 80-meter band for Technician class licensees who have passed a Morse code exam?

A. 3500–4000 kHz

B. 3675–3725 kHz

C. 7100–7150 kHz

D. 7000–7300 kHz

T1B09   What are the frequency limits of the 40-meter band in ITU Region 2 for Technician class licensees who have passed a Morse code exam?

A. 3500–4000 kHz

B. 3700–3750 kHz

C. 7100–7150 kHz

D. 7000–7300 kHz

T1B10   What are the frequency limits of the 15-meter band for Technician class licensees who have passed a Morse code exam?

A. 21.100–21.200 MHz

B. 21.000–21.450 MHz

C. 28.000–29.700 MHz

D. 28.100–28.200 MHz

T1B11   What are the frequency limits of the 10-meter band for Technician class licensees who have passed a Morse code exam?

A. 28.000–28.500 MHz

B. 28.100–29.500 MHz

C. 28.100–28.500 MHz

D. 29.100–29.500 MHz

T1B12   If you are a Technician licensee who has passed a Morse code exam, what is one document you can use to prove that you are authorized to use certain amateur frequencies below 30 MHz?

A. A certificate from the FCC showing that you have notified them that you will be using the HF bands

B. A certificate showing that you have attended a class in HF communications

C. A Certificate of Successful Completion of Examination showing that you have passed a Morse code exam

D. No special proof is required

T1C   Emission privileges authorized to the Technician control operator (VHF/UHF and HF)

T1C01   On what HF band may a Technician licensee use FM phone emission?

A. 10 meters

B. 15 meters

C. 75 meters

D. None

T1C02 On what frequencies within the 6-meter band may phone emissions be transmitted?

A. 50.0–54.0 MHz only

B. 50.1–54.0 MHz only

C. 51.0–54.0 MHz only

D. 52.0–54.0 MHz only

T1C03 On what frequencies within the 2-meter band may image emissions be transmitted?

A. 144.1–148.0 MHz only

B. 146.0–148.0 MHz only

C. 144.0–148.0 MHz only

D. 146.0–147.0 MHz only

T1C04 What frequencies within the 2-meter band are reserved exclusively for CW operations?

A. 146–147 MHz

B. 146.0–146.1 MHz

C. 145–148 MHz

D. 144.0–144.1 MHz

T1C05 What emission types are Technician control operators who have passed a Morse code exam allowed to use in the 80-meter band?

A. CW only

B. Data only

C. RTTY only

D. Phone only

T1C06 What emission types are Technician control operators who have passed a Morse code exam allowed to use from 7100 to 7150 kHz in ITU Region 2?

TECHNICIAN CLASS EXAMINATION QUESTION POOL

A. CW and data

B. Phone

C. Data only

D. CW only

T1C07 What emission types are Technician control operators who have passed a Morse code exam allowed to use on frequencies from 28.1 to 28.3 MHz?

A. All authorized amateur emission privileges

B. Data or phone

C. CW, RTTY and data

D. CW and phone

T1C08 What emission types are Technician control operators who have passed a Morse code exam allowed to use on frequencies from 28.3 to 28.5 MHz?

A. All authorized amateur emission privileges

B. CW and data

C. CW and single-sideband phone

D. Data and phone

T1C09 What emission types are Technician control operators allowed to use on the amateur 1.25-meter band in ITU Region 2?

A. Only CW and phone

B. Only CW and data

C. Only data and phone

D. All amateur emission privileges authorized for use on the band

T1C10 What emission types are Technician control operators allowed to use on the amateur 23-centimeter band?

A. Only data and phone

B. Only CW and data

C. Only CW and phone

D. All amateur emission privileges authorized for use on the band

T1C11   On what frequencies within the 70-centimeter band in ITU Region 2 may image emissions be transmitted?

A. 420.0–420.1 MHz only

B. 430.0–440.0 MHz only

C. 420.0–450.0 MHz only

D. 440.0–450.0 MHz only

T1D   Responsibility of licensee; station control; control operator requirements; station identification; points of communication and operation; business communications

T1D01   What is the control point of an amateur station?

A. The on/off switch of the transmitter

B. The input/output port of a packet controller

C. The variable frequency oscillator of a transmitter

D. The location at which the control operator function is performed

T1D02   Who is responsible for the proper operation of an amateur station?

A. Only the control operator

B. Only the station licensee

C. Both the control operator and the station licensee

D. The person who owns the station equipment

T1D03   What is your responsibility as a station licensee?

A. You must allow another amateur to operate your station upon request

B. You must be present whenever the station is operated

C. You must notify the FCC if another amateur acts as the control operator

D. You are responsible for the proper operation of the station in accordance with the FCC rules

T1D04   Who may be the control operator of an amateur station?

A. Any person over 21 years of age

B. Any person over 21 years of age with a General class license or higher

C. Any licensed amateur chosen by the station licensee

D. Any licensed amateur with a Technician class license or higher

T1D05   If you are the control operator at the station of another amateur who has a higher class license than yours, what operating privileges are you allowed?

A. Any privileges allowed by the higher license

B. Only the privileges allowed by your license

C. All the emission privileges of the higher license, but only the frequency privileges of your license

D. All the frequency privileges of the higher license, but only the emission privileges of your license

T1D06   When an amateur station is transmitting, where must its control operator be?

A. At the station's control point

B. Anywhere in the same building as the transmitter

C. At the station's entrance, to control entry to the room

D. Anywhere within 50 km of the station location

T1D07   How often must an amateur station be identified?

A. At the beginning of a contact and at least every ten minutes after that

B. At least once during each transmission

C. At least every 10 minutes during and at the end of a contact

D. At the beginning and end of each transmission

T1D08   What identification, if any, is required when two amateur stations begin communications?

A. No identification is required

B. One of the stations must give both stations' call signs

C. Each station must transmit its own call sign

D. Both stations must transmit both call signs

T1D09   What identification, if any, is required when two amateur stations end communications?

A. No identification is required

B. One of the stations must transmit both stations' call signs

C. Each station must transmit its own call sign

D. Both stations must transmit both call signs

T1D10   What is the longest period of time an amateur station can operate without transmitting its call sign?

A. 5 minutes

B. 10 minutes

C. 15 minutes

D. 30 minutes

T1D11   What emission type may always be used for station identification, regardless of the transmitting frequency?

A. CW

B. RTTY

C. MCW

D. Phone

T1D12   If you are a Technician licensee with a Certificate of Successful Completion of Examination (CSCE) for a Morse code exam, how should you identify your station when transmitting on the 10-meter band?

A. You must give your call sign followed by the words "plus plus"

B. You must give your call sign followed by the words "temporary plus"

C. No special form of identification is needed

D. You must give your call sign and the location of the VE examination where you obtained the CSCE

T1E   Third-party communication; authorized and prohibited transmissions; permissible one-way communication

T1E01   What kind of payment is allowed for third-party messages sent by an amateur station?

A. Any amount agreed upon in advance

B. Donation of repairs to amateur equipment

C. Donation of amateur equipment

D. No payment of any kind is allowed

T1E02   What is the definition of third-party communications?

A. A message sent between two amateur stations for someone else

B. Public service communications for a political party

C. Any messages sent by amateur stations

D. A three-minute transmission to another amateur

T1E03   What is a "third party" in amateur communications?

A. An amateur station that breaks in to talk

B. A person who is sent a message by amateur communications other than a control operator who handles the message

C. A shortwave listener who monitors amateur communications

D. An unlicensed control operator

T1E04   When are third-party messages allowed to be sent to a foreign country?

A. When sent by agreement of both control operators

B. When the third party speaks to a relative

C. They are not allowed under any circumstances

D. When the US has a third-party agreement with the foreign country or the third party is qualified to be a control operator

T1E05   If you let an unlicensed third party use your amateur station, what must you do at your station's control point?

A. You must continuously monitor and supervise the third-party's participation

B. You must monitor and supervise the communication only if contacts are made in countries that have no third-party communications agreement with the United States

C. You must monitor and supervise the communication only if contacts are made on frequencies below 30 MHz

D. You must key the transmitter and make the station identification

T1E06   Besides normal identification, what else must a U.S. station do when sending third-party communications internationally?

A. The U.S. station must transmit its own call sign at the beginning of each communication, and at least every ten minutes after that

B. The U.S. station must transmit both call signs at the end of each communication

C. The U.S. station must transmit its own call sign at the beginning of each communication, and at least every five minutes after that

D. Each station must transmit its own call sign at the end of each transmission, and at least every five minutes after that

T1E07   When is an amateur allowed to broadcast information to the general public?

A. Never

B. Only when the operator is being paid

C. Only when broadcasts last less than 1 hour

D. Only when broadcasts last longer than 15 minutes

T1E08 When is an amateur station permitted to transmit music?

A. Never, except incidental music during authorized rebroadcasts of space shuttle communications

B. Only if the transmitted music produces no spurious emissions

C. Only if it is used to jam an illegal transmission

D. Only if it is above 1280 MHz, and the music is a live performance

T1E09 When is the use of codes or ciphers allowed to hide the meaning of an amateur message?

A. Only during contests

B. Only during nationally declared emergencies

C. Never, except when special requirements are met

D. Only on frequencies above 1280 MHz

T1E10 Which of the following one-way communications may not be transmitted in the amateur service?

A. Telecommands to model craft

B. Broadcasts intended for the general public

C. Brief transmissions to make adjustments to the station

D. Morse code practice

T1E11 If you are allowing a non-amateur friend to use your station to talk to someone in the US, and a foreign station breaks in to talk to your friend, what should you do?

A. Have your friend wait until you find out if the US has a third-party agreement with the foreign station's government

B. Stop all discussions and quickly sign off

C. Since you can talk to any foreign amateurs, your friend may keep talking as long as you are the control operator

D. Report the incident to the foreign amateur's government

T1E12   When are you allowed to transmit a message to a station in a foreign country for a third party?

A. Anytime

B. Never

C. Anytime, unless there is a third-party agreement between the US and the foreign government

D. If there is a third-party agreement with the US government, or if the third party is eligible to be the control operator

T1F   Frequency selection and sharing; transmitter power; digital communications

T1F01   If the FCC rules say that the amateur service is a secondary user of a frequency band, and another service is a primary user, what does this mean?

A. Nothing special; all users of a frequency band have equal rights to operate

B. Amateurs are only allowed to use the frequency band during emergencies

C. Amateurs are allowed to use the frequency band only if they do not cause harmful interference to primary users

D. Amateurs must increase transmitter power to overcome any interference caused by primary users

T1F02   What rule applies if two amateur stations want to use the same frequency?

A. The station operator with a lesser class of license must yield the frequency to a higher-class licensee

B. The station operator with a lower power output must yield the frequency to the station with a higher power output

C. Both station operators have an equal right to operate on the frequency

D. Station operators in ITU Regions 1 and 3 must yield the frequency to stations in ITU Region 2

T1F03 If a repeater is causing harmful interference to another repeater and a frequency coordinator has recommended the operation of one repeater only, who is responsible for resolving the interference?

A. The licensee of the unrecommended repeater

B. Both repeater licensees

C. The licensee of the recommended repeater

D. The frequency coordinator

T1F04 If a repeater is causing harmful interference to another amateur repeater and a frequency coordinator has recommended the operation of both repeaters, who is responsible for resolving the interference?

A. The licensee of the repeater that has been recommended for the longest period of time

B. The licensee of the repeater that has been recommended the most recently

C. The frequency coordinator

D. Both repeater licensees

T1F05 What is the term for the average power supplied to an antenna transmission line during one RF cycle at the crest of the modulation envelope?

A. Peak transmitter power

B. Peak output power

C. Average radio-frequency power

D. Peak envelope power

T1F06 What is the maximum transmitting power permitted an amateur station on 146.52 MHz?

A. 200 watts pep output

B. 500 watts ERP

C. 1000 watts dc input

D. 1500 watts pep output

T1F07 On which band(s) may a Technician licensee who has passed a Morse code exam use up to 200 watts pep output power?

A. 80, 40, 15, and 10 meters

B. 80, 40, 20, and 10 meters

C. 1.25 meters

D. 23 centimeters

T1F08 What amount of transmitter power must amateur stations use at all times?

A. 25 watts pep output

B. 250 watts pep output

C. 1500 watts pep output

D. The minimum legal power necessary to communicate

T1F09 What name does the FCC use for telemetry, telecommand or computer communications emissions?

A. CW

B. Image

C. Data

D. RTTY

T1F10 What name does the FCC use for narrow-band direct-printing telegraphy emissions?

A. CW

B. Image

C. MCW

D. RTTY

T1F11 What is the maximum symbol rate permitted for packet transmissions on the 2-meter band?

A. 300 bauds

B. 1200 bauds

C. 19.6 kilobauds

D. 56 kilobauds

T1F12   What is the maximum symbol rate permitted for RTTY or data transmissions on the 6- and 2-meter bands?

A. 56 kilobauds

B. 19.6 kilobauds

C. 1200 bauds

D. 300 bauds

T1G   Satellite and space communications; false signals or unidentified communications; malicious interference

T1G01   What is an amateur space station?

A. An amateur station operated on an unused frequency

B. An amateur station awaiting its new call letters from the FCC

C. An amateur station located more than 50 kilometers above the Earth's surface

D. An amateur station that communicates with the International Space Station

T1G02   Who may be the licensee of an amateur space station?

A. An amateur holding an Amateur Extra class operator license

B. Any licensed amateur operator

C. Anyone designated by the commander of the spacecraft

D. No one unless specifically authorized by the government

T1G03   Which band may NOT be used by Earth stations for satellite communications?

A. 6 meters

B. 2 meters

C. 70 centimeters

D. 23 centimeters

T1G04   When may false or deceptive amateur signals or communications be transmitted?

A. Never

B. When operating a beacon transmitter in a "fox hunt" exercise

C. When playing a harmless "practical joke"

D. When you need to hide the meaning of a message for secrecy

T1G05   If an amateur pretends there is an emergency and transmits the word "MAYDAY," what is this called?

A. A traditional greeting in May

B. An emergency test transmission

C. False or deceptive signals

D. Nothing special; "MAYDAY" has no meaning in an emergency

T1G06   When may an amateur transmit unidentified communications?

A. Only for brief tests not meant as messages

B. Only if it does not interfere with others

C. Never, except transmissions from a space station or to control a model craft

D. Only for two-way or third-party communications

T1G07   What is an amateur communication called that does not have the required station identification?

A. Unidentified communications or signals

B. Reluctance modulation

C. Test emission

D. Tactical communication

T1G08   If an amateur transmits to test access to a repeater without giving any station identification, what type of communication is this called?

A. A test emission; no identification is required

B. An illegal unmodulated transmission

C. An illegal unidentified transmission

D. A non-communication; no voice is transmitted

T1G09 When may you deliberately interfere with another station's communications?

A. Only if the station is operating illegally

B. Only if the station begins transmitting on a frequency you are using

C. Never

D. You may expect, and cause, deliberate interference because it can't be helped during crowded band conditions

T1G10 If an amateur repeatedly transmits on a frequency already occupied by a group of amateurs in a net operation, what type of interference is this called?

A. Break-in interference

B. Harmful or malicious interference

C. Incidental interference

D. Intermittent interference

T1G11 What is a transmission called that disturbs other communications?

A. Interrupted CW

B. Harmful interference

C. Transponder signals

D. Unidentified transmissions

T1H Correct language; phonetics; beacons; radio control of model craft and vehicles

T1H01 If you are using a language besides English to make a contact, what language must you use when identifying your station?

A. The language being used for the contact

B. The language being used for the contact, provided the US has a third-party communications agreement with that country

C. English

D. Any language of a country that is a member of the International Telecommunication Union

T1H02 What do the FCC Rules suggest you use as an aid for correct station identification when using phone?

A. A speech compressor

B. Q signals

C. A phonetic alphabet

D. Unique words of your choice

T1H03 What is the advantage in using the International Telecommunication Union (ITU) phonetic alphabet when identifying your station?

A. The words are internationally recognized substitutes for letters

B. There is no advantage

C. The words have been chosen to be easily pronounced by Asian cultures

D. It preserves traditions begun in the early days of Amateur Radio

T1H04 What is one reason to avoid using "cute" phrases or word combinations to identify your station?

A. They are not easily understood by non-English-speaking amateurs

B. They might offend English-speaking amateurs

C. They do not meet FCC identification requirements

D. They might be interpreted as codes or ciphers intended to obscure the meaning of your identification

T1H05 What is an amateur station called that transmits communications for the purpose of observation of propagation and reception?

A. A beacon

B. A repeater

C. An auxiliary station

D. A radio control station

T1H06 What is the maximum transmitting power permitted an amateur station in beacon operation?

A. 10 watts pep output

B. 100 watts pep output

C. 500 watts pep output

D. 1500 watts pep output

T1H07 What minimum class of amateur license must you hold to operate a beacon or a repeater station?

A. Technician with credit for passing a Morse code exam

B. Technician

C. General

D. Amateur Extra

T1H08 What minimum information must be on a label affixed to a transmitter used for telecommand (control) of model craft?

A. Station call sign

B. Station call sign and the station licensee's name

C. Station call sign and the station licensee's name and address

D. Station call sign and the station licensee's class of license

T1H09 What is the maximum transmitter power an amateur station is allowed when used for telecommand (control) of model craft?

A. One milliwatt

B. One watt

C. 25 watts

D. 100 watts

T1I Emergency communications; broadcasting; indecent and obscene language

T1I01 If you hear a voice distress signal on a frequency outside of your license privileges, what are you allowed to do to help the station in distress?

A. You are NOT allowed to help because the frequency of the signal is outside your privileges

B. You are allowed to help only if you keep your signals within the nearest frequency band of your privileges

C. You are allowed to help on a frequency outside your privileges only if you use international Morse code

D. You are allowed to help on a frequency outside your privileges in any way possible

T1I02 When may you use your amateur station to transmit an "SOS" or "MAYDAY"?

A. Never

B. Only at specific times (at 15 and 30 minutes after the hour)

C. In a life- or property-threatening emergency

D. When the National Weather Service has announced a severe weather watch

T1I03 When may you send a distress signal on any frequency?

A. Never

B. In a life- or property-threatening emergency

C. Only at specific times (at 15 and 30 minutes after the hour)

D. When the National Weather Service has announced a severe weather watch

T1I04 If a disaster disrupts normal communication systems in an area where the amateur service is regulated by the FCC, what kinds of transmissions may stations make?

A. Those that are necessary to meet essential communication needs and facilitate relief actions

B. Those that allow a commercial business to continue to operate in the affected area

C. Those for which material compensation has been paid to the amateur operator for delivery into the affected area

D. Those that are to be used for program production or news gathering for broadcasting purposes

T1I05   What information is included in an FCC declaration of a temporary state of communication emergency?

A. A list of organizations authorized to use radio communications in the affected area

B. A list of amateur frequency bands to be used in the affected area

C. Any special conditions and special rules to be observed during the emergency

D. An operating schedule for authorized amateur emergency stations

T1I06   What is meant by the term broadcasting?

A. Transmissions intended for reception by the general public, either direct or relayed

B. Retransmission by automatic means of programs or signals from non-amateur stations

C. One-way radio communications, regardless of purpose or content

D. One-way or two-way radio communications between two or more stations

T1I07   When may you send obscene words from your amateur station?

A. Only when they do not cause interference to other communications

B. Never; obscene words are not allowed in amateur transmissions

C. Only when they are not retransmitted through a repeater

D. Any time, but there is an unwritten rule among amateurs that they should not be used on the air

T1I08 When may you send indecent words from your amateur station?

A. Only when they do not cause interference to other communications

B. Only when they are not retransmitted through a repeater

C. Any time, but there is an unwritten rule among amateurs that they should not be used on the air

D. Never; indecent words are not allowed in amateur transmissions

T1I09 Why is indecent and obscene language prohibited in the Amateur Service?

A. Because it is offensive to some individuals

B. Because young children may intercept amateur communications with readily available receiving equipment

C. Because such language is specifically prohibited by FCC Rules

D. All of these choices are correct

T1I10 Where can the official list of prohibited obscene and indecent words be found?

A. There is no public list of prohibited obscene and indecent words; if you believe a word is questionable, don't use it in your communications

B. The list is maintained by the Department of Commerce

C. The list is International, and is maintained by Industry Canada

D. The list is in the "public domain," and can be found in all amateur study guides

T1I11 Under what conditions may a Technician class operator use his or her station to broadcast information intended for reception by the general public?

A. Never, broadcasting is a privilege reserved for Extra and General class operators only

B. Only when operating in the FM broadcast band (88.1 to 107.9 MHz)

C. Only when operating in the AM broadcast band (530 to 1700 kHz)

D. Never, broadcasts intended for reception by the general public are not permitted in the Amateur Service

Subelement T2: Operating Procedures (5 Exam Questions—5 Groups)

T2A   Preparing to transmit; choosing a frequency for tune-up; operating or emergencies; Morse code; repeater operations and autopatch

T2A01   What should you do before you transmit on any frequency?

A. Listen to make sure others are not using the frequency

B. Listen to make sure that someone will be able to hear you

C. Check your antenna for resonance at the selected frequency

D. Make sure the SWR on your antenna feed line is high enough

T2A02   If you are in contact with another station and you hear an emergency call for help on your frequency, what should you do?

A. Tell the calling station that the frequency is in use

B. Direct the calling station to the nearest emergency net frequency

C. Call your local Civil Preparedness Office and inform them of the emergency

D. Stop your QSO immediately and take the emergency call

T2A03   Why should local amateur communications use VHF and UHF frequencies instead of HF frequencies?

A. To minimize interference on HF bands capable of long-distance communication

B. Because greater output power is permitted on VHF and UHF

C. Because HF transmissions are not propagated locally

D. Because signals are louder on VHF and UHF frequencies

T2A04   How can on-the-air interference be minimized during a lengthy transmitter testing or loading-up procedure?

A. Choose an unoccupied frequency

B. Use a dummy load

C. Use a nonresonant antenna

D. Use a resonant antenna that requires no loading-up procedure

T2A05   At what speed should a Morse code CQ call be transmitted?

A. Only speeds below 5 WPM

B. The highest speed your keyer will operate

C. Any speed at which you can reliably receive

D. The highest speed at which you can control the keyer

T2A06   What is an autopatch?

A. An automatic digital connection between a US and a foreign amateur

B. A digital connection used to transfer data between a hand-held radio and a computer

C. A device that allows radio users to access the public telephone system

D. A video interface allowing images to be patched into a digital data stream

T2A07   How do you call another station on a repeater if you know the station's call sign?

A. Say "break, break 79," then say the station's call sign

B. Say the station's call sign, then identify your own station

C. Say "CQ" three times, then say the station's call sign

D. Wait for the station to call "CQ," then answer it

T2A08   What is a courtesy tone (used in repeater operations)?

A. A sound used to identify the repeater

B. A sound used to indicate when a transmission is complete

C. A sound used to indicate that a message is waiting for someone

D. A sound used to activate a receiver in case of severe weather

T2A09   What is the meaning of the procedural signal "DE"?

A. "From" or "this is," as in "W0AIH DE KA9FOX"

B. "Directional Emissions" from your antenna

C. "Received all correctly"

D. "Calling any station"

T2A10   During commuting rush hours, which type of repeater operation should be discouraged?

A. Mobile stations

B. Low-power stations

C. Highway traffic information nets

D. Third-party communications nets

T2A11   What is the proper way to break into a conversation on a repeater?

A. Wait for the end of a transmission and start calling the desired party

B. Shout, "break, break!" to show that you're eager to join the conversation

C. Turn on an amplifier and override whoever is talking

D. Say your call sign during a break between transmissions

T2B   Definition and proper use; courteous operation; repeater frequency coordination; Morse code

T2B01   When using a repeater to communicate, which of the following do you need to know about the repeater?

A. Its input frequency and offset

B. Its call sign

C. Its power level

D. Whether or not it has an autopatch

T2B02   What is an autopatch?

A. Something that automatically selects the strongest signal to be repeated

B. A device that connects a mobile station to the next repeater if it moves out of range of the first

C. A device that allows repeater users to make telephone calls from their stations

D. A device that locks other stations out of a repeater when there is an important conversation in progress

T2B03   What is the purpose of a repeater time-out timer?

A. It lets a repeater have a rest period after heavy use

B. It logs repeater transmit time to predict when a repeater will fail

C. It tells how long someone has been using a repeater

D. It limits the amount of time someone can transmit on a repeater

T2B04   What is a CTCSS (or PL) tone?

A. A special signal used for telecommand control of model craft

B. A subaudible tone, added to a carrier, which may cause a receiver to accept a signal

C. A tone used by repeaters to mark the end of a transmission

D. A special signal used for telemetry between amateur space stations and Earth stations

T2B05   What is the usual input/output frequency separation for repeaters in the 2-meter band?

A. 600 kHz

B. 1.0 MHz

C. 1.6 MHz

D. 5.0 MHz

T2B06   What is the usual input/output frequency separation for repeaters in the 1.25-meter band?

A. 600 kHz

B. 1.0 MHz

C. 1.6 MHz

D. 5.0 MHz

T2B07   What is the usual input/output frequency separation for repeaters in the 70-centimeter band?

A. 600 kHz

B. 1.0 MHz

C. 1.6 MHz

D. 5.0 MHz

T2B08 What is the purpose of repeater operation?

A. To cut your power bill by using someone else's higher-power system

B. To help mobile and low-power stations extend their usable range

C. To transmit signals for observing propagation and reception

D. To communicate with stations in services other than amateur

T2B09 What is a repeater called that is available for anyone to use?

A. An open repeater

B. A closed repeater

C. An autopatch repeater

D. A private repeater

T2B10 Why should you pause briefly between transmissions when using a repeater?

A. To check the SWR of the repeater

B. To reach for pencil and paper for third-party communications

C. To listen for anyone wanting to break in

D. To dial up the repeater's autopatch

T2B11 Why should you keep transmissions short when using a repeater?

A. A long transmission may prevent someone with an emergency from using the repeater

B. To see if the receiving station operator is still awake

C. To give any listening non-hams a chance to respond

D. To keep long-distance charges down

T2C   Simplex operations; RST signal reporting; choice of equipment for desired communications; communications modes including amateur television (ATV), packet radio; Q signals, procedural signals and abbreviations

T2C01   What is simplex operation?

A. Transmitting and receiving on the same frequency

B. Transmitting and receiving over a wide area

C. Transmitting on one frequency and receiving on another

D. Transmitting one-way communications

T2C02   When should you use simplex operation instead of a repeater?

A. When the most reliable communications are needed

B. When a contact is possible without using a repeater

C. When an emergency telephone call is needed

D. When you are traveling and need some local information

T2C03   Why should simplex be used where possible, instead of using a repeater?

A. Signal range will be increased

B. Long distance toll charges will be avoided

C. The repeater will not be tied up unnecessarily

D. Your antenna's effectiveness will be better tested

T2C04   If you are talking to a station using a repeater, how would you find out if you could communicate using simplex instead?

A. See if you can clearly receive the station on the repeater's input frequency

B. See if you can clearly receive the station on a lower frequency band

C. See if you can clearly receive a more distant repeater

D. See if a third station can clearly receive both of you

T2C05   What does RST mean in a signal report?

A. Recovery, signal strength, tempo

B. Recovery, signal speed, tone

C. Readability, signal speed, tempo

D. Readability, signal strength, tone

T2C06   What is the meaning of: "Your signal report is five nine plus 20 dB..."?

A. Your signal strength has increased by a factor of 100

B. Repeat your transmission on a frequency 20 kHz higher

C. The bandwidth of your signal is 20 decibels above linearity

D. A relative signal-strength meter reading is 20 decibels greater than strength 9

T2C07   What is the meaning of the procedural signal "CQ"?

A. "Call on the quarter hour"

B. "New antenna is being tested" (no station should answer)

C. "Only the called station should transmit"

D. "Calling any station"

T2C08   What is a QSL card in the amateur service?

A. A letter or postcard from an amateur pen pal

B. A Notice of Violation from the FCC

C. A written acknowledgment of communications between two amateurs

D. A postcard reminding you when your license will expire

T2C09   What is the correct way to call CQ when using voice?

A. Say "CQ" once, followed by "this is," followed by your call sign spoken three times

B. Say "CQ" at least five times, followed by "this is," followed by your call sign spoken once

C. Say "CQ" three times, followed by "this is," followed by your call sign spoken three times

D. Say "CQ" at least ten times, followed by "this is," followed by your call sign spoken once

T2C10　How should you answer a voice CQ call?

A. Say the other station's call sign at least ten times, followed by "this is," then your call sign at least twice

B. Say the other station's call sign at least five times phonetically, followed by "this is," then your call sign at least once

C. Say the other station's call sign at least three times, followed by "this is," then your call sign at least five times phonetically

D. Say the other station's call sign once, followed by "this is," then your call sign given phonetically

T2C11　What is the meaning of: "Your signal is full quieting..."?

A. Your signal is strong enough to overcome all receiver noise

B. Your signal has no spurious sounds

C. Your signal is not strong enough to be received

D. Your signal is being received, but no audio is being heard

T2D　Distress calling and emergency drills and communications—operations and equipment; Radio Amateur Civil Emergency Service (RACES)

T2D01　What is the proper distress call to use when operating phone?

A. Say "MAYDAY" several times

B. Say "HELP" several times

C. Say "EMERGENCY" several times

D. Say "SOS" several times

T2D02　What is the proper distress call to use when operating CW?

A. MAYDAY

B. QRRR

C. QRZ

D. SOS

T2D03 What is the proper way to interrupt a repeater conversation to signal a distress call?

A. Say "BREAK" twice, then your call sign

B. Say "HELP" as many times as it takes to get someone to answer

C. Say "SOS," then your call sign

D. Say "EMERGENCY" three times

T2D04 What is one reason for using tactical call signs such as "command post" or "weather center" during an emergency?

A. They keep the general public informed about what is going on

B. They are more efficient and help coordinate public-service communications

C. They are required by the FCC

D. They increase goodwill between amateurs

T2D05 What type of messages concerning a person's well-being are sent into or out of a disaster area?

A. Routine traffic

B. Tactical traffic

C. Formal message traffic

D. Health and Welfare traffic

T2D06 What are messages called that are sent into or out of a disaster area concerning the immediate safety of human life?

A. Tactical traffic

B. Emergency traffic

C. Formal message traffic

D. Health and Welfare traffic

T2D07   Why is it a good idea to have a way to operate your amateur station without using commercial AC power lines?

A. So you may use your station while mobile

B. So you may provide communications in an emergency

C. So you may operate in contests where AC power is not allowed

D. So you will comply with the FCC rules

T2D08   What is the most important accessory to have for a hand-held radio in an emergency?

A. An extra antenna

B. A portable amplifier

C. Several sets of charged batteries

D. A microphone headset for hands-free operation

T2D09   Which type of antenna would be a good choice as part of a portable HF amateur station that could be set up in case of an emergency?

A. A three-element quad

B. A three-element Yagi

C. A dipole

D. A parabolic dish

T2D10   What is the maximum number of hours allowed per week for RACES drills?

A. One

B. Seven, but not more than one hour per day

C. Eight

D. As many hours as you want

T2D11   How must you identify messages sent during a RACES drill?

A. As emergency messages

B. As amateur traffic

C. As official government messages

D. As drill or test messages

T2E Voice communications and phonetics; SSB/CW weak-signal operations; radioteleprinting; packet; special operations

T2E01 To make your call sign better understood when using voice transmissions, what should you do?

A. Use Standard International Phonetics for each letter of your call

B. Use any words that start with the same letters as your call sign for each letter of your call

C. Talk louder

D. Turn up your microphone gain

T2E02 What does the abbreviation "RTTY" stand for?

A. "Returning to you," meaning "your turn to transmit"

B. Radioteletype

C. A general call to all digital stations

D. Morse code practice over the air

T2E03 What does "connected" mean in a packet-radio link?

A. A telephone link is working between two stations

B. A message has reached an amateur station for local delivery

C. A transmitting station is sending data to only one receiving station; it replies that the data is being received correctly

D. A transmitting and receiving station are using a digipeater, so no other contacts can take place until they are finished

T2E04 What does "monitoring" mean on a packet-radio frequency?

A. The FCC is copying all messages

B. A member of the Amateur Auxiliary to the FCC's Compliance and Information Bureau is copying all messages

C. A receiving station is displaying all messages sent to it, and replying that the messages are being received correctly

D. A receiving station is displaying all messages on the frequency, and is not replying to any messages

T2E05   What is a digipeater?

A. A packet-radio station that retransmits only data that is marked to be retransmitted

B. A packet-radio station that retransmits any data that it receives

C. A repeater that changes audio signals to digital data

D. A repeater built using only digital electronics parts

T2E06   What does "network" mean in packet radio?

A. A way of connecting terminal-node controllers by telephone so data can be sent over long distances

B. A way of connecting packet-radio stations so data can be sent over long distances

C. The wiring connections on a terminal-node controller board

D. The programming in a terminal-node controller that rejects other callers if a station is already connected

T2E07   When should digital transmissions be used on 2-meter simplex voice frequencies?

A. In between voice syllables

B. Digital operations should be avoided on simplex voice frequencies

C. Only in the evening

D. At any time, so as to encourage the best use of the band

T2E08   Which of the following modes of communication are NOT available to a Technician class operator?

A. CW and SSB on HF bands

B. Amateur television (ATV)

C. EME (Moon bounce)

D. VHF packet, CW, and SSB

T2E09   What speed should you use when answering a CQ call using RTTY?

A. Half the speed of the received signal

B. The same speed as the received signal

C. Twice the speed of the received signal

D. Any speed, since RTTY systems adjust to any signal speed

T2E10   When may you operate your amateur station aboard a commercial aircraft?

A. At any time

B. Only while the aircraft is not in flight

C. Only with the pilot's specific permission and not while the aircraft is operating under Instrument Flight Rules

D. Only if you have written permission from the commercial airline company and not during takeoff and landing

T2E11   When may you operate your amateur station somewhere in the US besides the address listed on your license?

A. Only during times of emergency

B. Only after giving proper notice to the FCC

C. During an emergency or an FCC-approved emergency practice

D. Whenever you want to

Subelement T3: Radio-Wave Propagation (3 Exam Questions—3 Groups)

T3A   Line of sight; reflection of VHF/UHF signals

T3A01   How are VHF signals propagated within the range of the visible horizon?

A. By sky wave

B. By line of sight

C. By plane wave

D. By geometric refraction

T3A02   When a signal travels in a straight line from one antenna to another, what is this called?

A. Line-of-sight propagation

B. Straight-line propagation

C. Knife-edge diffraction

D. Tunnel ducting

T3A03   How do VHF and UHF radio waves usually travel from a transmitting antenna to a receiving antenna?

A. They bend through the ionosphere

B. They go in a straight line

C. They wander in any direction

D. They move in a circle going either east or west from the transmitter

T3A04   What type of propagation usually occurs from one hand-held VHF transceiver to another nearby?

A. Tunnel propagation

B. Sky-wave propagation

C. Line-of-sight propagation

D. Auroral propagation

T3A05   What causes the ionosphere to form?

A. Solar radiation ionizing the outer atmosphere

B. Temperature changes ionizing the outer atmosphere

C. Lightning ionizing the outer atmosphere

D. Release of fluorocarbons into the atmosphere

T3A06   What type of solar radiation is most responsible for ionization in the outer atmosphere?

A. Thermal

B. Nonionized particle

C. Ultraviolet

D. Microwave

T3A07   Which two daytime ionospheric regions combine into one region at night?

A. E and F1

B. D and E

C. F1 and F2

D. E1 and E2

T3A08   Which ionospheric region becomes one region at night, but separates into two separate regions during the day?

A. D

B. E

C. F

D. All of these choices

T3A09   Ultraviolet solar radiation is most responsible for ionization in what part of the atmosphere?

A. Inner

B. Outer

C. All of these choices

D. None of these choices

T3A10   What part of our atmosphere is formed by solar radiation ionizing the outer atmosphere?

A. Ionosphere

B. Troposphere

C. Ecosphere

D. Stratosphere

T3A11   What can happen to VHF or UHF signals going towards a metal-framed building?

A. They will go around the building

B. They can be bent by the ionosphere

C. They can be easily reflected by the building

D. They are sometimes scattered in the ecosphere

T3B   Tropospheric ducting or bending; amateur satellite and EME operations

T3B01   Ducting occurs in which region of the atmosphere?

A. F2

B. Ecosphere

C. Troposphere

D. Stratosphere

T3B02   What effect does tropospheric bending have on 2-meter radio waves?

A. It lets you contact stations farther away

B. It causes them to travel shorter distances

C. It garbles the signal

D. It reverses the sideband of the signal

T3B03   What causes tropospheric ducting of radio waves?

A. A very-low-pressure area

B. An aurora to the north

C. Lightning between the transmitting and receiving stations

D. A temperature inversion

T3B04 What causes VHF radio waves to be propagated several hundred miles over oceans?

A. A polar air mass

B. A widespread temperature inversion

C. An overcast of cirriform clouds

D. A high-pressure zone

T3B05 In which of the following frequency ranges does tropospheric ducting most often occur?

A. UHF

B. MF

C. HF

D. VLF

T3B06 What weather condition may cause tropospheric ducting?

A. A stable high-pressure system

B. An unstable low-pressure system

C. A series of low-pressure waves

D. Periods of heavy rainfall

T3B07 How does the signal loss for a given path through the troposphere vary with frequency?

A. There is no relationship

B. The path loss decreases as the frequency increases

C. The path loss increases as the frequency increases

D. There is no path loss at all

T3B08 Why are high-gain antennas normally used for EME (moon-bounce) communications?

A. To reduce the scattering of the reflected signal as it returns to Earth

B. To overcome the extreme path losses of this mode

C. To reduce the effects of polarization changes in the received signal

D. To overcome the high levels of solar noise at the receiver

T3B09 Which of the following antenna systems would be the best choice for an EME (moonbounce) station?

A. A single-dipole antenna

B. An isotropic antenna

C. A ground-plane antenna

D. A high-gain array of Yagi antennas

T3B10 When is it necessary to use a higher transmitter power level when conducting satellite communications?

A. When the satellite is at its perigee

B. When the satellite is low to the horizon

C. When the satellite is fully illuminated by the sun

D. When the satellite is near directly overhead

T3B11 Which of the following conditions must be met before two stations can conduct real-time communications through a satellite?

A. Both stations must use circularly polarized antennas

B. The satellite must be illuminated by the sun during the communications

C. The satellite must be in view of both stations simultaneously

D. Both stations must use high-gain antenna systems

T3C Ionospheric propagation, causes and variation; maximum usable frequency; sporadic-E propagation; ground wave, HF propagation characteristics; sunspots and the sunspot cycle

T3C01 Which region of the ionosphere is mainly responsible for absorbing MF/HF radio signals during the daytime?

A. The F2 region

B. The F1 region

C. The E region

D. The D region

T3C02   If you are receiving a weak and distorted signal from a distant station on a frequency close to the maximum usable frequency, what type of propagation is probably occurring?

A. Ducting

B. Line-of-sight

C. Scatter

D. Ground-wave

T3C03   In relation to sky-wave propagation, what does the term "maximum usable frequency" (MUF) mean?

A. The highest frequency signal that will reach its intended destination

B. The lowest frequency signal that will reach its intended destination

C. The highest frequency signal that is most absorbed by the ionosphere

D. The lowest frequency signal that is most absorbed by the ionosphere

T3C04   When a signal travels along the surface of the Earth, what is this called?

A. Sky-wave propagation

B. Knife-edge diffraction

C. E-region propagation

D. Ground-wave propagation

T3C05   When a signal is returned to Earth by the ionosphere, what is this called?

A. Sky-wave propagation

B. Earth-Moon-Earth propagation

C. Ground-wave propagation

D. Tropospheric propagation

T3C06   What is a skip zone?

A. An area covered by ground-wave propagation

B. An area covered by sky-wave propagation

C. An area that is too far away for ground-wave propagation, but too close for sky-wave propagation

D. An area that is too far away for ground-wave or sky-wave propagation

T3C07   Which ionospheric region is closest to the Earth?

A. The A region

B. The D region

C. The E region

D. The F region

T3C08   Which region of the ionosphere is mainly responsible for long-distance sky-wave radio communications?

A. D region

B. E region

C. F1 region

D. F2 region

T3C09   Which of the ionospheric regions may split into two regions only during the daytime?

A. Troposphere

B. F

C. Electrostatic

D. D

T3C10   How does the number of sunspots relate to the amount of ionization in the ionosphere?

A. The more sunspots there are, the greater the ionization

B. The more sunspots there are, the less the ionization

C. Unless there are sunspots, the ionization is zero

D. Sunspots do not affect the ionosphere

T3C11   How long is an average sunspot cycle?

A. 2 years

B. 5 years

C. 11 years

D. 17 years

Subelement T4: Amateur Radio Practices (4 Exam Questions—4 Groups)

T4A   Lightning protection and station grounding; safety interlocks, antenna installation safety procedures; dummy antennas

T4A01   How can an antenna system best be protected from lightning damage?

A. Install a balun at the antenna feed point

B. Install an RF choke in the antenna feed line

C. Ground all antennas when they are not in use

D. Install a fuse in the antenna feed line

T4A02   How can amateur station equipment best be protected from lightning damage?

A. Use heavy insulation on the wiring

B. Never turn off the equipment

C. Disconnect the ground system from all radios

D. Disconnect all equipment from the power lines and antenna cables

T4A03   For best protection from electrical shock, what should be grounded in an amateur station?

A. The power supply primary

B. All station equipment

C. The antenna feed line

D. The ac power mains

T4A04   Why would there be an interlock switch in a high-voltage power supply to turn off the power if its cabinet is opened?

A. To keep dangerous RF radiation from leaking out through an open cabinet

B. To keep dangerous RF radiation from coming in through an open cabinet

C. To turn the power supply off when it is not being used

D. To keep anyone opening the cabinet from getting shocked by dangerous high voltages

T4A05   Why should you wear a hard hat and safety glasses if you are on the ground helping someone work on an antenna tower?

A. So you won't be hurt if the tower should accidentally fall

B. To keep RF energy away from your head during antenna testing

C. To protect your head from something dropped from the tower

D. So someone passing by will know that work is being done on the tower and will stay away

T4A06   What safety factors must you consider when using a bow and arrow or slingshot and weight to shoot an antenna-support line over a tree?

A. You must ensure that the line is strong enough to withstand the shock of shooting the weight

B. You must ensure that the arrow or weight has a safe flight path if the line breaks

C. You must ensure that the bow and arrow or slingshot is in good working condition

D. All of these choices are correct

T4A07   Which of the following is the best way to install your antenna in relation to overhead electric power lines?

A. Always be sure your antenna wire is higher than the power line, and crosses it at a 90-degree angle

B. Always be sure your antenna and feed line are well clear of any power lines

C. Always be sure your antenna is lower than the power line, and crosses it at a small angle

D. Only use vertical antennas within 100 feet of a power line

T4A08 What device is used in place of an antenna during transmitter tests so that no signal is radiated?

A. An antenna matcher

B. A dummy antenna

C. A low-pass filter

D. A decoupling resistor

T4A09 Why would you use a dummy antenna?

A. For off-the-air transmitter testing

B. To reduce output power

C. To give comparative signal reports

D. To allow antenna tuning without causing interference

T4A10 What minimum rating should a dummy antenna have for use with a 100-watt single-sideband phone transmitter?

A. 100 watts continuous

B. 141 watts continuous

C. 175 watts continuous

D. 200 watts continuous

T4A11 Would a 100-watt light bulb make a good dummy load for tuning a transceiver?

A. Yes; a light bulb behaves exactly like a dummy load

B. No; the impedance of the light bulb changes as the filament gets hot

C. No; the light bulb would act like an open circuit

D. No; the light bulb would act like a short circuit

T4B Electrical wiring, including switch location, dangerous voltages and currents; SWR meaning and measurements; SWR meters

T4B01 Where should the green wire in a three-wire ac line cord be connected in a power supply?

A. To the fuse

B. To the "hot" side of the power switch

C. To the chassis

D. To the white wire

T4B02 What is the minimum voltage that is usually dangerous to humans?

A. 30 volts

B. 100 volts

C. 1000 volts

D. 2000 volts

T4B03 How much electrical current flowing through the human body will probably be fatal?

A. As little as 1/10 of an ampere

B. Approximately 10 amperes

C. More than 20 amperes

D. Current through the human body is never fatal

T4B04 Which body organ can be fatally affected by a very small amount of electrical current?

A. The heart

B. The brain

C. The liver

D. The lungs

T4B05   What does an SWR reading of less than 1.5:1 mean?

A. An impedance match that is too low

B. An impedance mismatch; something may be wrong with the antenna system

C. A fairly good impedance match

D. An antenna gain of 1.5

T4B06   What does a very high SWR reading mean?

A. The antenna is the wrong length, or there may be an open or shorted connection somewhere in the feed line

B. The signals coming from the antenna are unusually strong, which means very good radio conditions

C. The transmitter is putting out more power than normal, showing that it is about to go bad

D. There is a large amount of solar radiation, which means very poor radio conditions

T4B07   If an SWR reading at the low-frequency end of an amateur band is 2.5:1, increasing to 5:1 at the high-frequency end of the same band, what does this tell you about your 1/2-wavelength dipole antenna?

A. The antenna is broadbanded

B. The antenna is too long for operation on the band

C. The antenna is too short for operation on the band

D. The antenna is just right for operation on the band

T4B08   If an SWR reading at the low-frequency end of an amateur band is 5:1, decreasing to 2.5:1 at the high-frequency end of the same band, what does this tell you about your 1/2-wavelength dipole antenna?

A. The antenna is broadbanded

B. The antenna is too long for operation on the band

C. The antenna is too short for operation on the band

D. The antenna is just right for operation on the band

T4B09 What instrument is used to measure the relative impedance match between an antenna and its feed line?

A. An ammeter

B. An ohmmeter

C. A voltmeter

D. An SWR meter

T4B10 If you use an SWR meter designed to operate on 3–30 MHz for VHF measurements, how accurate will its readings be?

A. They will not be accurate

B. They will be accurate enough to get by

C. If it properly calibrates to full scale in the set position, they may be accurate

D. They will be accurate providing the readings are multiplied by 4.5

T4B11 What does an SWR reading of 1:1 mean?

A. An antenna for another frequency band is probably connected

B. The best impedance match has been attained

C. No power is going to the antenna

D. The SWR meter is broken

T4C Meters and their placement in circuits, including volt, amp, multi, peak-reading and RF watt; ratings of fuses and switches
T4C01 How is a voltmeter usually connected to a circuit under test?

A. In series with the circuit

B. In parallel with the circuit

C. In quadrature with the circuit

D. In phase with the circuit

T4C02 How is an ammeter usually connected to a circuit under test?

A. In series with the circuit

B. In parallel with the circuit

C. In quadrature with the circuit

D. In phase with the circuit

T4C03 Where should an RF wattmeter be connected for the most accurate readings of transmitter output power?

A. At the transmitter output connector

B. At the antenna feed point

C. One-half wavelength from the transmitter output

D. One-half wavelength from the antenna feed point

T4C04 For which measurements would you normally use a multi-meter?

A. SWR and power

B. Resistance, capacitance, and inductance

C. Resistance and reactance

D. Voltage, current, and resistance

T4C05 What might happen if you switch a multimeter to measure resistance while you have it connected to measure voltage?

A. The multimeter would read half the actual voltage

B. It would probably destroy the meter circuitry

C. The multimeter would read twice the actual voltage

D. Nothing unusual would happen; the multimeter would measure the circuit's resistance

T4C06 If you switch a multimeter to read microamps and connect it into a circuit drawing 5 amps, what might happen?

A. The multimeter would read half the actual current

B. The multimeter would read twice the actual current

C. It would probably destroy the meter circuitry

D. The multimeter would read a very small value of current

T4C07 At what line impedance do most RF watt meters usually operate?

A. 25 ohms

B. 50 ohms

C. 100 ohms

D. 300 ohms

T4C08   What does a directional wattmeter measure?

A. Forward and reflected power

B. The directional pattern of an antenna

C. The energy used by a transmitter

D. Thermal heating in a load resistor

T4C09   If a directional RF wattmeter reads 90 watts forward power and 10 watts reflected power, what is the actual transmitter output power?

A. 10 watts

B. 80 watts

C. 90 watts

D. 100 watts

T4C10   Why might you use a peak-reading RF wattmeter at your station?

A. To make sure your transmitter's output power is not higher than that authorized by your license class

B. To make sure your transmitter is not drawing too much power from the ac line

C. To make sure all your transmitter's power is being radiated by your antenna

D. To measure transmitter input and output power at the same time

T4C11   What could happen to your transceiver if you replace its blown 5-amp ac line fuse with a 30-amp fuse?

A. The 30-amp fuse would better protect your transceiver from using too much current

B. The transceiver would run cooler

C. The transceiver could use more current than 5 amps and a fire could occur

D. The transceiver would not be able to produce as much RF output

T4D   RFI and its complications, resolution and responsibility

T4D01   What is meant by receiver overload?

A. Too much voltage from the power supply

B. Too much current from the power supply

C. Interference caused by strong signals from a nearby source

D. Interference caused by turning the volume up too high

T4D02   What is meant by harmonic radiation?

A. Unwanted signals at frequencies that are multiples of the fundamental (chosen) frequency

B. Unwanted signals that are combined with a 60-Hz hum

C. Unwanted signals caused by sympathetic vibrations from a nearby transmitter

D. Signals that cause skip propagation to occur

T4D03   What type of filter might be connected to an amateur HF transmitter to cut down on harmonic radiation?

A. A key-click filter

B. A low-pass filter

C. A high-pass filter

D. A CW filter

T4D04   If your neighbor reports television interference whenever you are transmitting from your amateur station, no matter what frequency band you use, what is probably the cause of the interference?

A. Too little transmitter harmonic suppression

B. Receiver VR tube discharge

C. Receiver overload

D. Incorrect antenna length

T4D05   If your neighbor reports television interference on one or two channels only when you are transmitting on the 15-meter band, what is probably the cause of the interference?

A. Too much low-pass filtering on the transmitter

B. De-ionization of the ionosphere near your neighbor's TV antenna

C. TV receiver front-end overload

D. Harmonic radiation from your transmitter

T4D06   What type of filter should be connected to a TV receiver as the first step in trying to prevent RF overload from an amateur HF station transmission?

A. Low-pass

B. High-pass

C. Band-pass

D. Notch

T4D07   What first step should be taken at a cable TV receiver when trying to prevent RF overload from an amateur HF station transmission?

A. Install a low-pass filter in the cable system transmission line

B. Tighten all connectors and inspect the cable system transmission line

C. Make sure the center conductor of the cable system transmission line is well grounded

D. Install a ceramic filter in the cable system transmission line

T4D08   What effect might a break in a cable television transmission line have on amateur communications?

A. Cable lines are shielded and a break cannot affect amateur communications

B. Harmonic radiation from the TV receiver may cause the amateur transmitter to transmit off-frequency

C. TV interference may result when the amateur station is transmitting, or interference may occur to the amateur receiver

D. The broken cable may pick up very high voltages when the amateur station is transmitting

T4D09   If you are told that your amateur station is causing television interference, what should you do?

A. First make sure that your station is operating properly, and that it does not cause interference to your own television

B. Immediately turn off your transmitter and contact the nearest FCC office for assistance

C. Connect a high-pass filter to the transmitter output and a low-pass filter to the antenna-input terminals of the television

D. Continue operating normally, because you have no reason to worry about the interference

T4D10   If harmonic radiation from your transmitter is causing interference to television receivers in your neighborhood, who is responsible for taking care of the interference?

A. The owners of the television receivers are responsible

B. Both you and the owners of the television receivers share the responsibility

C. You alone are responsible, since your transmitter is causing the problem

D. The FCC must decide if you or the owners of the television receivers are responsible

T4D11   If signals from your transmitter are causing front-end overload in your neighbor's television receiver, who is responsible for taking care of the interference?

A. You alone are responsible, since your transmitter is causing the problem

B. Both you and the owner of the television receiver share the responsibility

C. The FCC must decide if you or the owner of the television receiver are responsible

D. The owner of the television receiver is responsible

Subelement T5: Electrical Principles (3 Exam Questions—3 Groups)

T5A   Metric prefixes, e.g., pico, nano, micro, milli, centi, kilo, mega, giga; concepts, units and measurement of current, voltage; concept of conductor and insulator; concept of open and short circuits

T5A01   If a dial marked in kilohertz shows a reading of 28,450 kHz, what would it show if it were marked in hertz?

A. 284,500 Hz

B. 28,450,000 Hz

C. 284,500,000 Hz

D. 284,500,000,000 Hz

T5A02   If an ammeter marked in amperes is used to measure a 3000-milliampere current, what reading would it show?

A. 0.003 amperes

B. 0.3 amperes

C. 3 amperes

D. 3,000,000 amperes

T5A03   How many hertz are in a kilohertz?

A. 10

B. 100

C. 1000

D. 1,000,000

T5A04   What is the basic unit of electric current?

A. The volt

B. The watt

C. The ampere

D. The ohm

T5A05 Which instrument would you use to measure electric current?

A. An ohmmeter

B. A wavemeter

C. A voltmeter

D. An ammeter

T5A06 Which instrument would you use to measure electric potential or electromotive force?

A. An ammeter

B. A voltmeter

C. A wavemeter

D. An ohmmeter

T5A07 What is the basic unit of electromotive force (EMF)?

A. The volt

B. The watt

C. The ampere

D. The ohm

T5A08 What are three good electrical conductors?

A. Copper, gold, mica

B. Gold, silver, wood

C. Gold, silver, aluminum

D. Copper, aluminum, paper

T5A09 What are four good electrical insulators?

A. Glass, air, plastic, porcelain

B. Glass, wood, copper, porcelain

C. Paper, glass, air, aluminum

D. Plastic, rubber, wood, carbon

T5A10   Which electrical circuit can have no current?

A. A closed circuit

B. A short circuit

C. An open circuit

D. A complete circuit

T5A11   Which electrical circuit draws too much current?

A. An open circuit

B. A dead circuit

C. A closed circuit

D. A short circuit

T5B   Concepts, units, and calculation of resistance, inductance, and capacitance values in series and parallel circuits

T5B01   What does resistance do in an electric circuit?

A. It stores energy in a magnetic field

B. It stores energy in an electric field

C. It provides electrons by a chemical reaction

D. It opposes the flow of electrons

T5B02   What is the definition of 1 ohm?

A. The reactance of a circuit in which a 1-microfarad capacitor is resonant at 1 MHz

B. The resistance of a circuit in which a 1-amp current flows when 1 volt is applied

C. The resistance of a circuit in which a 1-milliamp current flows when 1 volt is applied

D. The reactance of a circuit in which a 1-millihenry inductor is resonant at 1 MHz

T5B03 What is the basic unit of resistance?

A. The farad

B. The watt

C. The ohm

D. The resistor

T5B04 What is one reason resistors are used in electronic circuits?

A. To block the flow of direct current while allowing alternating current to pass

B. To block the flow of alternating current while allowing direct current to pass

C. To increase the voltage of the circuit

D. To control the amount of current that flows for a particular applied voltage

T5B05 What is the ability to store energy in a magnetic field called?

A. Admittance

B. Capacitance

C. Resistance

D. Inductance

T5B06 What is one reason inductors are used in electronic circuits?

A. To block the flow of direct current while allowing alternating current to pass

B. To reduce the flow of ac while allowing dc to pass freely

C. To change the time constant of the applied voltage

D. To change alternating current to direct current

T5B07 What is the ability to store energy in an electric field called?

A. Inductance

B. Resistance

C. Tolerance

D. Capacitance

T5B08 What is one reason capacitors are used in electronic circuits?

A. To block the flow of direct current while allowing alternating current to pass

B. To block the flow of alternating current while allowing direct current to pass

C. To change the time constant of the applied voltage

D. To change alternating current to direct current

T5B09 If two resistors are connected in series, what is their total resistance?

A. The difference between the individual resistor values

B. Always less than the value of either resistor

C. The product of the individual resistor values

D. The sum of the individual resistor values

T5B10 If two equal-value inductors are connected in parallel, what is their total inductance?

A. Half the value of one inductor

B. Twice the value of one inductor

C. The same as the value of either inductor

D. The value of one inductor times the value of the other

T5B11 If two equal-value capacitors are connected in series, what is their total capacitance?

A. Twice the value of one capacitor

B. The same as the value of either capacitor

C. Half the value of either capacitor

D. The value of one capacitor times the value of the other

T5C   Ohm's Law (any calculations will be kept to a very low level— no fractions or decimals) and the concepts of energy and power; concepts of frequency, including ac vs. dc, frequency units, and wavelength

T5C01   How is the current in a dc circuit directly calculated when the voltage and resistance are known?

A. $I=R\times E$ [current equals resistance multiplied by voltage]

B. $I=R/E$ [current equals resistance divided by voltage]

C. $I=E/R$ [current equals voltage divided by resistance]

D. $I=E/P$ [current equals voltage divided by power]

T5C02   How is the resistance in a dc circuit calculated when the voltage and current are known?

A. $R=I/E$ [resistance equals current divided by voltage]

B. $R=E/I$ [resistance equals voltage divided by current]

C. $R=I\times E$ [resistance equals current multiplied by voltage]

D. $R=P/E$ [resistance equals power divided by voltage]

T5C03   How is the voltage in a dc circuit directly calculated when the current and resistance are known?

A. $E=I/R$ [voltage equals current divided by resistance]

B. $E=R/I$ [voltage equals resistance divided by current]

C. $E=I\times R$ [voltage equals current multiplied by resistance]

D. $E=I/P$ [voltage equals current divided by power]

T5C04   If a current of 2 amperes flows through a 50-ohm resistor, what is the voltage across the resistor?

A. 25 volts

B. 52 volts

C. 100 volts

D. 200 volts

T5C05   If a 100-ohm resistor is connected to 200 volts, what is the current through the resistor?

A. 1 ampere

B. 2 amperes

C. 300 amperes

D. 20,000 amperes

T5C06   If a current of 3 amperes flows through a resistor connected to 90 volts, what is the resistance?

A. 3 ohms

B. 30 ohms

C. 93 ohms

D. 270 ohms

T5C07   What term describes how fast electrical energy is used?

A. Resistance

B. Current

C. Power

D. Voltage

T5C08   What is the basic unit of electrical power?

A. The ohm

B. The watt

C. The volt

D. The ampere

T5C09   What happens to a signal's wavelength as its frequency increases?

A. It gets shorter

B. It gets longer

C. It stays the same

D. It disappears

T5C10 What is the name of a current that flows back and forth, first in one direction, then in the opposite direction?

A. An alternating current

B. A direct current

C. A rough current

D. A steady state current

T5C11 What is the name of a current that flows only in one direction?

A. An alternating current

B. A direct current

C. A normal current

D. A smooth current

Subelement T6—Circuit Components (2 Exam Questions—2 Groups)

T6A Electrical function and/or schematic representation of resistor, switch, fuse, or battery; resistor construction types, variable and fixed, color code, power ratings, schematic symbols

T6A01 What does a variable resistor or potentiometer do?

A. Its resistance changes when ac is applied to it

B. It transforms a variable voltage into a constant voltage

C. Its resistance changes when its slide or contact is moved

D. Its resistance changes when it is heated

T6A02 Which symbol of Figure T6-1 represents a fixed resistor?

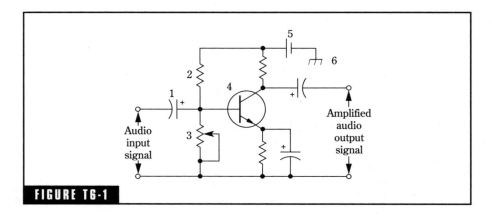

FIGURE T6-1

A. Symbol 2

B. Symbol 3

C. Symbol 4

D. Symbol 5

T6A03   Why would you use a double-pole, single-throw switch?

A. To switch one input to one output

B. To switch one input to either of two outputs

C. To switch two inputs at the same time, one input to either of two outputs, and the other input to either of two outputs

D. To switch two inputs at the same time, one input to one output, and the other input to the other output

T6A04   In Figure T6-2, which symbol represents a single-pole, single-throw switch?

A. Symbol 1

B. Symbol 2

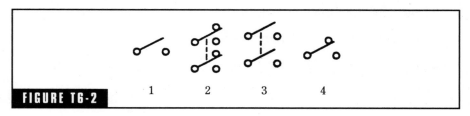

FIGURE T6-2

C. Symbol 3

D. Symbol 4

T6A05   Why would you use a fuse?

A. To create a short circuit when there is too much current in a circuit

B. To change direct current into alternating current

C. To change alternating current into direct current

D. To create an open circuit when there is too much current in a circuit

T6A06   In Figure T6-3, which symbol represents a fuse?

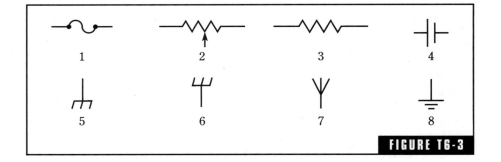

FIGURE T6-3

A. Symbol 1

B. Symbol 3

C. Symbol 5

D. Symbol 7

T6A07   Which of these components has a positive and a negative side?

A. A battery

B. A potentiometer

C. A fuse

D. A resistor

T6A08   In Figure T6-3, which symbol represents a single-cell battery?

A. Symbol 7

B. Symbol 5

C. Symbol 1

D. Symbol 4

T6A09   Why would a large-size resistor be used instead of a smaller one of the same resistance value?

A. For better response time

B. For a higher current gain

C. For greater power dissipation

D. For less impedance in the circuit

T6A10   What do the first three color bands on a resistor indicate?

A. The value of the resistor in ohms

B. The resistance tolerance in percent

C. The power rating in watts

D. The resistance material

T6A11   Which tolerance rating would indicate a high-precision resistor?

A. 0.1%

B. 5%

C. 10%

D. 20%

T6B   Electrical function and/or schematic representation of a ground, antenna, inductor, capacitor, transistor, integrated circuit; construction of variable and fixed inductors and capacitors; factors affecting inductance and capacitance

T6B01   Which component can amplify a small signal using low voltages?

A. A pnp transistor

B. A variable resistor

C. An electrolytic capacitor

D. A multiple-cell battery

T6B02   Which component is used to radiate radio energy?

A. An antenna

B. An earth ground

C. A chassis ground

D. A potentiometer

T6B03   In Figure T6-3, which symbol represents an earth ground?

A. Symbol 2

B. Symbol 5

C. Symbol 6

D. Symbol 8

T6B04   In Figure T6-3, which symbol represents an antenna?

A. Symbol 2

B. Symbol 3

C. Symbol 6

D. Symbol 7

T6B05   In Figure T6-4, which symbol represents an npn transistor?

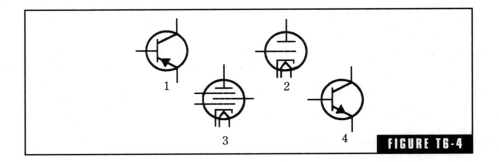

FIGURE T6-4

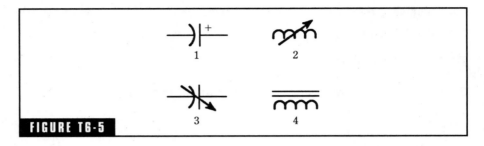

FIGURE T6-5

A. Symbol 1

B. Symbol 2

C. Symbol 3

D. Symbol 4

T6B06   Which symbol of Figure T6-5 represents a fixed-value capacitor?

A. Symbol 1

B. Symbol 2

C. Symbol 3

D. Symbol 4

T6B07   In Figure T6-5, which symbol represents a variable capacitor?

A. Symbol 1

B. Symbol 2

C. Symbol 3

D. Symbol 4

T6B08   What does an inductor do?

A. It stores energy electrostatically and opposes a change in voltage

B. It stores energy electrochemically and opposes a change in current

C. It stores energy electromagnetically and opposes a change in current

D. It stores energy electromechanically and opposes a change in voltage

T6B09   As an iron core is inserted in a coil, what happens to the coil's inductance?

A. It increases

B. It decreases

C. It stays the same

D. It disappears

T6B10   What does a capacitor do?

A. It stores energy electrochemically and opposes a change in current

B. It stores energy electrostatically and opposes a change in voltage

C. It stores energy electromagnetically and opposes a change in current

D. It stores energy electromechanically and opposes a change in voltage

T6B11   What determines the capacitance of a capacitor?

A. The material between the plates, the area of one side of one plate, the number of plates, and the spacing between the plates

B. The material between the plates, the number of plates, and the size of the wires connected to the plates

C. The number of plates, the spacing between the plates and whether the dielectric material is n type or p type

D. The material between the plates, the area of one plate, the number of plates, and the material used for the protective coating

Subelement T7: Practical Circuits (2 Exam Questions—2 Groups)

T7A   Functional layout of station components including transmitter, transceiver, receiver, power supply, antenna, antenna switch, antenna feed line, impedance-matching device, SWR meter; station layout and accessories for radiotelephone, radioteleprinter (RTTY), or packet

T7A01   What would you connect to your transceiver if you wanted to switch it between several antennas?

A. A terminal-node switch

B. An antenna switch

C. A telegraph key switch

D. A high-pass filter

T7A02   What connects your transceiver to your antenna?

A. A dummy load

B. A ground wire

C. The power cord

D. A feed line

T7A03   If your mobile transceiver works in your car but not in your home, what should you check first?

A. The power supply

B. The speaker

C. The microphone

D. The SWR meter

T7A04   What does an antenna tuner do?

A. It matches a transceiver output impedance to the antenna system impedance

B. It helps a receiver automatically tune in stations that are far away

C. It switches an antenna system to a transceiver when sending, and to a receiver when listening

D. It switches a transceiver between different kinds of antennas connected to one feed line

T7A05   In Figure T7-1, if block 1 is a transceiver and block 3 is a dummy antenna, what is block 2?

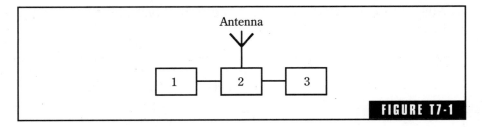

FIGURE T7-1

A. A terminal-node switch

B. An antenna switch

C. A telegraph key switch

D. A high-pass filter

T7A06　In Figure T7-1, if block 1 is a transceiver and block 2 is an antenna switch, what is block 3?

A. A terminal-node switch

B. An SWR meter

C. A telegraph key switch

D. A dummy antenna

T7A07　In Figure T7-2, if block 1 is a transceiver and block 3 is an antenna switch, what is block 2?

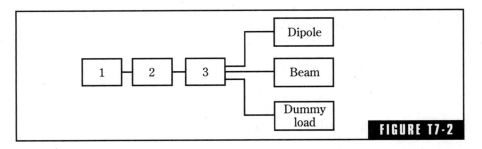

FIGURE T7-2

A. A terminal-node switch

B. A dipole antenna

C. An SWR meter

D. A high-pass filter

T7A08 In Figure T7-3, if block 1 is a transceiver and block 2 is an SWR meter, what is block 3?

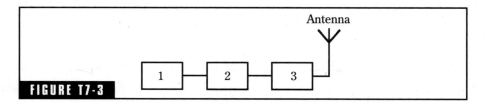

FIGURE T7-3

A. An antenna switch

B. An antenna tuner

C. A key-click filter

D. A terminal-node controller

T7A09 What would you connect to a transceiver for voice operation?

A. A splatter filter

B. A terminal-voice controller

C. A receiver audio filter

D. A microphone

T7A10 What would you connect to a transceiver for RTTY operation?

A. A modem and a teleprinter or computer system

B. A computer, a printer, and a RTTY refresh unit

C. A data-inverter controller

D. A modem, a monitor, and a DTMF keypad

T7A11 In packet-radio operation, what equipment connects to a terminal-node controller?

A. A transceiver and a modem

B. A transceiver and a terminal or computer system

C. A DTMF keypad, a monitor, and a transceiver

D. A DTMF microphone, a monitor, and a transceiver

T7B Transmitter and receiver block diagrams; purpose and operation of low-pass, high-pass, and band-pass filters

T7B01 What circuit uses a limiter and a frequency discriminator to produce an audio signal?

A. A double-conversion receiver

B. A variable-frequency oscillator

C. A superheterodyne receiver

D. An FM receiver

T7B02 What circuit is pictured in Figure T7-4 if block 1 is a variable-frequency oscillator?

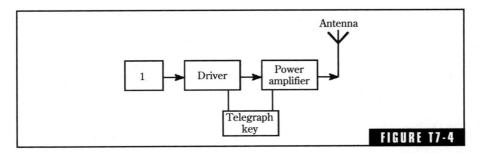

FIGURE T7-4

A. A packet-radio transmitter

B. A crystal-controlled transmitter

C. A single-sideband transmitter

D. A VFO-controlled transmitter

T7B03 What circuit is pictured in Figure T7-4 if block 1 is a crystal oscillator?

A. A crystal-controlled transmitter

B. A VFO-controlled transmitter

C. A single-sideband transmitter

D. A CW transceiver

T7B04 What type of circuit does Figure T7-5 represent if block 1 is a product detector?

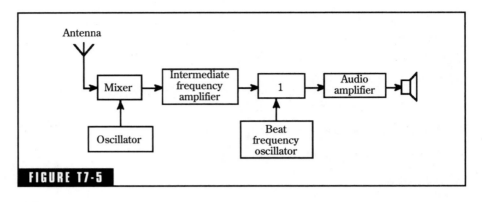

FIGURE T7-5

A. A simple phase modulation receiver

B. A simple FM receiver

C. A simple CW and SSB receiver

D. A double-conversion multiplier

T7B05 If Figure T7-5 is a diagram of a simple single-sideband receiver, what type of circuit should be shown in block 1?

A. A high-pass filter

B. A ratio detector

C. A low-pass filter

D. A product detector

T7B06 What circuit is pictured in Figure T7-6, if block 1 is a frequency discriminator?

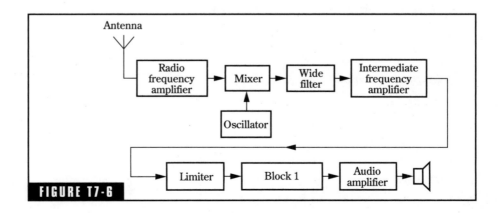

FIGURE T7-6

A. A double-conversion receiver

B. A variable-frequency oscillator

C. A superheterodyne receiver

D. An FM receiver

T7B07   Why do modern HF transmitters have a built-in low-pass filter in their RF output circuits?

A. To reduce RF energy below a cutoff point

B. To reduce low-frequency interference to other amateurs

C. To reduce harmonic radiation

D. To reduce fundamental radiation

T7B08   What circuit blocks RF energy above and below certain limits?

A. A band-pass filter

B. A high-pass filter

C. An input filter

D. A low-pass filter

T7B09   What type of filter is used in the IF section of receivers to block energy outside a certain frequency range?

A. A band-pass filter

B. A high-pass filter

C. An input filter

D. A low-pass filter

T7B10   What circuit function is found in all types of receivers?

A. An audio filter

B. A beat-frequency oscillator

C. A detector

D. An RF amplifier

T7B11   What would you use to connect a dual-band antenna to a mobile transceiver which has separate VHF and UHF outputs?

A. A dual-needle SWR meter

B. A full-duplex phone patch

C. Twin high-pass filters

D. A duplexer

Subelement T8: Signals and Emissions (2 Exam Questions—2 Groups)

T8A RF carrier, definition, and typical bandwidths; harmonics and unwanted signals; chirp; superimposed hum; equipment and adjustments to help reduce interference to others

T8A01 What is an RF carrier?

A. The part of a transmitter that carries the signal to the transmitter antenna

B. The part of a receiver that carries the signal from the antenna to the detector

C. A radio frequency signal that is modulated to produce a radiotelephone signal

D. A modulation that changes a radio frequency signal to produce a radiotelephone signal

T8A02 Which list of emission types is in order from the narrowest bandwidth to the widest bandwidth?

A. RTTY, CW, SSB voice, FM voice

B. CW, FM voice, RTTY, SSB voice

C. CW, RTTY, SSB voice, FM voice

D. CW, SSB voice, RTTY, FM voice

T8A03 What is the usual bandwidth of a single-sideband amateur signal?

A. 1 kHz

B. 2 kHz

C. Between 3 and 6 kHz

D. Between 2 and 3 kHz

T8A04   What is the usual bandwidth of a frequency-modulated amateur signal?

A. Less than 5 kHz

B. Between 5 and 10 kHz

C. Between 10 and 20 kHz

D. Greater than 20 kHz

T8A05   What is the name for emissions produced by switching a transmitter's output on and off?

A. Phone

B. Test

C. CW

D. RTTY

T8A06   What term describes the process of combining an information signal with a radio signal?

A. Superposition

B. Modulation

C. Demodulation

D. Phase-inversion

T8A07   What is the result of overdeviation in an FM transmitter?

A. Increased transmitter power

B. Out-of-channel emissions

C. Increased transmitter range

D. Poor carrier suppression

T8A08   What causes splatter interference?

A. Keying a transmitter too fast

B. Signals from a transmitter's output circuit are being sent back to its input circuit

C. Overmodulation of a transmitter

D. The transmitting antenna is the wrong length

T8A09 How does the frequency of a harmonic compare to the desired transmitting frequency?

A. It is slightly more than the desired frequency

B. It is slightly less than the desired frequency

C. It is exactly two, or three, or more times the desired frequency

D. It is much less than the desired frequency

T8A10 What should you check if you change your transceiver's microphone from a mobile type to a base-station type?

A. Check the CTCSS levels on the oscilloscope

B. Make an on-the-air radio check to ensure the quality of your signal

C. Check the amount of current the transceiver is now using

D. Check to make sure the frequency readout is now correct

T8A11 Why is good station grounding needed when connecting your computer to your transceiver to receive high-frequency data signals?

A. Good grounding raises the receiver's noise floor

B. Good grounding protects the computer from nearby lightning strikes

C. Good grounding will minimize stray noise on the receiver

D. FCC rules require all equipment to be grounded

T8B Concepts and types of modulation: CW, phone, RTTY, and data emission types; FM deviation

T8B01 What is the name for packet-radio emissions?

A. CW

B. Data

C. Phone

D. RTTY

T8B02   What is the name of the voice emission most used on VHF/UHF repeaters?

A. Single-sideband phone

B. Pulse-modulated phone

C. Slow-scan phone

D. Frequency-modulated phone

T8B03   What is meant by the upper sideband (USB)?

A. The part of a single-sideband signal that is above the carrier frequency

B. The part of a single-sideband signal that is below the carrier frequency

C. Any frequency above 10 MHz

D. The carrier frequency of a single-sideband signal

T8B04   What does the term "phone transmissions" usually mean?

A. The use of telephones to set up an amateur contact

B. A phone patch between amateur radio and the telephone system

C. AM, FM, or SSB voice transmissions by radiotelephony

D. Placing the telephone handset near a transceiver's microphone and speaker to relay a telephone call

T8B05   How is an HF RTTY signal usually produced?

A. By frequency-shift keying an RF signal

B. By on/off keying an RF signal

C. By digital pulse-code keying of an unmodulated carrier

D. By on/off keying an audio-frequency signal

T8B06   What are two advantages to using modern data-transmission techniques for communications?

A. Very simple and low-cost equipment

B. No parity-checking required and high transmission speed

C. Easy for mobile stations to use and no additional cabling required

D. High transmission speed and communications reliability

T8B07 Which sideband is commonly used for 10-meter phone operation?

A. Upper sideband

B. Lower sideband

C. Amplitude-compandored sideband

D. Double sideband

T8B08 What can you do if you are told your FM hand-held or mobile transceiver is overdeviating?

A. Talk louder into the microphone

B. Let the transceiver cool off

C. Change to a higher power level

D. Talk farther away from the microphone

T8B09 What does chirp mean?

A. An overload in a receiver's audio circuit whenever CW is received

B. A high-pitched tone that is received along with a CW signal

C. A small change in a transmitter's frequency each time it is keyed

D. A slow change in transmitter frequency as the circuit warms up

Subelement T9: Antennas and Feed Lines (2 Exam Questions—2 Groups)

T9A Wavelength vs. antenna length; 1/2 wavelength dipole and 1/4 wavelength vertical antennas; multiband antennas

T9A01 How do you calculate the length (in feet) of a half-wavelength dipole antenna?

A. Divide 150 by the antenna's operating frequency (in MHz) [150/f (in MHz)]

B. Divide 234 by the antenna's operating frequency (in MHz) [234/f (in MHz)]

C. Divide 300 by the antenna's operating frequency (in MHz) [300/f (in MHz)]

D. Divide 468 by the antenna's operating frequency (in MHz) [468/f (in MHz)]

T9A02 How do you calculate the length (in feet) of a quarter-wavelength vertical antenna?

A. Divide 150 by the antenna's operating frequency (in MHz) [150/f (in MHz)]

B. Divide 234 by the antenna's operating frequency (in MHz) [234/f (in MHz)]

C. Divide 300 by the antenna's operating frequency (in MHz) [300/f (in MHz)]

D. Divide 468 by the antenna's operating frequency (in MHz) [468/f (in MHz)]

T9A03 How long should you make a quarter-wavelength vertical antenna for 440 MHz (measured to the nearest inch)?

A. 12 inches

B. 9 inches

C. 6 inches

D. 3 inches

T9A04 How long should you make a quarter-wavelength vertical antenna for 28.450 MHz (measured to the nearest foot)?

A. 8 ft

B. 12 ft

C. 16 ft

D. 24 ft

T9A05   How long should you make a quarter-wavelength vertical antenna for 146 MHz (measured to the nearest inch)?

A. 112 inches

B. 50 inches

C. 19 inches

D. 12 inches

T9A06   If an antenna is made longer, what happens to its resonant frequency?

A. It decreases

B. It increases

C. It stays the same

D. It disappears

T9A07   If an antenna is made shorter, what happens to its resonant frequency?

A. It decreases

B. It increases

C. It stays the same

D. It disappears

T9A08   How could you decrease the resonant frequency of a dipole antenna?

A. Lengthen the antenna

B. Shorten the antenna

C. Use less feed line

D. Use a smaller size feed line

T9A09   How could you increase the resonant frequency of a dipole antenna?

A. Lengthen the antenna

B. Shorten the antenna

C. Use more feed line

D. Use a larger size feed line

T9A10   What is one advantage to using a multiband antenna?

A. You can operate on several bands with a single feed line

B. Multiband antennas always have high gain

C. You can transmit on several frequencies simultaneously

D. Multiband antennas offer poor harmonic suppression

T9A11   What is one disadvantage to using a multiband antenna?

A. It must always be used with a balun

B. It will always have low gain

C. It cannot handle high power

D. It can radiate unwanted harmonics

T9B   Parasitic beam directional antennas; polarization, impedance matching and SWR, feed lines, balanced vs. unbalanced (including baluns)

T9B01   What is a directional antenna?

A. An antenna that sends and receives radio energy equally well in all directions

B. An antenna that cannot send and receive radio energy by skywave or skip propagation

C. An antenna that sends and receives radio energy mainly in one direction

D. An antenna that uses a directional coupler to measure power transmitted

T9B02   How is a Yagi antenna constructed?

A. Two or more straight, parallel elements are fixed in line with each other

B. Two or more square or circular loops are fixed in line with each other

C. Two or more square or circular loops are stacked inside each other

D. A straight element is fixed in the center of three or more elements that angle toward the ground

T9B03   How many directly driven elements do most parasitic beam antennas have?

A. None

B. One

C. Two

D. Three

T9B04   What is a parasitic beam antenna?

A. An antenna in which some elements obtain their radio energy by induction or radiation from a driven element

B. An antenna in which wave traps are used to magnetically couple the elements

C. An antenna in which all elements are driven by direct connection to the feed line

D. An antenna in which the driven element obtains its radio energy by induction or radiation from director elements

T9B05   What are the parasitic elements of a Yagi antenna?

A. The driven element and any reflectors

B. The director and the driven element

C. Only the reflectors (if any)

D. Any directors or any reflectors

T9B06   What is a cubical quad antenna?

A. Four straight, parallel elements in line with each other, each approximately 1/2-electrical wavelength long

B. Two or more parallel four-sided wire loops, each approximately one-electrical wavelength long

C. A vertical conductor 1/4-electrical wavelength high, fed at the bottom

D. A center-fed wire 1/2-electrical wavelength long

T9B07　What type of nondirectional antenna is easy to make at home and works well outdoors?

A. A Yagi

B. A delta loop

C. A cubical quad

D. A ground plane

T9B08　What electromagnetic-wave polarization does most man-made electrical noise have in the HF and VHF spectrum?

A. Horizontal

B. Left-hand circular

C. Right-hand circular

D. Vertical

T9B09　What does standing-wave ratio mean?

A. The ratio of maximum to minimum inductances on a feed line

B. The ratio of maximum to minimum capacitances on a feed line

C. The ratio of maximum to minimum impedances on a feed line

D. The ratio of maximum to minimum voltages on a feed line

T9B10　Where would you install a balun to feed a dipole antenna with 50-ohm coaxial cable?

A. Between the coaxial cable and the antenna

B. Between the transmitter and the coaxial cable

C. Between the antenna and the ground

D. Between the coaxial cable and the ground

T9B11　Why does coaxial cable make a good antenna feed line?

A. You can make it at home, and its impedance matches most amateur antennas

B. It is weatherproof, and it can be used near metal objects

C. It is weatherproof, and its impedance is higher than that of most amateur antennas

D. It can be used near metal objects, and its impedance is higher than that of most amateur antennas

Subelement T0: RF Safety [3 Exam Questions— 3 Groups]

T0A   RF safety fundamentals, terms, and definitions

T0A01   Why is it a good idea to adhere to the FCC's Rules for using the minimum power needed when you are transmitting with your hand-held radio?

A. Large fines are always imposed on operators violating this rule

B. To reduce the level of RF radiation exposure to the operator's head

C. To reduce calcification of the NiCd battery pack

D. To eliminate self oscillation in the receiver RF amplifier

T0A02   Over what frequency range are the FCC Regulations most stringent for RF radiation exposure?

A. Frequencies below 300 kHz

B. Frequencies between 300 kHz and 3 MHz

C. Frequencies between 3 MHz and 30 MHz

D. Frequencies between 30 MHz and 300 MHz

T0A03   What is one biological effect to the eye that can result from RF exposure?

A. The strong magnetic fields can cause blurred vision

B. The strong magnetic fields can cause polarization lens

C. It can cause heating, which can result in the formation of cataracts

D. It can cause heating, which can result in astigmatism

T0A04   In the far field, as the distance from the source increases, how does power density vary?

A. The power density is proportional to the square of the distance

B. The power density is proportional to the square root of the distance

C. The power density is proportional to the inverse square of the distance

D. The power density is proportional to the inverse cube of the distance

T0A05   In the near field, how does the field strength vary with distance from the source?

A. It always increases with the cube of the distance

B. It always decreases with the cube of the distance

C. It varies as a sine wave with distance

D. It depends on the type of antenna being used

T0A06   Why should you never look into the open end of a microwave feed horn antenna while the transmitter is operating?

A. You may be exposing your eyes to more than the maximum permissible exposure of RF radiation

B. You may be exposing your eyes to more than the maximum permissible exposure level of infrared radiation

C. You may be exposing your eyes to more than the maximum permissible exposure level of ultraviolet radiation

D. All of these choices are correct

T0A07   Why are Amateur Radio operators required to meet the FCC RF radiation exposure limits?

A. The standards are applied equally to all radio services

B. To ensure that RF radiation occurs only in a desired direction

C. Because amateur station operations are more easily adjusted than those of commercial radio services

D. To ensure a safe operating environment for amateurs, their families, and neighbors

T0A08 Why are the maximum permissible exposure (MPE) levels not uniform throughout the radio spectrum?

A. The human body absorbs energy differently at various frequencies

B. Some frequency ranges have a cooling effect while others have a heating effect on the body

C. Some frequency ranges have no effect on the body

D. Radiation at some frequencies can have a catalytic effect on the body

T0A09 What does the term "specific absorption rate," or SAR, mean?

A. The degree of RF energy consumed by the ionosphere

B. The rate at which transmitter energy is lost because of a poor feed line

C. The rate at which RF energy is absorbed into the human body

D. The amount of signal weakening caused by atmospheric phenomena

T0A10 On what value are the maximum permissible exposure (MPE) limits based?

A. The square of the mass of the exposed body

B. The square root of the mass of the exposed body

C. The whole-body specific gravity (WBSG)

D. The whole-body specific absorption rate (SAR)

T0B RF safety rules and guidelines

T0B01 Where will you find the applicable FCC RF radiation maximum permissible exposure (MPE) limits defined?

A. FCC Part 97 Amateur Service Rules and Regulations

B. FCC Part 15 Radiation Exposure Rules and Regulations

C. FCC Part 1 and Office of Engineering and Technology (OET) Bulletin 65

D. Environmental Protection Agency Regulation 65

T0B02 What factors must you consider if your repeater station antenna will be located at a site that is occupied by antennas for transmitters in other services?

A. Your radiated signal must be considered as part of the total RF radiation from the site when determining RF radiation exposure levels

B. Each individual transmitting station at a multiple transmitter site must meet the RF radiation exposure levels

C. Each station at a multiple-transmitter site may add no more than 1% of the maximum permissible exposure (MPE) for that site

D. Amateur stations are categorically excluded from RF radiation exposure evaluation at multiple-transmitter sites

T0B03 To determine compliance with the maximum permitted exposure (MPE) levels, safe exposure levels for RF energy are averaged for an "uncontrolled" RF environment over what time period?

A. 6 minutes

B. 10 minutes

C. 15 minutes

D. 30 minutes

T0B04 To determine compliance with the maximum permitted exposure (MPE) levels, safe exposure levels for RF energy are averaged for a "controlled" RF environment over what time period?

A. 6 minutes

B. 10 minutes

C. 15 minutes

D. 30 minutes

T0B05   Which of the following categories describes most common amateur use of a hand-held transceiver?

A. Mobile devices

B. Portable devices

C. Fixed devices

D. None of these choices is correct

T0B06   How does an Amateur Radio operator demonstrate that he or she has read and understood the FCC rules about RF-radiation exposure?

A. By indicating his or her understanding of this requirement on an amateur license application form at the time of application

B. By posting a copy of Part 97 at the station

C. By completing an FCC Environmental Assessment Form

D. By completing an FCC Environmental Impact Statement

T0B07   What amateur stations must comply with the requirements for RF radiation exposure spelled out in Part 97?

A. Stations with antennas that exceed 10 dBi of gain

B. Stations that have a duty cycle greater than 50 percent

C. Stations that run more than 50 watts peak envelope power (pep)

D. All amateur stations regardless of power

T0B08   Who is responsible for ensuring that an amateur station complies with FCC Rules about RF radiation exposure?

A. The Federal Communications Commission

B. The Environmental Protection Agency

C. The licensee of the amateur station

D. The Food and Drug Administration

T0B09   Why do exposure limits vary with frequency?

A. Lower-frequency RF fields have more energy than higher-frequency fields

B. Lower-frequency RF fields penetrate deeper into the body than higher-frequency fields

C. The body's ability to absorb RF energy varies with frequency

D. It is impossible to measure specific absorption rates at some frequencies

T0B10   Why is the concept of "duty cycle" one factor used to determine safe RF radiation exposure levels?

A. It takes into account the amount of time the transmitter is operating at full power during a single transmission

B. It takes into account the transmitter power supply rating

C. It takes into account the antenna feed line loss

D. It takes into account the thermal effects of the final amplifier

T0B11   From an RF safety standpoint, what impact does the duty cycle have on the minimum safe distance separating an antenna and the neighboring environment?

A. The lower the duty cycle, the shorter the compliance distance

B. The compliance distance is increased with an increase in the duty cycle

C. Lower duty cycles subject the environment to lower radio-frequency radiation cycles

D. All of these answers are correct

T0C   Routine station evaluation (Practical applications for VHF/UHF and above operations)

T0C01   If you do not have the equipment to measure the RF power densities present at your station, what might you do to ensure compliance with the FCC RF radiation exposure limits?

A. Use one or more of the methods included in the amateur supplement to FCC OET Bulletin 65

B. Call an FCC-Certified Test Technician to perform the measurements for you

C. Reduce power from 200 watts pep to 100 watts pep

D. Operate only low-duty-cycle modes such as FM

T0C02   Is it necessary for you to perform mathematical calculations of the RF radiation exposure if your station transmits with more than 50 watts peak envelope power (pep)?

A. Yes, calculations are always required to ensure greatest accuracy

B. Calculations are required if your station is located in a densely populated neighborhood

C. No, calculations may not give accurate results, so measurements are always required

D. No, there are alternate means to determine if your station meets the RF radiation exposure limits

T0C03   Why should you make sure the antenna of a hand-held transceiver is not too close to your head when transmitting?

A. To help the antenna radiate energy equally in all directions

B. To reduce your exposure to the radio-frequency energy

C. To use your body to reflect the signal in one direction

D. To keep electrostatic charges from harming the operator

T0C04   What should you do for safety if you put up a UHF transmitting antenna?

A. Make sure the antenna will be in a place where no one can get near it when you are transmitting

B. Make sure that RF field screens are in place

C. Make sure the antenna is near the ground to keep its RF energy pointing in the correct direction

D. Make sure you connect an RF leakage filter at the antenna feed point

T0C05   How should you position the antenna of a hand-held transceiver while you are transmitting?

A. Away from your head and away from others

B. Towards the station you are contacting

C. Away from the station you are contacting

D. Down to bounce the signal off the ground

T0C06   Why should your antennas be located so that no one can touch them while you are transmitting?

A. Touching the antenna might cause television interference

B. Touching the antenna might cause RF burns

C. Touching the antenna might cause it to radiate harmonics

D. Touching the antenna might cause it to go into self-oscillation

T0C07   For the lowest RF radiation exposure to passengers, where would you mount your mobile antenna?

A. On the trunk lid

B. On the roof

C. On a front fender opposite the broadcast radio antenna

D. On one side of the rear bumper

T0C08   What should you do for safety before removing the shielding on a UHF power amplifier?

A. Make sure all RF screens are in place at the antenna feed line

B. Make sure the antenna feed line is properly grounded

C. Make sure the amplifier cannot accidentally be turned on

D. Make sure that RF leakage filters are connected

T0C09   Why might mobile transceivers produce less RF radiation exposure than hand-held transceivers in mobile operations?

A. They do not produce less exposure because they usually have higher power levels

B. They have a higher duty cycle

C. When mounted on a metal vehicle roof, mobile antennas are generally well shielded from vehicle occupants

D. Larger transmitters dissipate heat and energy more readily

T0C10   What are some reasons you should never operate a power amplifier unless its covers are in place?

A. To maintain the required high operating temperatures of the equipment and reduce RF radiation exposure

B. To reduce the risk of shock from high voltages and reduce RF radiation exposure

C. To ensure that the amplifier will go into self oscillation and to minimize the effects of stray capacitance

D. To minimize the effects of stray inductance and to reduce the risk of shock from high voltages

Answers

T1A01	B	T1D02	C	T1G02	B
T1A02	D	T1D03	D	T1G03	A
T1A03	D	T1D04	C	T1G04	A
T1A04	C	T1D05	B	T1G05	C
T1A05	C	T1D06	A	T1G06	C
T1A06	A	T1D07	C	T1G07	A
T1A07	D	T1D08	A	T1G08	C
T1A08	C	T1D09	C	T1G09	C
T1A09	A	T1D10	B	T1G10	B
T1A10	C	T1D11	A	T1G11	B
T1A11	A	T1D12	C	T1H01	C
T1A12	B	T1E01	D	T1H02	C
T1B01	B	T1E02	A	T1H03	A
T1B02	B	T1E03	B	T1H04	A
T1B03	B	T1E04	D	T1H05	A
T1B04	C	T1E05	A	T1H06	B
T1B05	D	T1E06	B	T1H07	B
T1B06	B	T1E07	A	T1H08	C
T1B07	A	T1E08	A	T1H09	B
T1B08	B	T1E09	C	T1I01	D
T1B09	C	T1E10	B	T1I02	C
T1B10	A	T1E11	A	T1I03	B
T1B11	C	T1E12	D	T1I04	A
T1B12	C	T1F01	C	T1I05	C
T1C01	D	T1F02	C	T1I06	A
T1C02	B	T1F03	A	T1I07	B
T1C03	A	T1F04	D	T1I08	D
T1C04	D	T1F05	D	T1I09	D
T1C05	A	T1F06	D	T1I10	A
T1C06	D	T1F07	A	T1I11	D
T1C07	C	T1F08	D	T2A01	A
T1C08	C	T1F09	C	T2A02	D
T1C09	D	T1F10	D	T2A03	A
T1C10	D	T1F11	C	T2A04	B
T1C11	C	T1F12	B	T2A05	C
T1D01	D	T1G01	C	T2A06	C

T2A07	B	T2E01	A	T3C06	C
T2A08	B	T2E02	B	T3C07	B
T2A09	A	T2E03	C	T3C08	D
T2A10	D	T2E04	D	T3C09	B
T2A11	D	T2E05	A	T3C10	A
T2B01	A	T2E06	B	T3C11	C
T2B02	C	T2E07	B	T4A01	C
T2B03	D	T2E08	A	T4A02	D
T2B04	B	T2E09	B	T4A03	B
T2B05	A	T2E10	C	T4A04	D
T2B06	C	T2E11	D	T4A05	C
T2B07	D	T3A01	B	T4A06	D
T2B08	B	T3A02	A	T4A07	B
T2B09	A	T3A03	B	T4A08	B
T2B10	C	T3A04	C	T4A09	A
T2B11	A	T3A05	A	T4A10	A
T2C01	A	T3A06	C	T4A11	B
T2C02	B	T3A07	C	T4B01	C
T2C03	C	T3A08	C	T4B02	A
T2C04	A	T3A09	B	T4B03	A
T2C05	D	T3A10	A	T4B04	A
T2C06	D	T3A11	C	T4B05	C
T2C07	D	T3B01	C	T4B06	A
T2C08	C	T3B02	A	T4B07	B
T2C09	C	T3B03	D	T4B08	C
T2C10	D	T3B04	B	T4B09	D
T2C11	A	T3B05	A	T4B10	C
T2D01	A	T3B06	A	T4B11	B
T2D02	D	T3B07	C	T4C01	B
T2D03	A	T3B08	B	T4C02	A
T2D04	B	T3B09	D	T4C03	A
T2D05	D	T3B10	B	T4C04	D
T2D06	B	T3B11	C	T4C05	B
T2D07	B	T3C01	D	T4C06	C
T2D08	C	T3C02	C	T4C07	B
T2D09	C	T3C03	A	T4C08	A
T2D10	A	T3C04	D	T4C09	B
T2D11	D	T3C05	A	T4C10	A

T4C11	C	T5C05	B	T7A10	A
T4D01	C	T5C06	B	T7A11	B
T4D02	A	T5C07	C	T7B01	D
T4D03	B	T5C08	B	T7B02	D
T4D04	C	T5C09	A	T7B03	A
T4D05	D	T5C10	A	T7B04	C
T4D06	B	T5C11	B	T7B05	D
T4D07	B	T6A01	C	T7B06	D
T4D08	C	T6A02	A	T7B07	C
T4D09	A	T6A03	D	T7B08	A
T4D10	C	T6A04	A	T7B09	A
T4D11	D	T6A05	D	T7B10	C
T5A01	B	T6A06	A	T7B11	D
T5A02	C	T6A07	A	T8A01	C
T5A03	C	T6A08	D	T8A02	C
T5A04	C	T6A09	C	T8A03	D
T5A05	D	T6A10	A	T8A04	C
T5A06	B	T6A11	A	T8A05	C
T5A07	A	T6B01	A	T8A06	B
T5A08	C	T6B02	A	T8A07	B
T5A09	A	T6B03	D	T8A08	C
T5A10	C	T6B04	D	T8A09	C
T5A11	D	T6B05	D	T8A10	B
T5B01	D	T6B06	A	T8A11	C
T5B02	B	T6B07	C	T8B01	B
T5B03	C	T6B08	C	T8B02	D
T5B04	D	T6B09	A	T8B03	A
T5B05	D	T6B10	B	T8B04	C
T5B06	B	T6B11	A	T8B05	A
T5B07	D	T7A01	B	T8B06	D
T5B08	A	T7A02	D	T8B07	A
T5B09	D	T7A03	A	T8B08	D
T5B10	A	T7A04	A	T8B09	C
T5B11	C	T7A05	B	T9A01	D
T5C01	C	T7A06	D	T9A02	B
T5C02	B	T7A07	C	T9A03	C
T5C03	C	T7A08	B	T9A04	A
T5C04	C	T7A09	D	T9A05	C

T9A06	A	T0A08	A
T9A07	B	T0A09	C
T9A08	A	T0A10	D
T9A09	B	T0B01	C
T9A10	A	T0B02	A
T9A11	D	T0B03	D
T9B01	C	T0B04	A
T9B02	A	T0B05	B
T9B03	B	T0B06	A
T9B04	A	T0B07	D
T9B05	D	T0B08	C
T9B06	B	T0B09	C
T9B07	D	T0B10	A
T9B08	D	T0B11	D
T9B09	D	T0C01	A
T9B10	A	T0C02	D
T9B11	B	T0C03	B
T0A01	B	T0C04	A
T0A02	D	T0C05	A
T0A03	C	T0C06	B
T0A04	C	T0C07	B
T0A05	D	T0C08	C
T0A06	A	T0C09	C
T0A07	D	T0C10	B

The W5YI RF Safety Tables*

There are two ways to determine whether your station's radio frequency signal radiation is within the maximum permissible exposure (MPE) guidelines established by the FCC for *controlled* and *uncontrolled* environments. One way is direct *measurement* of the RF fields. The second way is through *prediction* using various antenna modeling, equations, and calculation methods described in the FCC's *OET Bulletin 65* and *Supplement B*.

In general, most amateurs will not have access to the appropriate calibrated equipment to make precise field strength/power density measurements. The field-strength meters in common use by amateur operators and inexpensive, hand-peld field-strength meters do not provide the accuracy necessary for reliable measurements, especially when different frequencies may be encountered at a given measurement location. It is more practical for amateurs to determine their pep output power at the antenna and then look up the required distances to the controlled/uncontrolled environments using the following tables, which were developed using the prediction equations supplied by the FCC.

The FCC has determined that radio operators and their families are in the controlled environment and your neighbors and passers-by are in the uncontrolled environment. The estimated minimum compliance distances are in meters from the transmitting antenna to either the occupational/controlled exposure environment (Con) or the general population/uncontrolled exposure environment (Unc) using typical

*Courtesy of the W5YI Group and Master Publishing Inc.

antenna gains for the amateur service and assuming 100% duty cycle and maximum surface reflection. Therefore, these charts represent the worst-case scenario. They do not take into consideration compliance distance reductions that would be caused by:

1. Feed line losses, which reduce power output at the antenna, especially at the VHF and higher frequency levels.

2. Duty cycle caused by the emission type. The emission-type factor accounts for the fact that, for some modulated emission types that have a nonconstant envelope, the pep can be considerably larger than the average power. Multiply the distances by 0.4 if you are using CW Morse telegraphy, and by 0.2 for two-way single sideband (SSB) voice. There is no reduction for FM.

3. Duty cycle caused by on/off time or *time averaging.* The RF safety guidelines permit RF exposures to be averaged over certain periods of time with the average not to exceed the limit for continuous exposure. The averaging time for occupational/controlled exposure is 6 minutes, while the averaging time for general population/uncontrolled exposure is 30 minutes. For example, if the relevant time interval for time-averaging is 6 minutes, an amateur could be exposed to two times the applicable power density limit for three minutes as long as he or she were not exposed at all for the preceding or following three minutes.

A routine evaluation is not required for vehicular mobile or hand-held transceiver stations. Amateur radio operators should be aware, however, of the potential for exposure to RF electromagnetic fields from these stations, and take measures (such as reducing transmitting power to the minimum necessary, positioning the radiating antenna as far from humans as practical, and limiting continuous transmitting time) to protect themselves and the occupants of their vehicles.

Amateur radio operators should also be aware that the new FCC radio-frequency safety regulations address exposure to people—and not the strength of the signal. Amateurs may exceed the MPE limits as long as no one is exposed to the radiation.

How to read the chart: If you are radiating 500 watts from your 10-meter dipole (about 3 dB gain), there must be at least 4.5 meters (about 15 feet) between you (and your family) and the antenna—and a distance of 10 meters (about 33 feet) between the antenna and your neighbors.

TABLE B.1 Amateur Radio Frequency Bands

Freq. (MF/HF) (MHz/band)	Antenna gain (dBI)	Peak envelope power (watts)							
		100 watts		500 watts		1000 watts		1500 watts	
		Con	Unc	Con	Unc	Con	Unc	Con	Unc
2.0 (160 m)	0	0.1	0.2	0.3	0.5	0.5	0.7	0.6	0.8
2.0 (160 m)	3	0.2	0.3	0.5	0.7	0.6	1.06	0.8	1.2
4.0 (75/80 m)	0	0.2	0.4	0.4	1.0	0.6	1.3	0.7	1.6
4.0 (75/80 m)	3	0.3	0.6	0.6	1.3	0.9	1.9	1.0	2.3
7.3 (40 m)	0	0.3	0.8	0.8	1.7	1.1	2.5	1.3	3.0
7.3 (40 m)	3	0.5	1.1	1.1	2.5	1.6	3.5	1.9	4.2
7.3 (40 m)	6	0.7	1.5	1.5	3.5	2.2	4.9	2.7	6.0
10.15 (30 m)	0	0.5	1.1	1.1	2.4	1.5	3.4	1.9	4.2
10.15 (30 m)	3	0.7	1.5	1.5	3.4	2.2	4.8	2.6	5.9
10.15 (30 m)	6	1.0	2.2	2.2	4.8	3.0	6.8	3.7	8.3
14.35 (20 m)	0	0.7	1.5	1.5	3.4	2.2	4.8	2.6	5.9
14.35 (20 m)	3	1.0	2.2	2.2	4.8	3.0	6.8	3.7	8.4
14.35 (20 m)	6	1.4	3.0	3.0	6.8	4.3	9.6	5.3	11.8
14.35 (20 m)	9	1.9	4.3	4.3	9.6	6.1	13.6	7.5	16.7
18.168 (17 m)	0	0.9	1.9	1.9	4.3	2.7	6.1	3.3	7.5
18.168 (17 m)	3	1.2	2.7	2.7	6.1	3.9	8.6	4.7	10.6
18.168 (17 m)	6	1.7	3.9	3.9	8.6	5.5	12.2	6.7	14.9
18.168 (17 m)	9	2.4	5.4	5.4	12.2	7.7	17.2	9.4	21.1
21.145 (15 m)	0	1.0	2.3	2.3	5.1	3.2	7.2	4.0	8.8
21.145 (15 m)	3	1.4	3.2	3.2	7.2	4.6	10.2	5.6	12.5
21.145 (15 m)	6	2.0	4.6	4.6	10.2	6.4	14.4	7.9	17.6
21.145 (15 m)	9	2.9	6.4	6.4	14.4	9.1	20.3	11.1	24.9
24.99 (12 m)	0	1.2	2.7	2.7	5.9	3.8	8.4	4.6	10.3

TABLE B.1 Amateur Radio Frequency Bands (*Continued*)

Freq. (MF/HF) (MHz/band)	Antenna gain (dBI)	Peak envelope power (watts)							
		100 watts		500 watts		1000 watts		1500 watts	
		Con	Unc	Con	Unc	Con	Unc	Con	Unc
24.99 (12 m)	3	1.7	3.8	3.8	8.4	5.3	11.9	6.5	14.5
24.99 (12 m)	6	2.4	5.3	5.3	11.9	7.5	16.8	9.2	20.5
24.99 (12 m)	9	3.4	7.5	7.5	16.8	10.6	23.7	13.0	29.0
29.7 (10 m)	0	1.4	3.2	3.2	7.1	4.5	10.0	5.5	12.2
29.7 (10 m)	3	2.0	4.5	4.5	10.0	6.3	14.1	7.7	17.3
29.7 (10 m)	6	2.8	6.3	6.3	14.1	8.9	19.9	10.9	24.4
29.7 (10 m)	9	4.0	8.9	8.9	19.9	12.6	28.2	15.4	34.5
50 (6 m)	0	1.0	2.3	1.4	3.2	3.2	7.1	4.5	10.1
50 (6 m)	3	1.4	3.2	2.0	4.5	4.5	10.1	6.4	14.3
50 (6 m)	6	2.0	4.5	2.8	6.4	6.4	14.2	9.0	20.1
50 (6 m)	9	2.8	6.4	4.0	9.0	9.0	20.1	12.7	28.4
50 (6 m)	12	4.0	9.0	5.7	12.7	12.7	28.4	18.0	40.2
50 (6 m)	15	5.7	12.7	8.0	18.0	18.0	40.2	25.4	56.8
144 (2 m)	0	1.0	2.3	1.4	3.2	3.2	7.1	4.5	10.1
144 (2 m)	3	1.4	3.2	2.0	4.5	4.5	10.1	6.4	14.3
144 (2 m)	6	2.0	4.5	2.8	6.4	6.4	14.2	9.0	20.1
144 (2 m)	9	2.8	6.4	4.0	9.0	9.0	20.1	12.7	28.4
144 (2 m)	12	4.0	9.0	5.7	12.7	12.7	28.4	18.0	40.2
144 (2 m)	15	5.7	12.7	8.0	18.0	18.0	40.2	25.4	56.8
144 (2 m)	20	10.1	22.6	14.3	32.0	32.0	71.4	45.1	101.0
222 (1.25 m)	0	1.0	2.3	1.4	3.2	3.2	7.1	4.5	10.1
222 (1.25 m)	3	1.4	3.2	2.0	4.5	4.5	10.1	6.4	14.3
222 (1.25 m)	6	2.0	4.5	2.8	6.4	6.4	14.2	9.0	20.1

TABLE B.1 Amateur Radio Frequency Bands (*Continued*)

Freq. (MF/HF) (MHz/band)	Antenna gain (dBI)	Peak envelope poser (watts)							
		100 watts		500 watts		1000 watts		1500 watts	
		Con	Unc	Con	Unc	Con	Unc	Con	Unc
222 (1.25 m)	9	2.8	6.4	4.0	9.0	9.0	20.1	12.7	28.4
222 (1.25 m)	12	4.0	9.0	5.7	12.7	12.7	28.4	18.0	40.2
222 (1.25 m)	15	5.7	12.7	8.0	18.0	18.0	40.2	25.4	56.8
450 (70 cm)	0	0.8	1.8	1.2	2.6	2.6	5.8	3.7	8.2
450 (70 cm)	3	1.2	2.6	1.6	3.7	3.7	8.2	5.2	11.6
450 (70 cm)	6	1.6	3.7	2.3	5.2	5.2	11.6	7.4	16.4
450 (70 cm)	9	2.3	5.2	3.3	7.3	7.3	16.4	10.4	23.2
450 (70 cm)	12	3.3	7.3	4.6	10.4	10.4	23.2	14.7	32.8
902 (33 cm)	0	0.6	1.3	0.8	1.8	1.8	4.1	2.6	5.8
902 (33 cm)	3	0.8	1.8	1.2	2.6	2.6	5.8	3.7	8.2
902 (33 cm)	6	1.2	2.6	1.6	3.7	3.7	8.2	5.2	11.6
902 (33 cm)	9	1.6	3.7	2.3	5.2	5.2	11.6	7.4	16.4
902 (33 cm)	12	2.3	5.2	3.3	7.3	7.3	16.4	10.4	23.2
1240 (23 cm)	0	0.5	1.1	0.7	1.6	1.6	3.5	2.2	5.0
1240 (23 cm)	3	0.7	1.6	1.0	2.2	2.2	5.0	3.1	7.0
1240 (23 cm)	6	1.0	2.2	1.4	3.1	3.1	7.0	4.4	9.9
1240 (23 cm)	9	1.4	3.1	2.0	4.4	4.4	9.9	6.3	14.0
1240 (23 cm)	12	2.0	4.4	2.8	6.2	6.2	14.0	8.8	19.8

All distances are in meters. To convert from meters to feet multiply meters by 3.28. Distance indicated is shortest line-of-sight distance to point where MPE limit for apprpriate exposure tier is predicted to occur.

ABOUT THE AUTHOR

An electronics engineer and college instructor, Clay Laster has been an avid ham radio operator for more than 40 years. He holds the Advanced Class License and the FCC General Radiotelephone Operator's License.